T4-AJS-322

The Influence of Cooperative Bacteria on Animal Host Biology

Ninety percent of the cells in the human body are bacteria, and humans may be host to many thousands of different species of bacteria. These striking statistics are part of a new paradigm in microbiology in which bacteria are no longer viewed as disease-causing killers but more as lifelong partners that are often essential for the survival of their host. This book brings together a group of diverse scientists – evolutionary biologists, immunologists, molecular biologists, microbiologists, pathologists, and mathematicians – to discuss the evolution and mechanisms of bacteria–host interactions at all levels of complexity, ranging from associations of one bacterium with its host to the many hundreds of bacteria normally associated with mammals. Chapters deal with the evolution of bacteria–host interactions over the last 60 years (since the introduction of antibiotics) to a period of 3.8 billion years (since the evolution of single-celled life) and discuss bacterial interactions with multicellular life forms such as coral reefs, insects, mice, and men. This book should be of interest to the widest range of biological scientists.

Margaret J. McFall-Ngai is Professor of Medical Microbiology and Immunology at the University of Wisconsin School of Medicine.

Brian Henderson is Professor of Cell Biology and head of the Cellular Microbiology Research Group in the Division of Microbial Diseases, Eastman Dental Institute, University College London. He is the co-editor of *Molecular Chaperones and Cell Signalling* (2005), *Bacterial Evasion of Host Immune Responses* (2003), and co-author of *Bacterial Disease Mechanisms* (2002).

Edward G. Ruby is Professor of Medical Microbiology and Immunology at the University of Wisconsin School of Medicine.

Published Titles

1. *Bacterial Adhesion to Host Tissues.* Edited by Michael Wilson 0521801079
2. *Bacterial Evasion of Host Immune Responses.* Edited by Brian Henderson & Petra Oyston 0521801737
3. *Dormancy in Microbial Diseases.* Edited by Anthony Coates 0521809401
4. *Susceptibility to Infectious Diseases.* Edited by Richard Bellamy 0521815258
5. *Bacterial Invasion of Host Cells.* Edited by Richard Lamont 0521809541
6. *Mammalian Host Defense Peptides.* Edited by Deirdre Devine & Robert Hancock 0521822203
7. *Bacterial Protein Toxins.* Edited by Alistair Lax 052182091X
8. *The Dynamic Bacterial Genome.* Edited by Peter Mullany 0521821576
9. *Salmonella Infections.* Edited by Pietro Mastroeni & Duncan Maskell 0521835046

Forthcoming Titles in the Series

Quorum Sensing and Bacterial Cell-to-Cell Communication. Edited by Donald Demuth & Richard Lamont 0521846382

Phagocytosis of Bacteria and Bacterial Pathogenicity. Edited by Joel Ernst & Olle Stendahl 0521845696

Over the past decade, the rapid development of an array of techniques in the fields of cellular and molecular biology have transformed whole areas of research across the biological sciences. Microbiology has perhaps been influenced most of all. Our understanding of microbial diversity and evolutionary biology, and of how pathogenic bacteria and viruses interact with their animal and plant hosts at the molecular level, for example, has been revolutionized. Perhaps the most exciting recent advance in microbiology has been the development of the interface discipline of Cellular Microbiology, a fusion of classic microbiology, microbial molecular biology, and eukaryotic cellular and molecular biology. Cellular Microbiology is revealing how pathogenic bacteria interact with host cells in what is turning out to be a complex evolutionary battle of competing gene products. Molecular and cellular biology are no longer discrete subject areas but vital tools and an integrated part of current microbiological research. As part of this revolution in molecular biology, the genomes of a growing number of pathogenic and model bacteria have been fully sequenced, with immense implications for our future understanding of microorganisms at the molecular level.

Advances in Molecular and Cellular Microbiology is a series edited by researchers active in these exciting and rapidly expanding fields. Each volume focuses on a particular aspect of cellular or molecular microbiology and provides an overview of the area, as well as examines current research. This series will enable graduate students and researchers to keep up with the rapidly diversifying literature in current microbiological research.

MCM

ADVANCES IN MOLECULAR AND CELLULAR MICROBIOLOGY

Advances in Molecular and Cellular Microbiology 10

The Influence of Cooperative Bacteria on Animal Host Biology

EDITED BY
MARGARET J. McFALL-NGAI
University of Wisconsin, Madison

BRIAN HENDERSON
University College London

EDWARD G. RUBY
University of Wisconsin, Madison

CAMBRIDGE
UNIVERSITY PRESS

CAMBRIDGE UNIVERSITY PRESS
Cambridge, New York, Melbourne, Madrid, Cape Town, Singapore, São Paulo

Cambridge University Press
40 West 20th Street, New York, NY 10011-4211, USA

www.cambridge.org
Information on this title: www.cambridge.org/9780521834650

First published 2005

Printed in the United States of America

A catalog record for this publication is available from the British Library.

Library of Congress Cataloging in Publication Data

The influence of cooperative bacteria on animal host biology / edited by
Margaret J. McFall-Ngai, Brian Henderson, Edward G. Ruby.
p. ; cm. – (Advances in molecular and cellular microbiology ; 10)
Includes bibliographical references and index.
ISBN 0-521-83465-1 (hardback)
1. Host-bacteria relationships. 2. Symbiosis.
[DNLM: 1. Immunity, Cellular. 2. Bacteria – genetics. 3. Bacteria – immunology.
4. Bacteria – pathogenicity. 5. Host-Parasite Relations. 6. Models, Animal.
7. Symbiosis. QW 568 I43 2005] I. McFall-Ngai, Margaret Jean. II. Henderson,
Brian, PhD. III. Ruby, Edward G. IV. Title. V. Series.
QR100.8.S9I546 2005
579.3′17852 – dc22 2005006653

ISBN-13 978-0-521-83465-0 hardback
ISBN-10 0-521-83465-1 hardback

Contents

Preface

Bacteriology can be traced back to Anton van Leuwenhooek in the late seventeenth century, who first saw the magnitude of the bacterial colonisation of the planet Earth. However, it was in the nineteenth century that bacteriology, as we know it, was created by the towering figures of scientists such as Louis Pasteur and Robert Koch of bacterial disease fame and Serge Winogradsky, who pioneered the characterisation of the activity of bacteria in natural habitats. Bacteriology begat immunology, and the twentieth century saw the development of antibacterial vaccines and then the discovery of antibiotics. These naturally occurring drugs created the false illusion that we had "beaten the bugs," and bacteriology as a scientific discipline went into decline. Nevertheless, the study of bacteria, such as *Escherichia coli*, and of bacteriophage was at the foundation of modern molecular biology. The emergence of widespread antibiotic resistance in the late twentieth century was the spur to reactivate bacteriology, which is now a flourishing discipline in its many guises, such as molecular microbiology, cellular microbiology, and environmental microbiology.

Throughout the twentieth century, there was a growing realisation that bacteria did more than simply cause disease, and evidence began to mount that most animals contained their own populations of bacteria that were variously termed: commensal bacteria, indigenous microbiota, microflora, and so on. It was not until the advent of techniques allowing molecular phylogenetic analysis of the life forms in environmental samples (including those from animals) that the magnitude of the diversity of the prokaryotic world became realised. We now understand that microbial life forms, particularly bacteria, are the predominant organisms on our planet, with a staggering diversity. What has been even more astonishing is the discovery that the majority of multicellular creatures on our planet live with many specific bacterial partners. In this volume, we describe these organisms as cooperative bacteria. Humans, most of whom fear bacterial infection, may be the acme of bacterial cooperation as ninety percent of the cells in the average human are bacteria, and the number of bacterial species living happily with us is estimated to be between 1,000 and 3,000. Compare this with the 50 or so bacteria (many of which form part of the cooperative assembly) that have the potential to cause human disease. This suggests that for the last century or more, we have been looking at bacteria through the wrong end of the telescope.

This volume arose from a meeting held in the beautiful lake resort of Bellagio, Italy, and funded by the Rockefeller Foundation. Scientists from diverse disciplines but with a common interest in bacteria–host interactions came together to discuss the biology of animal–bacterial cooperation and what it means for both partners.

This volume is divided into four sections. Part I discusses the evolution of cooperation and addresses key questions about the role of cooperation as an evolutionary pressure on both prokaryotes and eukaryotes, as well as the influence of bacterial cooperation in the evolution of the complex vertebrate acquired immune response. Biological evolution is generally viewed on the scale of millions of years. However, as the last two chapters in this section describe, rapid changes in phenotype can arise in bacterial populations in response to selection factors in their environment, such as antibiotics. In Part II attention focuses on the ecological interactions between bacteria and their multicellular hosts, and the reader is introduced to the roles of bacteria in such diverse activities as coral bleaching and the control of insect reproduction. A major problem in the study of cooperative bacteria is the complexity of the "system" that is created between bacteria and their hosts. The final chapters in this section deal with methods of analysis of cooperative systems using metagenomics and mathematical modelling. Part III discusses our emerging knowledge of how cooperative bacteria interact with their hosts at the molecular level in a variety of systems ranging from intracellular bacteria in insects to the complex systems found in mammalian tissues such as the gut, genitourinary system, and respiratory tract. The final section, Part IV, considers how cooperative bacteria interact with innate and acquired immunity, and discusses our rapidly advancing knowledge of the innate immune responses of invertebrates and the mechanisms by which vertebrates immunologically recognise and cope with bacteria, including their cooperative partners. The final chapter reviews how much we have gained control over our cooperative bacteria in the last 50 years.

This volume will be of interest to a wide range of biological scientists, including basic microbiologists, medical microbiologists, epidemiologists, cell biologists, virologists, parasitologists, ecologists, and zoologists, and should be useful to undergraduate, postgraduate, and postdoctoral scientists working at the frontier of a new understanding of the role of cooperative bacteria in animal host biology.

Contributors

Caroline Anselme
Laboratoire de Biologie Fonctionnelle Insectes et Interactions
URM INRA/INSA de Lyon
Bât. Louis Pasteur
20 Avenue Albert Einstein
69621 Villeurbanne Cedex
France
anselme@insa-lyon.fr

Göran Bergsten
Department of Microbiology, Immunology and Glycobiology
Institute of Laboratory Medicine
Lund University
S-223 62 Lund
Sweden
Goran.Bergsten@mig.lu.se

Clare E. Bryant
Department of Clinical Veterinary Medicine
University of Cambridge
Madingley Road
Cambridge CB3 OES
United Kingdom
Ceb27@cus.cam.ac.uk

Sylvain Charlat
Department of Biology
University College London
4 Stephenson Way
London NW1 2HE
United Kingdom
s.charlat@ucl.ac.uk

Julian Davies
Department of Microbiology and Immunology
University of British Columbia
Vancouver
British Columbia
Canada V6T 1Z3
JED@UNIXG.UBC.CA

Fernando de la Cruz
Departmento de Biologia Molecular
(Unidad Asociada al CIB)
Universidad de Cantabria
Spain
delacruz@medi.unican.es

Ulrich Dobrindt
Institut für Molekulare
Infektionsbiologie
Röntgenring 11
Würzburg D-97070
Germany
u.dobrindt@mail.uni-wuerzburg.de

Jörg Hacker
Institut für Molekulare
Infektionsbiologie
Röntgenring 11
Würzburg D-97070
Germany
j.hacker@mail.uni-wuerzburg.de

Jo Handelsman
Department of Plant Pathology
University of Wisconsin
1630 Linden Dr.
Madison
Wisconsin 53706-1598
USA
JOH@PLANTPATH.WISC.EDU

Abdelaziz Heddi
Laboratoire de Biologie
Fonctionnelle Insectes et
Interactions
URM INRA/INSA de Lyon
Bât. Louis Pasteur
20 Avenue Albert Einstein
69621 Villeurbanne Cedex.
France
Abdelaziz.heddi@insa-lyon.fr

Brian Henderson
Division of Microbial Diseases
Eastman Dental Institute
University College London
256 Gray's Inn Road
London WC1X 8LD
United Kingdom
b.henderson@eastman.ucl.ac.uk

Ute Hentschel
Institut für Molekulare
Infektionsbiologie
Röntgenring 11
Würzburg D-97070
Germany
u.hentschel@mail.uni-wuerzburg.de

Carole S. Hickman
Department of Integrative Biology
University of California Berkeley
Berkeley
California 94720-3140
USA
caroleh@socrates.berkeley.edu

Lora V. Hooper
Center for Immunology
The University of Texas
Southwestern Medical Center
at Dallas
5323 Harry Hines Blvd.
Dallas
Texas 75390
USA
Lora.hooper@utsouthwestern.edu

Emily A. Hornett
Department of Biology
University College London
4 Stephenson Way
London NW1 2HE
United Kingdom
e.hornett@ucl.ac.uk

Gregory D. Hurst
Department of Biology
University College London
4 Stephenson Way
London NW1 2HE
United Kingdom
g.hurst@ucl.ac.uk

Seema Mattoo
Department of Pharmacology
University of California, San Diego
School of Medicine
Leich tag Biomedical Research
Building
Rm 249, Bay M
9500 Gilman Drive
Dept 0721
La Jolla
California 92093-0721
smattoo@ucas.edu.

Margaret McFall-Ngai
Department of Medical
Microbiology and Immunology
University of Wisconsin
1300 University Avenue
Madison
Wisconsin 53706
USA
mjmcfallngai@wisc.edu

Hilde Merkert
Institut für Molekulare
Infektionsbiologie
Röntgenring 11
Würzburg D-97070
Germany
h.merkert@mail.uni-wuerzburg.de

Jeffrey F. Miller
UCLA School of Medicine
10833 LeConte Avenue
43-46CHS
Los Angeles
California 90095
USA
jfmiller@ucla.edu

Horacio Montenegro
Department of Biology
University College London
4 Stephenson Way
London NW1 2HE
United Kingdom
h.montenegro@ucl.ac.uk

Andrew S. Neish
Department of Pathology
Emory University School of
Medicine
105-F Whitehead
615 Michaels St.
Atlanta
Georgia 30322
USA
aneish@emory.edu

Kenneth F. Raffa
Department of Entomology
University of Wisconsin
1630 Linden Dr.
Madison
Wisconsin 53706-1598
USA
raffa@entomology.wisc.edu

Paul B. Rainey
School of Biological Sciences
University of Auckland
Private Bag 92019
Auckland
New Zealand
p.rainey@auckland.ac.nz

Rino Rappuoli
Chiron Vaccines
Via Fiorentina 1
53100 Siena
Italy
RINO.RAPPUOLI@CHIRON.COM

Max Reuter
Department of Biology
University College London
4 Stephenson Way
London NW1 2HE
United Kingdom
m.reuter@ucl.ac.uk

Courtney J. Robinson
Department of Plant Pathology
University of Wisconsin
1630 Linden Dr.
Madison
Wisconsin 53706-1598
USA
cjr@plantpath.wisc.edu

Eugene Rosenberg
Department of Molecular
Microbiology and Biotechnology
Tel Aviv University
Ramat Aviv 69978
Israel
eros@post.tau.ac.il

Edward N. Ruby
Department of Medical
Microbiology and Immunology
University of Wisconsin
1300 University Avenue
Madison
Wisconsin 53706
USA
egruby@wisc.edu

Robert M. Seymour
Department of Mathematics
University College London
Gower Street
London WC1E 6BT
United Kingdom
r.seymour@math.ucl.ac.uk

L. Courtney Smith
Department of Biological Sciences
George Washington University
Lisner Hall 334
2023 G St. NW
Washington, DC 20052
USA
csmith@gwu.edu

Michael Steinert
Institut für Molekulare
Infektionsbiologie
Röntgenring 11
Würzburg D-97070
Germany
m.steinert@mail.uni-wuerzburg.de

Catharina Svanborg
Department of Microbiology, Immunology and Glycobiology
Institute of Laboratory Medicine
Lund University
S-223 62 Lund
Sweden
Catharina.svanborg@mig.lu.se

Sabine Tötemeyer
Department of Clinical Veterinary Medicine
University of Cambridge
Madingley Road
Cambridge CB3 OES
United Kingdom
st2b2@cam.ac.uk

Zoe Veneti
Department of Biology
University College London
4 Stephenson Way
London NW1 2HE
United Kingdom
z.veneti@ucl.ac.uk

Björn Wullt
Department of Urology
Lund University
22185 Lund
Sweden
bjorn.wullt@urokir.lu.se

1 Evolutionary biology of animal host–bacteria interactions

CHAPTER 1

How have bacteria contributed to the evolution of multicellular animals?

Carole S. Hickman

1.1. INTRODUCTION

Mutualistic interactions between animals and microbes have a long evolutionary history and a profound influence on the continuously changing shape of life on Earth. The impact of animal–microbial symbioses extends far beyond the ecological and evolutionary effects on the interacting partners to include fundamental modifications of the biosphere and geosphere. A new discipline uniting the divergent perspectives of zoology and microbiology on animal–microbe interactions must extend beyond the life sciences to include geology and the predominantly physical sciences of biogeochemistry and geomicrobiology. Astrobologists, exploring the properties of microbial consortia in extreme ecosystems (Cady 1998), and paleontologists, discovering new worlds of microbial complexity in ancient ecosystems (Hagadorn et al. 1999), are converging on principles of evolving partnerships and evidence of the same interconnectivity that are being sought by workers who envision a new field of beneficial animal–microbe interactions (Ruby et al. 2004).

This chapter establishes the geological context and examines the geologic and fossil evidence for three evolutionary propositions: (1) bacteria prepared the biosphere and played a major role in the origin of animals, (2) bacteria played a role in the subsequent diversification and shaping of animal life and its response to biotic crises, and (3) bacteria have influenced the evolution of animal development. The chapter also outlines a set of specific contributions that animal–microbe symbioses have made to the history and shape of life and presents a set of challenges for future research.

1.2. THE ROLE OF BACTERIA IN THE EVOLUTIONARY ORIGINS OF ANIMAL LIFE

1.2.1. A Perspective from Deep Time and Geology

The most striking fact of the fossil record of bacteria and animals is the strongly asymmetric timing of their origins (Fig. 1.1). Earth formed 4.5 billion years ago (Ga), and the first geochemical evidence (isotope ratios indicative of biological fractionation of carbon) for microbial activity appears in the rock record between 3.8 and 3.5 Ga (Mojzsis et al. 1996; Falkowski and Raven 1997; Rosing 1999). Prokaryotic fossils have been reported from rocks as old as 3.47

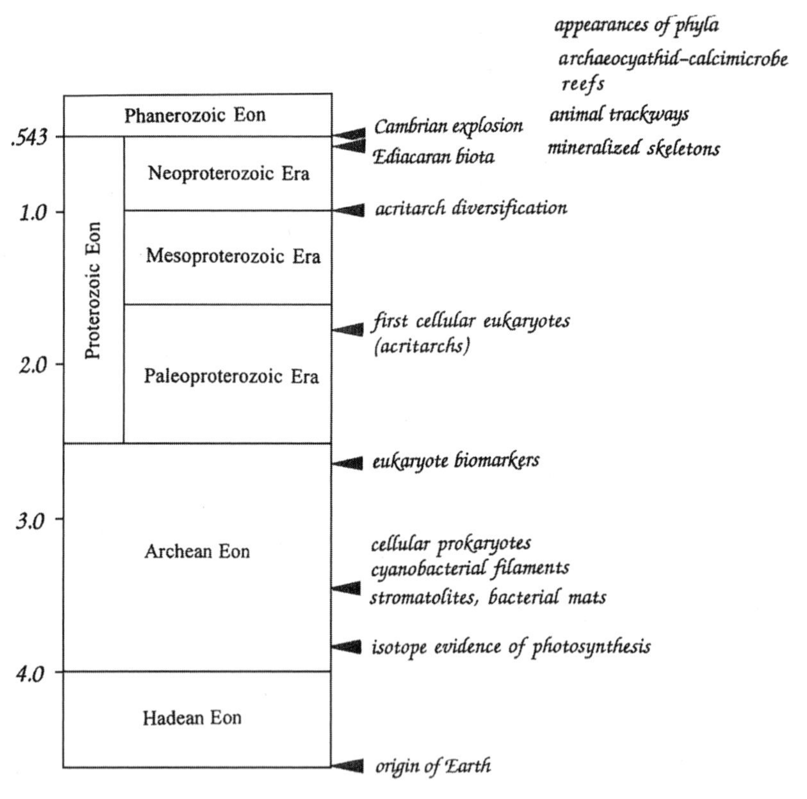

Figure 1.1. Major dated events and first appearances in the Precambrian geologic record. Evidence is based on fossils and sedimentary structures. Chemical and physical data are in the form of isotopes, biomarker molecules, biominerals, sedimentary structures, trace fossils, and body fossils. Hypothetical events, for which there is no physical or chemical evidence, are not included.

billion years (Schopf, 1993), and structural evidence of prokaryotic aggregations in the form of films, mats, and stromatolites dates to 3.5 Ga (Awramik 1984; Schopf 1992). Debates over the biogenicity of the very first putative cellular prokaryotes are peripheral to the central observation that prokaryotic life arose, diversified, and underwent more than 2 billion years of evolutionary change before the emergence of animals. Microbes have been the only form of life on Earth throughout most of its history. Animals first appear in the fossil record approximately 500 million years ago (Ma). By this time, microbes had profoundly modified the physical and chemical environments of the atmosphere, hydrosphere, and lithosphere, establishing the habitats in which multicellular life made its debut. Animals arose in a world teeming with microbes that had already established complex symbiotic interactions (both extra- and intracellular) with one another.

Stromatolites and microbial mat textures are the most important features of Precambrian sedimentary rocks, providing evidence of the interaction of communities with the geosphere, both in terms of sediment trapping and binding, and in terms of biologically induced precipitation of calcium carbonate. Although modern stromatolite ecosystems contain some elements that were not present in the Precambrian, they provide models in both marine settings (Fig. 1.2a) and lakes (Fig. 1.2b) for the study of microbial interactions and collaborations. However, it is the 3.5 billion–year fossil record of stromatolites and microbial mats that holds the key to understanding the history and evolution of these miniature ecosystems as accumulations of successive symbiotic microbial structures.

1.2.2. A Paleo-environmental Perspective

Even though there is abundant and widespread fossil evidence of stromatolitic communities dominated by cyanobacteria by 3 Ga, it is puzzling that the first eukaryotic cells do not appear in the rock record for another billion years. The lag time in the emergence of eukaryotes is less perplexing in light of independent geologic evidence that a stable aerobic global environment was not established until approximately 2 Ga. What happened to the oxygen? The rock record offers a compelling suggestion of a major oxygen sink in the form of all the dissolved iron that accumulated in the oceans during early Earth history. The evidence that oxygen combined with soluble iron is manifest in the banded iron formations (BIFs), massive accumulations of iron oxides that formed during the early part of the Proterozoic. Study of the evolutionary roles of microbes in promoting mineral formation and in establishing control of geochemical transformations and geochemical cycles

Figure 1.2. Living stromatolites provide diverse and complex examples of layered microbial ecosystems that can serve as a basis for study and modeling of microbial symbioses. **a**. Marine stromatolites in Shark Bay (Hamelin Pool), Western Australia. **b**. Non-marine stromatolites in Lake Clifton, Western Australia.

is in its infancy (Newman and Banfield 2002). Although microbial partnerships with the geosphere do not co-evolve in the biological sense of microbial partnerships with multicellular organisms, it may be useful to explore the relationships between these two kinds of interactions.

Land and sea-surface temperatures during most of Earth history have been significantly warmer than they are today. Biogeochemical data as well as molecular genetic data suggest that prokaryotic life arose in the extreme environments of hydrothermal vents (Pace 1997). Thermally altered rocks provide the hard evidence that these are the most ancient ecosystems on Earth, whereas the genomes of extant thermophiles place them at the base of the bacterial phylogenetic tree (Reysenbach and Shock 2002).

At the other extreme of temperature, there were two significant glacial episodes during the Precambrian, the first approximately 2.4–2.2 Ga and the second approximately 820–550 Ma. In the most extreme interpretation of these glaciations, the entire surface of the earth was covered in ice – the "Snowball Earth" hypothesis (Kirschvink 1992; Hoffman et al. 1998). Regardless of the controversy surrounding the extent of extreme environments on early Earth, the evidence is clear that they were more widespread than they have been in the more recent past, existing for long intervals during the emergence and early evolution of microbial ecosystems.

1.2.3. Microbial Diversification

Not only have microbes been evolving for approximately 3.5 billion years, the domains of Bacteria and Archaea have established themselves as the major reservoir of diversity of life on Earth. Prokaryotic life has remained small and relatively lacking in structural and behavioral diversity, but by other measures the diversity far exceeds that of animals. Estimates of taxonomic diversity are currently in the millions (Torsvik et al. 2002), although use of the species as the traditional unit of biodiversity does not work well for prokaryotes. Until recently, bacterial species have been characterized through pure culture in the laboratory. The approximately 4,500 species characterized to date are primarily a reflection of the fact that most microbes cannot be cultured under standard laboratory conditions. Amplified sequences from organisms that are unable to be isolated provide a new measure of diversity and a clear indication that most genetic diversity resides in the microbial world (Pace 1997; Nealson and Stahl 1997).

Even the lack of behavioral diversity of prokaryotes is subject to question. On a fine scale, there is great spatial heterogeneity and complexity within the water column and on the solid surfaces where bacterial films and mats form.

There is growing evidence that bacteria respond rapidly to orient themselves and maintain positions in steep chemical concentration gradients and that in bacterial communities, symbiotic pairs of bacteria can arrange themselves in characteristic geometric configurations (Fenchel 2002). Responses to physical environmental cues (light, mechanical disturbance, gravity, and magnetic fields) further extend the behavioral diversity of microbial life (Fenchel 1987).

Another measure of prokaryotic diversity is the range of "energy substrates" and metabolic pathways that have evolved. Microbial metabolic activity affects the formation and dissolution of minerals, and in the case of the most important chemical species with which they interact, they become the driving force of biogeochemical cycles (Nealson and Stahl 1997). New and unexpected microbial metabolic pathways are still being discovered, especially in extreme environments.

A further measure of the remarkable diversity of prokaryotes is the range of environments they occupy and their sphere of influence. In addition to populating every habitable environment on Earth (Newman and Banfield 2001), they occupy a vastly broader range of the atmosphere, hydrosphere, and lithosphere than do animals. Microbes have changed Earth beyond the boundaries of the known biosphere via processes that include oxygenic photosynthesis, nitrogen fixation, and carbon sequestration, as well as participating in fundamental abiotic interactions that fall within the scope of geomicrobiology.

1.2.4. Microbes and the Origins of the Eukaryotic Cell

There is general agreement for the proposition that eukaryotes arose through symbiotic incorporation of bacterial chloroplasts and mitochondria (Margulis 1970, 1993) and that symbiosis was an important source of evolutionary innovation in the early history of life (Margulis and Fester 1991). The fossil record is, of course, silent on cellular structure and cannot be used to evaluate alternative hypotheses for the order of assembly of elements of the eukaryotic cell. These range from a hypothetical stepwise incorporation of chloroplasts from cyanobacteria and subsequent incorporation of mitochondria from proteobacteria to a simultaneous merger of symbiotic partners in a microbial mat consortium (Nisbet and Sleep 2001). Evidence of eukaryote origins in the geologic record consists of molecular biomarkers and the morphology of cellular microfossils (e.g., cell shapes and colony forms not seen in bacteria). Rocks dated at 2.7 Ga contain molecular signatures of eukaryote metabolism (Brock et al. 1999), but those sterane geochemical biomarkers record only one eukaryotic property – the ability to make sterol compounds. It is not until 1.8–1.5 Ga that large microfossils, called acritarchs, begin to appear in the fossil record. Acritarchs have spiny spore coats and diverse cell shapes, implying the presence of a dynamic internal cytoskeleton. The "big bang" of eukaryote evolution did not begin until 1 Ga (Knoll 2003).

1.2.5. Microbes and the Oldest Animal Fossils – the Ediacaran Biota

The evolutionary transition from unicellular to multicellular life has occurred multiple times, and there are many theoretical considerations of how and why it might have happened. However, there is no evidence for the environments, selective pressures, or pathways involved in this pivotal step in achieving higher orders of organismal complexity and few empirical studies (but see Boraas et al. 1998; Rainey and Rainey 2003).

Explaining the emergence of animals (metazoans) is perhaps the most familiar problem in the set of unicellular–multicellular transitions. Although there are sedimentary structures more than 1 billion years old that have been interpreted as metazoan trace fossils (Seilacher et al. 1998), convincing traces and body fossils do not make their debut until late in the Proterozoic Eon.

The most striking historical evidence of a fundamental change in the shape of life appears worldwide in the rock record at or soon after 600 Ma in the form of diverse body fossils of the Ediacaran biota (Narbonne 1998). The flat, effectively two-dimensional body plan and unique modular construction of many Ediacaran organisms (Fig. 1.3a) have evoked images of "alien beings here on earth" (Lewin 1983). Although they initially were interpreted as animals and allocated to existing metazoan phyla (Glaessner 1984), there have been numerous subsequent interpretations, including an extinct sister group to metazoans (Buss and Seilacher 1994) and a separate kingdom, the Vendozoa (Seilacher 1989) or Vendobionta (Buss and Seilacher 1994).

Regardless of their placement in the tree of life, the Ediacaran organisms raise the possibility of a very interesting initial experiment in multicellular evolution via symbiosis. Were the first "animals" autotrophs, depending on either photosymbiotic or chemosymbiotic microbial symbionts for their nutrition? McMenamin (1986) hypothesized that Ediacaran organisms housed photosymbionts based on their two-dimensional construction and lack of morphology that could be interpreted as mouth, anus, food-collecting structures, or gut. Seilacher (1989) proposed a chemosymbiotic mode of nutrition, characterizing the body plan as a two-dimensional reaction chamber between oxic and anoxic environments. Both hypotheses are difficult, but perhaps not impossible to test. Runnegar (1992) observed that oxic–anoxic

Figure 1.3. Late Precambrian and Early Cambrian earliest animal fossils are associated with ancient microbial mats. **a**. The two-dimensional body plan and modular construction of the enigmatic *Dickinsonia*, Late Precambrian Ediacaran biota of South Australia, is adapted to life on microbial mats and possibly chemosymbiotic nutritional strategy at ancient oxic–anoxic interfaces. **b**. Meandering crawling tracks of an unknown animal in Lower Cambrian rocks in the White-Inyo Mountains, California.

boundaries could well have existed at the sediment–water interface if extensive late Precambrian microbial mats effectively sealed the sediment surface. New studies of enigmatic sedimentary structures in Late Proterozoic rocks show that resistant microbial mats (matground structures) were ubiquitous at the sediment–water interface at this time (Gehling 1999; Hagadorn and Bottjer 1999; Seilacher 1999).

The circumstantial evidence in favor of an initial animal–microbe experiment in symbiosis, producing an ecological assemblage of "autotrophic animals," has an empirical basis in the fossil record and sedimentary record. The hypothesis is testable, and the results are more concrete than the speculative expectation that symbioses must have been involved in the origins of multicellular animal life. The first macroscopic animal life had to adapt to substrates that were structured by layered microbial communities (Hagadorn et al. 1999).

Following the disappearance of the Ediacaran biota, there is evidence of a new, more active lifestyle that also had to adapt to marine substrates dominated and stabilized by microbial mat ecosystems. Lower Cambrian rocks record increasing disturbance of microbial mats in the form of horizontal trackways (Fig. 1.3b) and traces of organisms that did not fossilize and were probably soft-bodied (Bottjer et al. 2000).

1.2.6. Microbes and the Most Ancient Extant Animals – Sponges

Sponges are the basal metazoans in the tree of extant eukaryotes, and their present and past associations with complex microbial communities suggest a different kind of microbial role in the emergence of animals.

Sponges as Three-Dimensional Biofilms

In many respects, sponges represent a transitional grade in the emergence of true multicellularity and have many of the properties of biofilms, which are thin, submillimetric veneers of microbial populations and communities in an adhesive matrix of extracellular polymeric substances (Riding 2000). Because biofilms may be structurally heterogeneous and complexly organized, with conduits for flow of nutrients, oxygen, and metabolites, and because they may involve complex metabolic interactions among constituent bacterial species, they have been compared with metazoan tissue (Costerton et al. 1995). Although sponges lack tissues and organs, they construct a larger three-dimensional environmental matrix, the mesohyl. Within the mesohyl, their own motile cells, along with microbial consortia, cooperate to build a community structure – a mixed population of cells with some of the advantages but none of the vulnerability of a multicellular individual. Although sponges have differentiated cell types and clearly are animals, their microbial symbionts are a prominent part of the body plan. This speculative view of the sponge, as a chimeric multicellular organism, finds support in data from both living sponges and the sponge fossil record.

Microbial Symbioses with Modern Sponges

Modern sponges are well known for housing large, complex, symbiotic microbial communities. Bacteria occur both intracellularly and within the intercellular mesohyl matrix (Wilkinson 1978). Bacteria may account for as much as sixty percent of the sponge biomass (Reitner and Schumann-Kindel 1997) and exceed the concentrations in seawater by several orders of magnitude (Friedrich et al. 2001). Microbes within the mesohyl are physically isolated from seawater by contiguous host membranes. They form communities that are phylogenetically diverse, with a uniform phylogenetic signature that is distinct from that of both marine plankton and marine sediments (Hentschel et al. 2002). Most are uncultured or unculturable and perform postulated symbiotic functions that include nutrient acquisition, stabilization of the sponge skeleton, processing of metabolic waste, and production of useful secondary metabolites. Molecular sequences and phylogenetic analyses have revealed high microbial diversity and have confirmed the presence of at least five major bacterial subdivisions in a single sponge species (Lopez et al. 1999), and at least fourteen monophyletic sequence clusters that are sponge specific (Hentschel et al. 2002).

Sponges may be paraphyletic. One of the major sponge groups, the Calcarea, appears to be more closely allied to cnidarians and lacks the communities of mesohyle bacteria characteristic of Demospongiae and Hexactinellida, the two other major sponge groups (Borchiellini et al. 2001). The symbiotic communities of demosponges are dominated by Eubacteria (Schumann-Kindel et al. 1997) whereas the communities in hexactinellids are dominated by Archaea (Thiel et al. 2002). These sponge–bacterial associations are ancient, since demosponges and hexactinellids are both recognized unequivocally in Cambrian rocks, with strong inference of their presence in rocks of the late Proterozoic age. Consideration of the sponge fossil record and its close association with microbial carbonate rocks (sedimentary rocks produced by bacterially induced precipitation of calcium carbonate) provides a further suggestion of a symbiotic sponge origin.

Microbial Associations with Ancient Sponges

The Precambrian record of putative sponges includes Paleoproterozoic biomarkers considered characteristic of demosponges (Moldowan et al. 1994), Neoproterozoic phosphatized sponge tissue and larvae (Li et al. 1998), and Late Proterozoic spicules and body fossils of demosponge affinities (Reitner and Wörheide 2002). Rigby (1986) noted that most of the Precambrian reports of sponge spicules can be rejected on closer examination.

However, the Phanerozoic fossil record contains a series of major developments in the form ancient reef building ecosystems that suggest an intimate relationship between hypercalcified sponges and microbes. The sponge–microbe connection has not been explored systematically, but a few examples are briefly described to suggest that there is a recurring phenomenon worthy of interdisciplinary investigation.

Early Cambrian archaeocyathid-calcimicrobe reefs represent the first major metazoan hypercalcified reef building in the history of life (Fig. 1.4). The extensive literature devoted to this extinct group over the past 150 years is predominantly taxonomic and devoted to arguing its phylogenetic affinities. They have been treated alternatively as a class of sponges, a separate phylum, and a separate nonmetazoan kingdom (see Rowland 2001 for a review), but

Figure 1.4. Lower Cambrian archaeocyathid reef. The most ancient collaboration in reef building is between an extinct group of sponges or sponge grade organisms and calcimicrobes. Esmerelda County, Nevada.

there is a growing consensus, based on functional and constructional analyses, that they were aspiculate calcareous sponges (Debrenne and Reitner 2001; Debrenne et al. 2002). Archaeocyathids diversified rapidly during the Early Cambrian, formed extensive reef systems, and were effectively extinct by the end of the Early Cambrian. The microbial connection that makes these reefs particularly interesting is that the framework produced by the sponge is relatively open and calcified microbes are more significant in their contribution to the construction of Early Cambrian reefs (Copper 1974; Riding 2001). The repetition of this theme of sponge-calcimicrobe collaborations is particularly striking because different major sponge groups were involved in the collaboration during Phanerozoic history. The ecology of coral reef–bacterial interactions in damage to modern coral reefs is the subject of Chapter 6.

In the Middle and Late Cambrian, following the demise of archaeocyathids as frame-builders, reefs were constructed almost exclusively by microbes, with no significant animal input (Pratt et al. 2001; Riding 2001). However, there are intriguing suggestions of repeated collaborations that developed between calcimicrobes and sponges. Kruse (1999) describes Middle Cambrian reefs in Iran constructed by a consortium of microbes and anthaspidellid sponges. In Silurian reefs in Alaska, there is evidence of the first collaboration between microbially constructed stromatolites and sphinctozoan sponges (Soja 1994). Stromatoporoid sponges (extinct and possibly a polyphyletic grade of sponge construction) were the dominant hypercalcified reef-builders from the Late Silurian to the Middle Devonian (Cook 2002). Late Permian reefs are anomalously structured by a combination of massive inorganic and microbially precipitated calcium carbonate frameworks (Grotzinger and Knoll 1995) in which coralline demosponges were a major component (Wood et al. 1994; Reitner and Wörheide 2002). The dominant sponge taxa in microbial–sponge reefs were not all calcareous, and there are many different relationships between sponge and microbial components. For example, Late Jurassic microbe–sponge reef tracts are dominated by siliceous sponges (Olóriz et al. 2003), and siliceous sponge–microbe associations are a recurrent theme in the Phanerozoic fossil record (Brunton and Dixon 1994).

1.2.7. Microbes and the Cambrian Explosion of Skeletons

The appearance of abundant large body fossils of familiar-looking marine metazoans began 543 Ma corresponding with the beginning of the Cambrian Period and Phanerozoic Eon. Popularly known as the Cambrian "explosion," it is one of the most dramatic features of the fossil record. There is no convincing evidence of macroscopic body fossils prior to the Ediacaran biota (500 Ma), and no evidence at all in support of hypothesized cryptic origins of animals as interstitial (Fortey et al. 1997) or as larval forms (Peterson et al. 1997). Although taxonomic allocation of the Neoproterozoic fossil tissue and "embryos" in the phosphates of the Doushantuo Formation in China (Xiao et al. 1998) is controversial, these fossils provide evidence of multicellularity as far back as 570 Ma (Li et al. 1998). The Doushantuo fossils reinforce the inference that something held the lid on the emergence and diversification of conspicuous multicellular life and that the body plans that appear suddenly in the record 543 Ma reflect a silent record of the evolution of the developmental regulatory networks that direct pattern formation of the major animal groups (Knoll and Carroll 1999).

What exploded in the Cambrian explosion? The difficulty of "explaining" the Cambrian explosion is compounded by the number of distinct changes that appear simultaneously in the shape of life. Signor and Lipps (1992) reviewed 20 hypotheses proposed to "account for the origin and Precambrian–Cambrian radiation of Metazoans." The hypotheses treat a diverse range of theoretical questions ranging from "why" and "where" to questions of what actually proliferated. Was it a taxonomic explosion of animal phyla? A structural (and corresponding functional) explosion of body plans? Or was it primarily an explosion in the number of fossils – an artifact of preservation?

Perhaps the most intriguing element of the Cambrian explosion – the element that implicates bacteria in the origins and evolutionary radiation of animals – is the sudden appearance of skeletons and the origin of diverse modes of biomineralization. Biomineralization in all of the metazoan phyla involves some degree of regulation of crystal nucleation and crystal growth, ranging from weak in "biologically induced" (Lowenstam and Weiner 1983) mineralization to strongly controlled in "organic matrix-mediated" (Lowenstam 1981) mineralization.

Considerable emphasis has been placed on an ecological explanation for the sudden evolutionary innovation of calcified skeletons: the need for protection from predators. Alternative explanations invoking an environmental trigger fail to produce any convincing geological evidence of a corresponding climatic or environmental event. Although both of these "explanations" may be involved, they do not add up to a satisfying reconstruction of evolutionary history. Two lines of evidence suggest a role for microbes. The first is the environment and ecosystem in which the earliest mineralized animal fossils occur. In Late Proterozoic microbial reefs of the Nama Group in Namibia, formed by calcifying cyanobacteria (between 550 and 543 Ma), lightly mineralized tubes of two species of the genus *Cloudina* provide evidence of animal calcification in a microbially dominated environment (Grotzinger et al. 2000). The Nama Group also contains the earliest-known fully biomineralized metazoan, *Namapoikia*, which is reported to attain sizes of up to 1 meter in diameter (Wood et al. 2002).

The association of the earliest mineralized animals with microbially constructed calcium carbonate reefs provides an environmental setting, but the emergence of the biological control of mineralization implies a much deeper origin of mechanisms of intracellular calcium ion control. The idea that calcium physiology predated the origin of multicellularity was proposed originally by Lowenstam and Margulis (1980a, b). Based on the essential roles of intracellular calcium ion regulation in physiology and information transfer in the eukaryotic cell, they argued that biomineralized skeletons were a consequence of the co-opting of machinery that evolved much earlier. How much earlier, and might unicellular eukaryotes have acquired it from an endosymbiotic prokaryote source? We may never know the answer to these questions, but it is plausible given 2 billion years of evolution of microbial physiology.

There are at least seven rare biominerals that are known only in living prokaryotes (Lowenstam and Weiner 1983), and Proterozoic manganese-encrusting bacteria are known from deposits dated at approximately 1.6 Ga (Muir 1978). Pyrite, another biomineral previously known only from bacteria (Lowenstam and Weiner 1983), has been reported recently in the form of elaborate imbricate scales on the foot of a hydrothermal vent gastropod from 2,440 meters in the Indian Ocean (Warén et al. 2003). Within the imbrications, the sclerite surfaces are coated with bacteria, raising the intriguing possibility that they play a symbiotic role in mediating sulfide deposition.

1.2.8. Microbial Interactions with the Dead

Bacteria also "cooperate" with one another and with dead animal hosts in the processes that form the fossil record of life on Earth, and perhaps the fossil record of life elsewhere in the universe. Although bacterial cooperation with the dead is not exactly mutualistic and certainly does not result in coevolution, it is a biogeochemical interaction with some surprisingly important implications for the understanding of the fossil record and the history of life.

Although it seems counterintuitive, microbes may play a positive role in fossilization. Exceptional fossil preservation is frequently mediated by bacteria (Duncan et al. 1998; Harding and Chant 2000). Microbial mats and biofilms control the precipitation of a range of minerals that preserve soft tissues, and paleontologists have become increasingly aware of the importance of the physical and chemical environments in which postmortem bacterially mediated biomineralization occurs (Duane 2003). Early and rapid mineralization is a key to the preservation of soft tissue of both animals (Briggs and Kear 1993) and plants (Dunn et al. 1997).

A broad range of microbes has been implicated in the progressive biomineralization leading to fossilization (Duane 2003), and biostratonomic and early diagenetic systems currently under study, both in the fossil record and in the laboratory, are considered models in the search for extraterrestrial life (Westall 1999).

Microbial interaction with the dead also produces taphonomic feedback in ecosystems. Reefs are perhaps the best example of the constructive effect of this feedback. The massive biomineralized structure of living and fossil reefs, regardless of the animal species responsible for the reef framework, are mineralized primarily by the calcimicrobe consortium that encrusts and binds the reef. The encrusting and binding microbial calcification is often much greater than the metazoan input and involves a range of live–dead constructional interactions.

1.3. THE ROLE OF BACTERIA IN THE EVOLUTIONARY DIVERSIFICATION AND PERSISTENCE OF ANIMALS IN EXTREME ENVIRONMENTS

Mutually beneficial associations between bacteria and animals are common and widespread in the modern world. Some of the most spectacular examples are in extreme environments. Although extremophilic designs for survival are associated with the "toughness, tenacity, and metabolic diversity" of prokaryotes (Nealson and Conrad 1999), and microbes have been characterized as able to "solve every conceivable biochemical problem" (Price 1991), animals show a remarkable ability to prosper when they collaborate with microbes in unusual settings (Hickman 2003). The proposition developed here is that animals and microbes repeatedly have evolved collaborations that extend the survival potential of both partners. Symbioses have enabled certain groups of animals to survive and prosper during times of global biotic crises, to survive mass extinctions, and to repopulate the biosphere during intervals of recovery from mass extinction.

The sponge–microbe example, developed above, illustrates a repeated pattern observed in the recovery of Phanerozoic tropical reef ecosystems following mass extinction events. The prevalence of stromatolites and microbial fabrics in rocks immediately following these events led Schubert and Bottjer (1992) to refer to them as "disaster forms." Although it is possible that the relaxation of normal metazoan-dominated ecology in the wake of massive loss of animal diversity is what permitted microbial communities to expand, it does not follow that microbial life was forced to contract again with the rise of novel metazoan ecosystems without contributing significantly to the recovery. The repeated reformation of sponge–microbe reef ecosystems during the Paleozoic and Mesozoic is consistent with the hypothesis of an essential and positive role for symbioses in ecosystem recovery. The reader is referred to Chapter 6 for a current perspective on reef–microbe interactions.

A second form of evidence for geological persistence of metazoan–microbial symbioses occurs in the Phanerozoic record of extreme environments. Environments with extreme physical and chemical features leave strong signatures in the rock record, reflected in names of sedimentary rock types such as *evaporite* and *microbialite*. Astrobiology and the search for evidence of life elsewhere in the universe have intensified the study of fossil microbial life and microbially generated sedimentary structures, biomarkers, and biogeochemical signals in the rock record. Two ancient environments of particular interest are hypersaline (evaporite) settings and seafloor vents and seeps emitting unusual geochemical fluids (Hickman 2003).

1.3.1. Animal–Microbial Symbioses in Hypersaline Ecosystems: An Example from Shark Bay

Modern evaporite systems include salt marshes, hypersaline estuaries, anchialine ponds, silled basins, and littoral lagoons. These systems are particularly common on the fringes of arid coastlines, where evaporation exceeds precipitation and runoff. Hypersaline environments have occupied vast regions of Earth at times in the geologic past and are relatively uncommon today. The rocks that form in these systems not only have distinctive physical and chemical features, they also have distinctive microbially mediated sedimentary features and distinctive fossil biotas.

Molluscs, particularly bivalves, offer some of the most dramatic examples of exuberant success in hypersaline ecosystems, where they dominate in terms of abundance, density, biomass, and productivity. Although Shark Bay in Western Australia is best known for the spectacular development of microbial mats and stromatolites (Fig. 1.2a), it also is home to dense populations of an extraordinary small cockle, fueled nutritionally by its photosymbiotic microbes (Fig. 1.5).

Fragum erugatum occurs only in the hypersaline reaches of Shark Bay (Slack-Smith 1990). It was first known from the vast beaches and beach ridges (up to 1 kilometer wide and 4 meters deep in some places) of pure *Fragum* shells along the shores of the bay. The shell beaches are one of the tourist

Figure 1.5. Living mat of the photosymbiotic bivalve *Fragum erugatum*, Shark Bay, Western Australia.

attractions of Shark Bay as well as a phenomenon of considerable geological interest (Berry and Playford 1997, 1998).

What supports this incredible productivity? Shell production alone is estimated to be on the order of 50 grams per square meter per year (Berry and Playford 1998). The living animals have provided some interesting answers. They occur at densities of 4,000 per square meter (Morton 2000), reaching even greater densities (Hickman 2003) where individuals are massed together in thick mucous-bound mats of live animals and dead shells (Fig. 1.5). Animals open their shells to expose mantle tissue containing dense populations of photosynthetic zooxanthellae (Berry and Playford 1997). Zooxanthellae in the mantle also are exposed to sunlight that passes through the thin, translucent shell, and Morton (2000) reports dense concentrations of zooxanthellae in the gills, which can be exposed to sunlight through the siphons. Although the symbiont is a unicellular eukaryote, *Symbiodinium*, its light-harvesting genes are bacterial in origin and the mode of energy acquisition is driven by ancient metabolic machinery.

The nutritional advantages of this symbiosis are obvious in the case of the bivalve. The considerable advantages to the microbial partner are revealed in the evolutionary modifications of structure and behavior of the bivalve host. *Fragum erugatum* is not a typical sedentary infaunal bivalve. The *Fragum* mat is in constant motion, with individuals moving between exposed positions in the surface of the mat to deeper in the mat by thrusting and probing of the long, muscular foot. Through this behavior, an animal can adjust the exposure of its symbionts to sunlight. Adjustment also occurs by regulating the degree of gape (opening) of the shell valves and by contracting or expanding the exposed mantle tissue.

In addition to physical mechanisms for providing symbionts with exposure to ambient illumination, some bivalves have developed light-filtering pigments that can modify the quality of light reaching their symbionts (Isaji et al. 2001). *Fragum erugatum* also may protect its symbionts from harmful ultraviolet radiation with mycosporine-like amino acids (MAAs), which absorb UV-A and -B light. In an analysis of mantle tissue of zooxanthellate bivalves that included one species of *Fragum*, Ishikura et al. (1997) found significant amounts of MAAs in the outermost layer of tissue where the symbionts were concentrated. They found no MAAs in symbionts isolated from mantle tissue. While UV-B irradiation of the clam did not suppress photosynthesis, it was suppressed in isolated symbionts.

1.3.2. Animal–Microbe Collaborations at Hydrothermal Vents and Cold Seeps

The discovery of hydrothermal vent and cold seep communities of invertebrates fueled by chemosynthesis was one of the most exciting biological discoveries of the twentieth century. The geologic record of vents and seeps and the fossil record of animal–bacterial chemosymbioses document the antiquity and deep roots of these collaborations. Hydrothermal vents are the oldest known ecosystems on earth (Reysenbach and Shock 2002), influencing both geochemical and prokaryote evolution for more than 3 billion years. Modern vent communities are postulated refugia for a fauna that has evolved in isolation from other marine faunas since the dawn of animal life (Tunnicliffe 1992). Many indirect lines of evidence point to the antiquity of chemosymbioses, including the obligacy of the relationships and the intracellular housing of the bacteria, often involving modifications of host morphology, physiology, and behavior suggestive of a long history of selection (Distel 1998).

Hydrothermal vents and seeps emitting chemically hostile fluids have been more common and widespread at other times in Earth history. Intensified interest in these environments has led to reinterpretation of many large-scale geologic features as the products of chemosynthesis. These include Jurassic pseudobioherms in southern France (Gaillard et al. 1972),

Carboniferous carbonate mounds in Newfoundland (von Bitter et al. 1992), Cretaceous carbonate mounds in the Canadian Arctic (Beauchamp and Savard 1992), and the Cretaceous Tepee Buttes in Colorado (Kauffman et al. 1996). The Tepee Buttes occur over a striking spatial scale, with hundreds of carbonate mounds aligned along a Laramide fracture zone, where fluid venting occurred for more than 1 million years Chemosymbiotic bivalves are a dominant component of sharply zoned biofaces of the Tepee Buttes.

Although many chemosymbiotic lineages have long geologic histories, there also is evidence that the composition and dominance of animals in ancient seep-and-vent communities have changed over the past 600 million years For example, Paleozoic seep-and-vent faunas were dominated by brachiopods rather than bivalves (Campbell and Botttjer 1995), and other Palaeozoic vent systems preserve mounds with bryozoan–microbial associations in which microbial activity had a strong influence on bryozoan development and colony form (Morris et al. 2002).

1.3.3. Collaborations between Lucinid Bivalves and Chemosymbiotic Bacteria

Lucinid bivalve species form dense populations in a variety of physiologically hostile marine environments where there are few other metazoan species (Hickman 1994). These environments include seafloor vents discharging fluids of unusual physicochemical compositions, including petroleum, sulfide, methane, and brine seeps. Lucinids also may be abundant in sulfide-rich anoxic sediments beneath sea-grass beds and at sewage outfalls, where their productivity is linked to a microbial symbiosis. In this instance, the symbionts are harnessing chemical energy rather than light energy. The enlarged gills of all lucinids studied to date are packed with specialized cells (bacteriocytes) containing chemolithotrophic bacteria that use sulfide as an energy substrate. In each case, the bivalves live in transition zones at oxic–anoxic boundaries.

Lucinids are not the only bivalves with chemosymbionts. Chemosymbiosis is documented in five bivalve families (Distel 1998), representing at least five independent evolutionary radiations (Barnes and Hickman 2001). Lucinids are developed as an example because approximately half of the described bivalve species with chemoautotrophic endosymbionts are lucinids (Fisher 1990) and the relationship has been studied most intensively and comparatively in this family.

As in the symbiotic association of *Fragum erugatum* and its photosymbionts, the nutritional advantage to the lucinid host is relatively obvious. Nutritional benefit is documented by at least eight independent lines of evidence (see Hickman 1994 for a summary and references). Although it is not clear that they receive all of their nutrition from their symbionts, the lucinid digestive system is reduced in size and complexity. It is the benefit to the symbionts that requires explanation, and again it is postulated that the large animal host extends microbial capability to bridge larger interfaces.

Lucinid bivalves position themselves at oxic–anoxic boundaries, where they obtain oxygen via an inhalant siphon from oxygenated water above the sediment–water interface. At the same time, they tap reduced sulfur from underlying anoxic substrates via a long probing foot.

The hypothesis that chemosymbiotic associations span interfaces is particularly attractive in view of the nature of the steep physical and chemical gradients that occur at hydrothermal vents and cold seeps. Microbial life is known to concentrate at physical and chemical interfaces, where energy and nutrients fluctuate most dramatically (Ash et al. 2002). Microbiologists have emphasized the importance of motility and chemosensory adaptations as keys to success in a heterogeneous world in which microbes track optimal chemical and physical conditions (Fenchel 2002). However, for a very small organism, maintaining an optimal position in a spatially or temporally fluctuating physical or chemical gradient is no trivial matter.

Symbioses that bridge interfaces behaviorally may be more common and widespread in eukaryotic organisms that are covered in bacterial overcoats. Ott et al. (1998) described an extraordinary symbiosis between a macroscopic, sessile, colonial ciliate and the sulfur-oxidizing bacteria that encase the exterior of the colony. The ciliate lives in steep sulfide–oxygen gradients at the surface of mangrove peat, where the colonies alternately expand to expose their bacteria to oxygenated water and contract into anoxic sulfidic water. This ingenious spatial and temporal decoupling of sulfide and oxygen uptake is not restricted to sessile organisms. In other gradient environments, Polz et al. (1994) described nematodes encased in epigrowths of filamentous bacteria that migrate up and down an oxygen–sulfide gradient at an oxic–anoxic interface in sediment. On a larger spatial scale, Polz et al. (1999) described aggregations of shrimp with sulfur-oxidizing epibacteria that swim in and out of sulfide plumes at a hydrothermal vent site. The probability of finding additional examples of behavioral adaptations for bridging environmental gradients is good and can be directed by knowledge of where such gradient environments are likely to exist.

1.4. THE ROLE OF BACTERIA IN BIOSPHERE RECOVERY FROM MASS EXTINCTIONS

Does animal life reorganize after global disasters with help from microbial friends? During episodes of severe global stress and mass extinction, major ecosystems repeatedly have collapsed, and the highly successful animal groups that dominated them have been extirpated. The causes of mass extinction have been much more thoroughly investigated than the recovery, reorganization, and repopulation of the biosphere following these events.

As noted in the previous section, biofilms, microbial mats, stromatolites, and sponges are often the first forms of life to appear in the rocks following a mass extinction event. As ancient grades of symbiotic organization, they seem to fit the characterization of "disaster forms" (Schubert and Bottjer 1992) as long-ranging opportunistic taxa that owe their geologic longevity to periodic retreats into refugial environments, expanding only during times of ecological relaxation in the wake of mass extinctions. Microbial carbonate sequences were particularly abundant following the great mass extinction at the end of the Permian Period (Ezaki et al. 2003). Rocks deposited during intervals following major mass extinctions typically contain few fossils. However, the forms that are present are often unusual, abundant, and widespread. They disappear again after a low-diversity recovery interval in which physically or chemically abnormal oceanic conditions may have persisted.

What is wrong with this picture of biosphere recovery? It assumes that entire microbial ecosystems and their component species advanced into empty ecospace and then retreated to their refugia without participating in the evolution of new metazoan taxa and ecosystems. Should we not expect that microbes would have evolved and assembled into novel consortia as they expanded into stressed or chemically unusual postcrisis environments? And should we not expect that microbes would have established novel symbiotic relationships with metazoans during recovery intervals? It may be particularly productive to reexamine the fossil record of the unusual metazoan taxa that exploded during recovery intervals. Do they represent experiments in cooperation with microbes? For example, reefs during the early Jurassic interval following the end of Triassic mass extinction were constructed by lithiotids, a short-lived and bizarre group of bivalves with features suggesting the presence of photosynthetic symbionts (Fraser et al. 2004).

Microbial carbonates in the geologic record are the mineralized product of microstructurally and metabolically complex ecosystems. They are more than relicts of a premetazoan biosphere. It is difficult to imagine the reestablishment of metazoan ecosystems in a microbial milieu without evolving novel forms of "détente among the domains" (*sensu* McFall-Ngai 1998) and exploring new stable and beneficial associations.

1.5. THE ROLE OF SYMBIOTIC BACTERIA IN THE EVOLUTION OF ANIMAL DEVELOPMENT

One of the most exciting questions concerning co-evolved animal–bacterial partnerships is their influence on animal developmental programs (McFall-Ngai 2002). To what extent have bacteria served as selective pressures that have modified animal tissues and created specialized cell types and novel organs? Although the fossil record is not likely to serve as a primary source of data for exploring these questions, it is the only historical record of what has actually happened. Hypotheses can be tested with morphological data from fossils and stratigraphic data identifying the times of first appearance of taxa and structures. Some hypotheses are tested most effectively when biological and geological data are combined.

1.5.1. Did the Bivalve Gill Originate as an Organ Shaped by Intracellular Chemosynthetic Bacterial Symbionts?

Distel (1998) proposed that chemosymbiosis may have played an important role in the early diversification of Bivalvia. Chemosymbiotic bivalve lineages are ancient. Although there is no record of the symbionts, the morphological features of Early Paleozoic lucinid and solemyid bivalves and their relationship to enclosing sediments are interpreted as adaptations for chemosymbiosis in their living relatives.

The characteristic molluscan gill or ctenidium is a unique molluscan feature that is reputed to have evolved as a respiratory organ and only secondarily co-opted for ciliary suspension feeding. However, the architecture of the ctenidium also meets an engineering paradigm for housing chemosymbiotic bacteria. Subfilamentar gill tissue is an ideal location for the specialized bacteriocyte cells of chemosymbiotic bivalves. Cells are directly exposed both to seawater as a source of dissolved nutrients and oxygen, and to the blood supply of the host that delivers sulfide. The gills of chemosymbiotic bivalves tend to be hypertrophied, but is the large gill a primitive rather than a derived state? Or is gill tissue easily invaded and easily modified by bacteria?

The same questions may be asked of the tissue of the mantle epithelium. Is mantle hypertrophy easily achieved by manipulative photosynthetic symbionts? In the case of molluscan taxa such as *Fragum* and *Tridacna*, the

Symbiodinium symbionts are housed extracellularly in the blood sinuses of the mantle. What makes certain tissues more susceptible than others?

1.6. CONCLUSIONS

Data from geology, the fossil record, and living marine invertebrates provide an evolutionary perspective on the direct and indirect benefits of microbial interactions with animals. Propositions that are consistent with multiple lines of evidence and that merit continued investigation include:

1. Microbial ecosystems formed more than 3 billion years ago and evolved metabolic collaborations for more than 2 billion years prior to the dawn of animal life.
2. Microbial interactions with the geosphere prepared and shaped the ancient biosphere in which multicellularity and animal life emerged.
3. Microbes contributed genes that continue to serve animals in their ancestral roles of harvesting energy as well as in co-opted roles that range from cell signaling and adhesion to building mineralized skeletons.
4. Microbial associations and symbioses with animals in extreme environments (hydrothermal vents, cold seeps, and hypersaline systems) are ancient. Symbioses provide unusual benefits to both host and symbiont, promoting their macroevolutionary persistence.
5. Microbial communities and animals with microbial symbionts have played an important role in recovery of the biosphere following the global crises and periodic mass extinction events that have extirpated major groups of organisms and entire ecosystems.
6. Microbial symbionts have developmentally and evolutionarily modified the tissues and organs of their diverse marine invertebrate hosts and have played a role in the origin of novel organs in some invertebrates.

Understanding the evolution of beneficial interactions between microbes and animals requires greater exchange among geologists, paleontologists, and biologists. The discoveries of “unexplored microbial worlds” (Hagadorn et al. 1999) in the geologic record and discoveries of unexpected microbial metabolisms and molecular-scale interactions of microbes with the geosphere (Newman and Banfield 2002) have galvanized new fields of microbial paleontology and geomicrobiology. Paleobiology has recognized that chemosymbiotic relationships are so deeply rooted in Earth history as to have played a role in the origins of evolutionary novelty, and perhaps the origin of life itself (Beauchamp and von Bitter 1992). The field of astrobiology has encouraged scientists from disparate disciplines to explore the potential for life elsewhere in the universe, and this effort extends beyond microbial life to the potential for evolutionary increases in properties such as size, complexity, and diversity. This is a good time, technologically, to develop a new understanding of the positive interactions of microbes and microbial consortia with animals in both ecological and evolutionary time.

ACKNOWLEDGMENTS

I am grateful to the conference organizers, the Rockefeller Foundation, and the National Science Foundation for the opportunity to develop a paleobiological and evolutionary perspective on beneficial animal–microbe partnerships. The National Science Foundation supported my earlier field research in Australia.

REFERENCES

Ash, C., Hanson, B., and Norman, C. (2002). Earth, air, fire, and water. *Science* **296**, 1055.

Awramik, S. M. (1984). Ancient stromatolites and microbial mats. In *Microbial Mats: Stromatolites,* ed. Y. Cohen, R.W. Castenholtz, and H. O. Halvorson, pp. 1–22. New York: Alan R. Liss.

Barnes, P. A. G. and Hickman, C. S. (2001). Biogeography and ecology of shallow-water chemoautotrophic bivalves. In *Abstracts, World Congress of Malacology 2001,* ed. L. Salvini-Plawen, J. Voltzow, H. Sattmann, and G. Steiner, p. 22. Vienna: Unitas Malacologica.

Beauchamp, B. and Savard, M. (1992). Cretaceous chemosynthetic carbonate mounds in the Canadian Arctic. *Palaios* **7**, 434–450.

Beauchamp, B. and von Bitter, P. (1992). Chemo what? *Palaios* **7**, 337.

Berry, P. F. and Playford, P. E. (1997). Biology of modern *Fragum erugatum* (Mollusca, Bivalvia, Cardiidae) in relation to deposition of the Hamelin Coquina, Shark Bay, Western Australia. *Marine and Freshwater Research* **48**, 415–420.

Berry, P. F. and Playford, P. (1998). Unearthing shell beach secrets. *Landscope* Spring, 49–52.

Boraas, M. E., Seale, D. B., and Boxhorn, J. E. (1998). Phagotrophy by a flagellate selects for colonial prey: A possible origin of multicellularity. *Evolutionary Ecology* **12**, 153–164.

Borchiellini, C., Manuel, M, Alivon, E., Boury-Esnault, N., Vachelet, J., and Le Parco, Y. (2001). Sponge paraphyly and the origin of Metazoa. *Journal of Evolutionary Biology* **14**, 171–179.

Bottjer, D. J., Hagadorn, J. W., and Dornbois, S. Q. (2000). The Cambrian substrate revolution. *GSA Today* **10**, 1–7.

Briggs, D. E. G. and Kear, A. J. (1993). Fossilization of soft tissue in the laboratory. *Science* **259**, 1439–1442.

Brock, J. J., Logan, G. A., Buick, R., and Summons, R. E. (1999). Archean molecular fossils and the rise of Eukaryotes. *Science* **285**, 1033–1036.

Brunton, F. R. and Dixon, O. A. (1994). Siliceous sponge-microbe biotic associations and their recurrence through the Phanerozoic as reef mound constructors. *Palaios* **9**, 370–387.

Buss, L. W. and Seilacher, A. (1994). The phylum Vendobionta: a sister group of the Eumetazoa? *Paleobiology* **20**, 1–4.

Cady, S. (1998). Astrobiology: A new frontier for 21st century paleontologists. *Palaios* **13**, 95–97.

Campbell, K. A. and Bottjer, D. J. (1995). Brachiopods and chemosymbiotic bivalves in Phanerozoic hydrothermal vent and cold seep environments. *Geology* **23**, 321–324.

Cook, A. (2002). 'Class Stromatoporoidea' Nicholson & Murie, 1878: Stromatoporoids. In *Systema Porifera: A Guide to the Classification of Sponges*, ed. J. N. A. Hooper, and R. W. M. Van Soest, pp. 69–70. New York: Kluwer Academic/Plenum Publishers.

Copper, P. (1974). Structure and development of Early Paleozoic Reefs. *Proceedings of the Second International Coral Reef Symposium* **1**, 746–754.

Costerton, J. W., Lewandowski, Z., Caldwell, D. E., Korber, D. R., and Lappin-Scott, H. M. (1995). Microbial biofilms. *Annual Review of Microbiology* **49**, 711–745.

Debrenne, F. and Reitner, J. (2001). Sponges, cnidarians, and cetnophores. In *The Ecology of the Cambrian Radiation*, ed. A. Yu. Zhuravlev and R. Riding, pp. 301–325. New York: Columbia University Press.

Debrenne, F., Zhuravlev, A. Yu., and Kruse, P. D. (2002). Class Archaeocyathida Bornemann. In *Systema Porifera: A Guide to the Classificationn of Sponges*, ed. J. N. A. Hooper and R. W. M. Van Soest, pp. 1539–1592. New York: Kluwer Academic/Plenum Publishers.

Distel, D. L. (1998). Evolution of chemoautotrophic endosymbioses in bivalves. *BioScience* **48**, 277–286.

Duane, M. J. (2003). Unusual preservation of crustaceans and microbial colonies in a vadose zone, northwest Morocco. *Lethaia* **36**, 21–32.

Duncan, I. J., Briggs, D. E. G., and Archer, M. (1998). Three-dimensionally mineralized insects and millipedes from the Tertiary of Riversleigh, Queensland, Australia. *Palaeontology* **41**, 835–851.

Dunn, K. A., McLean, R. J. C., Upchurch, G. R., and Folk. R. L. (1997). Enhancement of leaf fossilization potential by bacterial biofilms. *Geology* **25**, 1119–1122.

Ezaki, Y., Jianbo, L., and Adachi, N. (2003). Earliest Triassic microbialite micro- to megastructures in the Huaying area of Sichuan Province, South China: Implications for the nature of oceanic conditions after the end-Permian extinction. *Palaios* **18**, 388–402.

Falkowski, P. G. and Raven, J. A. (1997). *Aquatic Photosynthesis.* Malden, MA: Blackwell Science.

Fenchel, T. (1987). *The Ecology of Protozoa.* Berlin: Springer-Verlag.

Fenchel, T. (2002). Microbial behavior in a heterogeneous world. *Science* **296**, 1068–1070.

Fisher, C. R. (1990). Chemoautotrophic and methanotrophic symbioses in marine invertebrates. *Reviews in Aquatic Sciences* **2**, 339–436.

Fortey, R. F., Briggs, D. E. G., and Wills, M. A. (1997). The Cambrian evolutionary "explosion" recalibrated. *BioEssays* **19**, 429–434.

Fraser, N. M., Bottjer, D. J., and Fischer, A. G. (2004). Dissecting "*Lithiotis*" bivalves: implications for the early Jurassic reef eclipse. *Palaios* **19**, 51–67.

Friedrich, A. B., Fischer, I., Proksch, P., Hacker, J., and Hentschel, U. (2001). Temporal variation of the microbial community associated with the Mediterranean sponge *Aplysina aerophoba. FEMS Microbiology Ecology* **38**, 105–113.

Galliard, C., Rio, M., Rolin, Y., and Roux, M. (1992). Fossil chemosynthetic communities related to vents or seeps in sedimentary basins: the pseudobioherms of southeastern France compared with other world examples. *Palaios* **7**, 451–465.

Gehling, J. G. (1999). Microbial mats in terminal Proterozoic siliciclastics: Ediacaran death masks. *Palaios* **14**, 40–57.

Glaessner, M. F. (1984). *The Dawn of Animal Life.* Cambridge: Cambridge University Press.

Grotzinger, J. P. and Knoll, A. H. (1995). Anomalous carbonate precipitates: Is the Precambrian the key to the Permian? *Palaios* **10**, 578–596.

Grotzinger, J. P., Watters, W. A., and Knoll, A. H. (2000). Calcified metazoans in thrombolite-stromatolite reefs of the terminal Proterozoic Nama Group, Namibia. *Paleobiology* **26**, 334–359.

Hagadorn, J. W. and Bottjer, D. J. (1999). Restriction of a Late Neoproterozoic biotope: Suspect-microbial structures and trace fossils at the Vendian-Cambrian transition. *Palaios* **14**, 73–85.

Hagadorn, J. W., Pflüger, F., and Bottjer, D. J. (1999). Unexplored microbial worlds. *Palaios* **14**, 1–2.

Harding, I. C. and Chant, L. S. (2000). Self-sedimented diatom mats as agents of exceptional fossil preservation in the Oligocene Florissant lake beds, Colorado, United States. *Geology* **28**, 195–198.

Hentschel, U., Hopke, J., Horn, M., Friedrich, A. B., Wagner, M., Hacker, J., and Moore, B. S. (2002). Molecular evidence for a uniform microbial community in sponges from different oceans. *Applied Environmental Microbiology* **68**, 4431–4440.

Hickman, C. S. (1994). The genus *Parvilucina* in the Eastern Pacific: Making evolutionary sense of a chemosymbiotic species complex. *The Veliger* **37**, 43–61.

Hickman, C. S. (2003). Mollusc–microbe mutualisms extend the potential for life in hypersaline systems. *Astrobiology* **3**, 631–644.

Hoffman, P. F., Kaufman, A. J., Halverson, G. P., and Schrag, D. P. (1998). A Neoproterozoic snowball earth. *Science* **281**, 1342–1346.

Isaji, S., Ohno, T., and Nishi, E. (2001). Fine structure and distribution of iridophores in the photosymbiotic bivalve subfamily Fraginae (Cardioidea). *The Veliger* **44**, 54–65.

Ishikura, M., Kato, C., and Maruyama, T. (1997). UV-absorbing substances in zooxanthellate and azooxanthellate clams. *Marine Biology* **128**, 649–655.

Kauffman, E. G., Arthur, M. A., Howe, B., and Scholle, P. A. (1996). Widespread venting of methane-rich fluids in Late Cretaceous (Campanian) submarine springs (Tepee Buttes), Western Interior seaway, USA *Geology* **24**, 799–802.

Kirschvink, J. (1992). Late Proterozoic low latitude glaciation: The snowball Earth. In *The Proterozoic Biosphere: A Multidisciplinary Study*, ed. J. W. Schopf and C. Cline, pp. 51–52. Cambridge: Cambridge University Press.

Knoll, A. H. (2003). *Life on a Young Planet*. Princeton, NJ: Princeton University Press.

Knoll, A. H. and Carroll, S. B. (1999). Early animal evolution: Emerging views from comparative biology and geology. *Science* **284**, 2129–2137.

Kruse, P. D. (1999). Distinctive Middle Cambrian sponge-calcimicrobe reefs in Iran. *Memoirs of the Queensland Museum* **44**, 298.

Lewin, R. (1983). Alien beings here on Earth? *Science* **223**, 39.

Li, C. W., Chen, J. Y., and Hua, T. E. (1998). Precambrian sponges with cellular structures. *Science* **279**, 879–882.

Lopez, J. V., McCarthy, P. J., Janda, K. E., Willoughby, R., and Pomponi, S. A. (1999). Molecular techniques reveal wide phyletic diversity of heterotrophic microbes associated with *Discodermia* spp. (Porifera: Demospongiae). *Memoirs of the Queensland Museum* **44**, 329–341.

Lowenstam, H. A. (1981). Minerals formed by organisms. *Science* **211**, 1126–1131.

Lowenstam, H. A. and Margulis, L. (1980a). Calcium regulation and the appearance of calcareous skeletons in the fossil record. In *The Mechanisms of Biomineralization in Animals and Plants*, ed. M. Omori and N. Watabe, pp. 289–300. Tokyo, Tokai University Press.

Lowenstam, H. A. and Margulis, L. (1980b). Evolutionary prerequisites for early Phanerozoic calcareous skeletons. *BioSystems* **121**, 27–41.

Lowenstam, H. A. and Weiner, S. (1983). Mineralization by organisms and the evolution of biomineralization. In *Biomineralization and Biological Metal Accumulation*, ed. P. Westbroek and E. W. de Jong, pp. 191–203. Dordrecht, Holland: Riedel.

Margulis, L. (1970). *Origin of Eukaryotic Cells.* New Haven, CT: Yale University Press.

Margulis L. (1993). *Symbiosis in Cell Evolution.* 2nd ed. New York: W. H. Freeman.

Margulis, L. and Fester, R. (eds.) (1991). *Symbiosis as a Source of Evolutionary Innovation.* Cambridge, MA: MIT Press.

McFall-Ngai, M. J. (1998). The development of cooperative associations between animals and bacteria: Establishing détente among domains. *American Zoologist* **38**, 593–608.

McFall-Ngai, M. J. (2002). Unseen forces: The influence of bacteria on animal development. *Developmental Biology* **242**, 1–14.

McMenamin, M. A. S. (1986). The garden of Ediacara. *Palaios* **1**, 178–182.

Mojzsis, S. J., Arrhenius, G., McKeegan, K. D., Harrison, T. M., Nutman, A. P., and Friend, C. R. L. (1996). Evidence for life on Earth before 3,800 million years ago. *Nature* **384**, 55–59.

Moldowan, J. M., Dahl, J., Jacobson, S. R., Huizinga, B. J., McCaffrey, M. A., and Summons, R. E. (1994). Molecular fossil evidence for late Proterozoic–Early Paleozoic environments. *Terra Nova* **6**, 4.

Morris, P. A., von Bitter, P. H., Schenk, P. E., and Wentworth, S. J. (2002). Interactions of bryozoans and microbes in a chemosynthetic hydrothermal vent system: Big Cove Formation (Lower Codroy Group, Lower Carboniferous, Middle Viséan/Arundian), Port au Port Peninsula, West Newfoundland, Canada. In *Bryozoan Studies 2001*, ed. W. Jackson, B. Jones, and S. Jones, pp. 221–227. Lisse, Netherlands: Swets & Zeitlinger.

Morton, B. (2000). The biology and functional morphology of *Fragum erugatum* (Bivalvia: Cardiidae) from Shark Bay, Western Australia: the significance of its relationship with entrained zooxanthellae. *Journal of Zoology, London* **251**, 39–52.

Muir, M. D. (1978). Microenvironments of some modern and fossil iron- and manganese-oxidizing bacteria. In *Environmental Biogeochemistry and*

Geomicrobiology, ed. W. E. Krumbein, pp. 937–944. Ann Arbor, MI: Ann Arbor Scientific Publications.

Narbonne, G. M. (1998). The Edicaran biota: A terminal Neoproterozoic experiment in the evolution of life. *GSA Today* **8**, 1–6.

Nealson, K. H. and Conrad, P. G. (1999). Life: past, present and future. *Philosophical Transactions of the Royal Society of London B*, **354**, 1923–1939.

Nealson, K. H. and Stahl, D. A. (1997). Microorganisms and biogeochemical cycles: what can we learn from layered microbial communities? In *Geomicrobiology: Interactions between Microbes and Minerals*, ed. J. F. Banfield and K. H. Nealson, pp. 5–34. Washington, DC: Mineralogical Society of America.

Newman, D. K. and Banfield, J. F. (2002). Geomicrobiology: How molecular-scale interactions underpin biogeochemical systems. Science **296**, 1071–1077.

Nisbet, E. G. and Sleep, N. H. (2001). The habitat and nature of early life. *Nature* **409**, 1083–1091.

Olóriz, F., Reolid, M., and Rodríguez-Tovar, F. J. (2003). A Late Jurassic carbonate ramp colonized by sponges and benthic microbial communities (External Prebetic, Southern Spain). *Palaios* **18**, 528–545.

Ott, J. A., Bright, M., and Schiemer, F. (1998). The ecology of a novel symbiosis between a marine peritrich ciliate and chemoautotrophic bacteria. *P.S.Z.N.: Marine Ecology* **19**, 229–243.

Pace, N. R. (1997). A molecular view of microbial diversity and the biosphere. *Science* **276**, 734–740.

Peterson, K. J., Cameron, R. A., and Davidson, E. H. (1997). Set-aside cells in maximal indirect development: Evolutionary and developmental significance. *BioEssays* **19**, 623–631.

Polz, M. F., Distel, D. L., Zarda, B., Amann, R., Felbeck, H., Ott, J. A., and Cavanaugh, C. M. (1994). A highly specific association between ectosymbiotic, sulphur-oxidizing bacteria and a marine nematode and its phylogenetic relationship to endosymbionts and free-living bacteria. *Applied Environmental Microbiology* **60**, 4461–4467.

Polz, M. F., Robinson, J. J., Cavanaugh, C. M., and Van Dover, C. L. (1999). Trophic ecology of massive shrimp aggregations at a Mid-Atlantic Ridge hydrothermal vent site. *Limnology and Oceanography* **43**, 1631–1638.

Pratt, B. B., Spincer, B. R., Wood, R. A., and Zhuravlev, A. Yu. (2001). Ecology and evolution of Cambrian reefs. In *The Ecology of the Cambrian Radiation*, ed. A. Yu. Zhuravlev, and R. Riding, pp. 254–274. New York: Columbia University Press,

Price, P. W. (1991). The web of life: development over 3.8 billion years of trophic relationships. In *Symbiosis as a Source of Evolutionary Innovation*, ed. L. Margulis and R. Fester, pp. 262–287. Cambridge, MA: MIT Press.

Rainey, P. B. and Rainey, K. (2003). Evolution of cooperation and conflict in experimental bacterial populations. *Nature* **425**, 72–74.

Reitner, J. and Schumann-Kindel, G. (1997). Pyrite in mineralized sponge tissue–product of sulfate reducing sponge related bacteria? *Facies* **36**, 272–276.

Reitner, J. and Wörheide, G. (2002). Non-lithistid fossil Demospongiae–origins of their palaeobiodiversity and highlights in history of preservation. In *Systema Porifera: A Guide to the Classification of Sponges*, ed. J. N. A. Hooper and R. W. M. Van Soest, pp. 52–68. New York: Kluwer Academic/ Plenum.

Reysenbach, A. L. and Shock, E. (2002). Merging genomes with geochemistry in hydrothermal ecosystems. *Science* **296**, 1077–1082.

Riding, R. (2000). Microbial carbonates: the geological record of calcified bacterial–algal mats and biofilms. *Sedimentology* **47** (Suppl. 1), 179–214.

Riding, R. (2001). Calcified algae and bacteria. In *The Ecology of the Cambrian Radiation*, ed. A. Yu. Zhuravlev and R. Riding, pp. 445–473. New York: Columbia University Press.

Rigby, J. K. (1986). Sponges of the Burgess Shale (Middle Cambrian), British Columbia. *Palaeontographica Canadiana* **2**, 1–105.

Rosing, M. T. (1999). C-13-depleted carbon microparticles in >3700-Ma sea-floor sedimentary rocks from west Greenland. *Science* **283**, 674–676.

Rowland, S. M. (2001). Archaeocyaths – a history of phylogenetic interpretation. *Journal of Paleontology* **75**, 1065–1078.

Ruby, E., Henderson, B., and McFall-Ngai, M. 2004. We get by with a little help from our (little) friends. *Science* **303**, 1305–1307.

Runnegar, B. (1992). Evolution of the earliest animals. In *Major Events in the History of Life*, ed. J. W. Schopf, pp. 65–93. Boston: Jones and Bartlett.

Schopf, J. W. (1992). Paleobiology of the Archaean. In *The Proterozoic Biosphere*, ed. J. W. Schopf and C. Klein, pp. 25–39. Cambridge: Cambridge University Press.

Schopf, J. W. (1993). Microfossils of the Early Archaen Apex chert: New evidence of the antiquity of life. *Science* **260**, 640–646.

Schubert, J. K. and Bottjer, D. J. (1992). Early Triassic stromatolites as post-mass extinction disaster forms. *Geology* **20**, 883–886.

Schumann-Kindel, G., Bergbauer, M., Manz W., Szewzyk, U. and Reitner, J. (1997). Aerobic and anaerobic microorganisms in modern sponges: a possible relationship to fossilization processes. *Facies* **36**, 268–272.

Seilacher, A. (1989). Vendozoa: Organismic construction in the Proterozoic biosphere. *Lethaia* **22**, 229–239.

Seilacher, A. (1999). Biomat-related lifestyles in the Precambrian. *Palaios* **14**, 86–93.

Seilacher, A., Bose, P. K., and Pflüger, F. (1998). Triploblastic animals more than 1 billion years ago: Trace fossil evidence from India. *Science* **282**, 80–83.

Signor, P. W. and Lipps, J. H. (1992). Origin and early radiation of the Metazoa. In *Origin and Early Evolution of the Metazoa*, ed. J. H. Lipps and P. W. Signor, pp. 3–32. New York: Plenum.

Slack-Smith, S. M. (1990). The bivalves of Shark Bay, Western Australia. In *Research in Shark Bay. Report of the France-Australe Bicentenary Expedition Committee*, ed. P. F. Berry, S. D. Bradshaw, and B. R. Wilson, pp. 129–157. Perth, Australia: Western Australian Museum.

Soja, C. M. (1994). Significance of Silurian stromatolite-sphinctozoan reefs. *Geology* **22**, 355–358.

Thiel, V., Blumenberg, M., Hefter, J., Pape, T., Pomponi, S. A., Reed, J., Reitner, J., Wörheide, G., and Michaelis, W. (2002). A chemical view of the most ancestral Metazoa – biomarker chemotaxonomy of hexactinellid sponges. *Naturwissenschaften* **89**, 60–66.

Torsvik, V., Ovreas, L., and Thingstad, T. F. (2002). Prokaryotic diversity – magnitude, dynamics, and controlling factors. *Science* **296**, 1064–1066.

Tunnicliffe, V. (1992). The nature and origin of the modern hydrothermal vent fauna. *Palaios* **7**, 338–350.

von Bitter, P., Scott, S. D., and Schenk, P. E. (1992). Chemosynthesis: an alternate hypothesis for Carboniferous biotas in bryozoan/microbial mounds, Newfoundland, Canada. *Palaios* **7**, 466–484.

Warén, A., Bengston, S., Goffredi, S. K., and Van Dover, C. L. (2003). A hot-vent gastropod with iron sulphide dermal sclerites. *Science* **30**, 1007.

Westall, F. (1999). The nature of fossil bacteria: a guide to the search for extraterrestrial life. *Journal of Geophysical Research* **104**, 16437–16451.

Wilkinson, C. R. (1978). Microbial associations in sponges. I. Ecology, physiology, and microbial populations of coral reef sponges. *Marine Biology* **49**, 161–167.

Wood, R. A., Dickson, J. A. D., and Kirkland-George, B. (1994). Turning the Capitan Reef upside down: A new appraisal of the ecology of the Permian Capitan Reef, Guadalupe Mountains, Texas and New Mexico. *Palaios* **9**, 422–427.

Wood, R. A., Grotzinger, J. P., and Dickson, J. A. D. (2002). Proterozoic modular biomineralized metazoan from the Nama Group, Namibia. *Science* **296**, 2383–2386.

Xiao, S., Zhang, Y., and Knoll, A. H. (1998). Three-dimensional preservation of algae and animal embryos in a Neoproterozoic phosphorite. *Nature* **391**, 553–558.

CHAPTER 2

The interface of microbiology and immunology: A comparative analysis of the animal kingdom

Margaret McFall-Ngai

We shall not cease from exploration
And the end of all our exploring
Will be to arrive where we started
And know the place for the first time.

– T. S. Eliot, *Little Gidding*

2.1. INTRODUCTION

An ever-increasing awareness of the pivotal role of microorganisms in ecology, the growing threat of antibiotic resistance, and fears of bacteria as bioweapons have resulted in a renaissance in the study of environmental and pathogenic microbiology. The study of microorganisms became codified into these two subdisciplines in the late nineteenth century. This dichotomy, which dominates the field to the present day, was the result of the pioneering efforts of such leaders as Robert Koch and Louis Pasteur with the study of microbial diseases and Sergei Winogradsky with the characterization of the activity of bacteria in natural habitats. These subdisciplines meet when the natural environment of animal or plant tissues is the specific niche of a microbial population or community, that is in co-evolved cooperative symbioses. The extent to which microbes have been important in the biology of their hosts has been difficult to determine because such associations are often highly complex, with many unculturable species. However, the development of new technology has made the study of such diverse systems a current reality. The reader should refer to Chapters 1 and 8 for more details of these topics. One area likely to be heavily impacted by the development of this frontier is our conceptual view of the immune system, the study and intellectual framework of which grew out of pathogenic microbiology. The most widely accepted views of the immune system characterize it as a non-self recognition system, with *self* defined as the cellular products of the single eukaryotic animal genome; the associated microbial partners are viewed as "tolerated" passengers. Thus, when traditional views of microbiology and immunology intersect with the phenomenon of cooperative symbioses, a critical question arises: In the context of the immune system, how do animals and plants establish associations with complex communities of microorganisms that are stable throughout the life of an individual and over evolutionary time? The present chapter explores this question in the light of new findings in animal host biology, microbiology, and immunology. The reader should also refer to Chapters 9 and 14 for consideration of the relationship between cooperative bacteria and the evolution of innate and acquired immunity.

2.2. THE INVERTEBRATE WORLD

2.2.1. The Paradox: Hale and Hearty, More Than Ninety-Five Percent of all Animals Thrive without a 'Combinatorial' Immune System

All of the thirty-plus animal phyla arose within the world's oceans. More than ninety-nine percent of the organic constituents are, presently and most likely historically, in the form of "dissolved organic material" (DOM) rather than particulate material (Hedges 1987). In addition, these environments are, and were, microbe rich, with 10^5 to 10^6 bacteria per milliliter of seawater (Knoll 2003). To take advantage of the DOM in the water column, many invertebrate body plans evolved to present large surface areas of unprotected epithelia to the environment (Wright and Manahan 1989). Thus, these animals were then, and are now, challenged with the seemingly opposite goals of taking up the nutrients but preventing the overgrowth in their tissues of the microorganisms that share this nutrient pool. To inhibit colonization by environmental bacteria, invertebrates rely solely on the innate immune system, which consists of a germline-encoded receptor system either associated with cells (e.g., macrophage-like blood cells or epithelia) or in the extracellular milieu (e.g., in the haemolymph) (Janeway et al. 2001). Our current understanding of how animal hosts employ a battery of receptors to recognize bacteria is reviewed in Chapter 15.

In view of the apparent constraints of their immune systems, the question has often been posed: as invertebrates appear no less challenged by pathogens than vertebrates, and are no less healthy nor suffer higher morbidity, how do they manage without the sophisticated "armory" of the combinatorial immune system? Leaders in the field of immunology have suggested a variety of theories to address this issue (e.g., Medzhitov and Janeway 1997; Rinkevich 1999), but none has been satisfactory. Most theories deal with perceived differences in life-history strategy between vertebrates and invertebrates, for instance, characterizing invertebrates as generally "r selected" (Pianka 1970). An animal that is typically r selected has a short life span, with its mortality being independent of its population density; reproduces early in its life span, producing a large number of young that develop quickly; is semelparous, that is, has one bout of reproduction and then dies; and, has a small body. A survey of invertebrate species, which constitute more than ninety-five percent of the animal diversity, demonstrates that these generalizations are not accurate (Table 2.1; Finch 1990). For example, mollusks and arthropods have ranges for life history traits that are comparable to those of the vertebrates. Thus, this most prevalent idea of how invertebrates can cope without the "protection" of a combinatorial immune system does not hold up to careful scrutiny.

2.2.2. The Associations of Invertebrates with the Microbial World – Keeping It Simple?

How does this vast array of invertebrates manage to resist bacterially induced pathogenesis with only the benefit of the innate immune system? One possibility is that they do so by *restricting* their long-term interactions with bacteria, that is, by evolving strategies through which they associate persistently with only a few microbial partners, and dissuading long-term or stable colonization by all others. Although an exhaustive survey of interactions remains to be developed, invertebrate hosts are known to exhibit an impressive panoply of symbiotic associations that occur in various configurations – binary (one host and one microbial species) to consortial (one host with dozens to hundreds of microbial species) (Buchner 1965; Saffo 1992; Douglas 1994). In these relationships, microbes may associate with host cells either intracellularly or extracellularly. However, the vast majority of well-characterized invertebrate associations are intracellular symbioses with only one or two microbial species, which are often vertically passed between generations. Thus they are passed in or on the egg and participate in all stages of the host's life cycle. Examples include bacteria that occur in the "bacteriomes" of at least eleven percent of the insect species (Douglas 1989; Baumann et al. 1995; O'Neill et al. 1997); *Wolbachia*-type bacteria that colonize the majority of insects, and many nematodes in particular geographic locations (see Chapter 7 for more details); chemosynthetic bacteria associated with a number of marine invertebrate phyla (see Chapter 1 and McMullin et al. 2000); and the

Table 2.1. *Life history strategy of the two most speciose invertebrate phyla in comparison to the vertebrate chordates.**

Character of r/K selection	Arthropoda	Mollusca	Chordata
Length of life	Minutes → decades (mayflies → lobsters)	Months → 100s of yrs (nudibranchs → quahog)	1 yr → 100+ yrs (*Betta* spp. → tortoise)
Fecundity	Single → millions (Onychophora → crabs)	Single → millions (land snails → nudibranchs)	Single → millions (bird/mammals → fish)
Selection favors: – body size (grams)	$10^{-6} \rightarrow 10^{5}$ (ants → lobsters)	$10^{-3} \rightarrow 1.5 \times 10^{5}$ (vermiform snails → giant clams)	$10^{-3} \rightarrow 10^{8}$ (fish → blue whale)
– reproduction % semelparity	~1%	~1%	~0.5% → ~5% (fish/marsupials)

*From Finch 1990.

algal zooxanthellae that occur within the endodermal cells of a wide variety of marine invertebrates, including reef-building corals (see Chapter 6 for more information). The phyla Arthropoda, Mollusca, Chordata, Nematoda, Platyhelminthes, Cnidaria, Annelida, Echinodermata, Porifera, and Bryozoa (the top ten animal phyla in terms of numbers of described species) have members in which binary intracellular symbioses have been reported (Buchner 1965; Saffo 1992; Douglas 1994). Such associations have been found in many of the minor phyla as well.

How common is it for invertebrates to associate with complex, *co-evolved* microbial consortia? Microbial communities have been reported in the bodies of sponges (see e.g., Hentschel et al. 2002; see Chapter 1) and cnidarians (Rohwer et al. 2002), in guts of a number of species in the phyla Arthropoda (for reviews, see Buchner 1965 and Dillon and Dillon 2004) and Echinodermata (see e.g., Prim and Lawrence 1975; Guerinot and Patriquin 1981; Thorsen 1999), and in the reproductive tissues of squids (Mollusca: Cephalopoda) (Kaufman et al. 1998; Barbieri et al. 2001). These microhabitats are all nutrient rich, so it is not surprising that microbial growth is fostered. The issue is not one of whether invertebrates have communities in association with them, but rather how the invertebrate hosts relate to these communities. In other words, do they form lasting associations?

A consideration of what is presently known suggests that co-evolved microbial communities may be rare among invertebrates. Some studies report situations in which no permanent bacterial community occurs in association with a particular invertebrate. In other words, when a microbial community is present, it is transient (Boyle and Mitchell 1978; Garland et al. 1982). Where consortia have been reported in invertebrates, it appears that they often have a different relationship to the host tissues than they do in vertebrates. The invertebrate consortia described thus far are usually (i) of a limited number of phylotypes (a few dozen at most), (ii) separated from host tissues by a physical barrier that precludes direct host cell–bacterial interaction, and/or (iii) a transient rather than a persistent, co-evolved community.

These features are well demonstrated in the arthropods, which have the best-studied microbial consortia among the invertebrates. A recent study has suggested that, at least in some insects, most of the microbial community of the gut is composed of "tourists," in other words, microbes that aid in the "economy" of the gut, but are just passing through. Specifically, a study of the gypsy moth gut microbiota showed that, when they are maintained on a particular food source, two to three dozen phylotypes are reproducibly present in the gut (Broderick et al. 2004; see Chapter 8). However, only two of these phylotypes persisted with all food types presented, suggesting that the co-evolved partnerships are highly exclusive. Although the gut microbiota of other insects has been studied, none has been experimentally manipulated as extensively as the gypsy moth. In some instances, relatively diverse communities have been reported in various arthropod groups, particularly in the insects and millipedes, but the stability of these consortia is undefined as yet.

Among the invertebrates, the most complex consortia reported occur in the hindguts of the termites and their relatives (Cruden and Markovetz 1987; Lilburn et al. 1999; Schmitt-Wagner et al. 2003). Recent molecular studies of termites have revealed that their co-evolved consortium is composed of dozens, if not hundreds, of microbial phylotypes. However, this community does not interact directly with host cells, but is separated from the host epithelia by a chitinous layer, and the entire community is shed each time the termite molts (Buchner 1965). In addition, in termites and other insects with gut symbionts, the membranes separating the animal epithelium from the microbes preclude a sampling of the microbial community by host cells, a behavior that is characteristic of the M cells, which are associated with the gut-associated lymphoid tissue and mucosal dendritic cells of mammals (Trier 1991; Bell et al. 1999).

Other than gut-associated microbial consortia, the only well-described consortium in the invertebrates occurs in the accessory nidamental gland (ANG) of the females of particular squid species (Kaufman et al. 1998; Barbieri et al. 2001). Although the precise function of these glands is not defined, they most likely participate in elaboration of the egg capsules. The ANG bacterial community, which consists of a couple dozen phylotypes, resides in a series of tubules. Often correlating with bacteria in the ANG is the presence of bacteria in the outer layer of the egg capsules. A function for this character remains unclear, although it has been suggested that the bacteria may provide a type of "probiotic" protection for the egg; that is, guarding the developing embryo from bacterial or fungal infection. This particular host-consortial system provides interesting opportunities for studying the stability of such communities, because the complexities of food-associated "tourists" do not affect this situation.

2.2.3. The Innate Immune System and the Active Discouragement by Most Invertebrates of Persistent Partnership with Diverse Microbial Species

The innate immune system of invertebrates appears to be well suited to mediate interactions with both pathogens and the limited array of beneficial bacteria, and this may characterize the vast majority of invertebrate species.

In the last few years, through the pioneering work of a large number of laboratories, characterizations of the innate immune system have revealed that the bacterial ligands, host receptors, and activated biochemical pathways associated with their interactions are highly conserved across the animal kingdom (for review, see Janeway and Medzhitov 2002 and also Chapter 15). Specifically, bacteria-specific molecules, or PAMPs (pathogen-associated molecular patterns – more correctly MAMPs – microbial-associated molecular patterns, as they are also possessed by cooperative bacteria), such as lipopolysaccharide (LPS), peptidoglycan, and CpG DNA, have been identified as the ligands for an array of host membrane, intracellular, and humoral receptors. Best studied are the members of the Toll and Toll-like receptor (TLR) family. The MAMP–TLR interaction results in the activation of response pathways, most notably the NF-κB pathway, which change transcription of genes associated with response to microorganisms. The reader should refer to Chapters 15 and 17 for a fuller description of the mechanism of the NF-κB system and its signaling activity. Because MAMPs are specific to microbes, animals use these molecules to sense and react to their presence. The host response results in elimination of pathogens, but in beneficial associations MAMPs may promote host-microbe interactions.

To avoid this generalized response, a microbial species could be physically separated from the host's immune system by being intracellular or behind an impermeable barrier, which appears to be the strategy in the insect bacteriome and gut, respectively. Alternatively, the symbionts could specifically modify the generalized response, a difficult proposition when it would have to do so without compromising the ability of the innate immune system to sense pathogens through P/M/AMPs recognition. Recent studies of the innate immune system of invertebrates have demonstrated that it is much more complex than previously appreciated. For example, evidence is mounting to suggest that variations in the Toll and the other pattern-recognition receptors confer certain specificity and diversity of response. Although this diversity has a function in controlling pathogenesis, within this context of an unexpected complexity, it seems plausible for animals to have formed co-evolved associations with a few extracellular microorganisms, wherein the innate immune system serves the dual function of management of these co-evolved microbial partners and defense against potentially pathogenic interlopers.

The mechanisms by which invertebrates may harbor a limited number of microbes may be an example of true "tolerance"; that is, the system has evolved so that the immune system is not challenged by the co-evolved microbe – a sort of "live and let live" situation. In contrast, as discussed below, with our current knowledge of the vertebrates, the term *tolerance* may no longer be the best descriptor of the dynamic process of interchange between beneficial microbes and their vertebrate hosts. Further aspects of the complexity of interactions between bacteria and host cells is described in Chapters 15 and 17.

2.3. A NOVEL AND OPPOSING STRATEGY OF INTERACTING WITH MICROBES MAY HAVE APPEARED WITH THE EVOLUTION OF THE GNATHOSTOMES

2.3.1. General Features of Vertebrate–Microbe Interactions

Although an understanding of animal–microbial relationships is still in its infancy, what we know suggests that the way vertebrates associate with microbes is very different from that of invertebrates (Fig. 2.1). In contrast to the invertebrates, binary associations with intracellular microbial species rarely if ever exist in vertebrates except as pathogenic interactions (O'Riordan and Portnoy 2002; Young et al. 2002; see Chapter 10). Remarkably, and perhaps significantly, no long-term beneficial associations with microbes have been described in which the microbial symbionts evade the immune system by taking up an intracellular lifestyle. Some fishes do have binary associations with *extracellular* bacteria, that is, a population of one bacterial species with a host fish. For example, a few dozen fish families have members with symbioses involving monospecific cultures of luminous bacteria (McFall-Ngai and Toller 1991). Outside of the fishes, the author knows of no other vertebrates that have binary associations.

Mammals, and perhaps all vertebrates, appear to interact more commonly with diverse sets of defined microbial communities, most notably along mucosa, that are picked up anew each generation and persist throughout the life of the host (McFall-Ngai 2002; Xu and Gordon 2003; see Chapter 12). The advent of large-scale sequencing capability has allowed relatively quick and reliable determination of the number of culturable and nonculturable microbial phylotypes that occur within a given niche. In addition, using bacterial-artificial-chromosome (BAC) libraries, biologists are able to gain extensive genomic sequence information of nonculturable microbes to gain insight into their possible functions within the community (Diaz-Torres et al. 2003; Schloss and Handelsman 2003; see Chapters 8 and 14).

With these and other tools, biologists are defining the diversity of these consortia in vertebrate hosts and characterizing their activities. The limited data available suggest that, unlike the invertebrates, the associated microbial partners of vertebrates (i) are diverse and numerous, (ii) often interact directly

	Binary, Intracellular	Binary, Extracellular	Consortial
Invertebrate phylum:			
Arthropoda	+	+	+
Mollusca	+	+	+
Chordata (invert.)	+	+	-
Nematoda	+	+	-
Platyhelminthes	+	+	-
Cnidaria	+	+	+
Annelida	+	+	-
Echinodermata	+	+	+
Porifera	+	+	+
Bryozoa	+	+	-
Vertebrate class:			
Agnatha	-	-	?
Chondrichthyes	-	-	+
Osteichthyes	-	+	+
Amphibia	-	-	+
Reptilia	-	-	+
Aves	-	-	+
Mammalia	-	-	+

Figure 2.1. Relationships between microbes and their animal hosts. Review of the available data and searches of literature databases reveal that all ten of the most speciose invertebrate phyla have representatives with beneficial binary, intracellular symbioses, whereas there have been no reports of beneficial intracellular symbionts in any class of the vertebrates. All of these invertebrate phyla also have representatives with binary, extracellular symbioses; such symbioses are absent in the vertebrates, with the exception of certain species of bony fish (Osteichthyes). Complex gut consortia have been reported in members of all vertebrate classes except the jawless fishes (Agnatha), whereas consortia are relatively rare among the invertebrates. +, present in some members; −, not reported in any member.

with host cells, and (iii) are persistent. Recent analyses have demonstrated that more than 600 phylotypes of bacteria reside in the healthy human mouth (Paster et al. 2001; Kazor et al. 2003), and it is estimated that the human gastrointestinal tract has several hundred phylotypes (Wilson 1997). Detailed characterizations of the oral microbiota have shown that these communities occur in the forms of highly structured biofilms (Kolenbrander 2000). Coupled with the numbers, known or suspected, for the consortia in other portions of the body (e.g., skin, urogenital tract, nasopharynx), current evidence suggests that mammals are persistently colonized by a consistent, defined assemblage of perhaps more than 2,000 bacterial phylotypes. In some areas, such as portions of the intestine, most of the resident microbiota are held at "arm's length" from the epithelium by thick layers of mucus (Deplancke and Gaskins 2001). However, in other areas, such as the subgingival spaces of the oral cavity, the microbial community interacts intimately with the epithelium (Roberts and Darveau 2002). As with invertebrates, a proportion of the microbiota will be tourists, but current data suggest that it is a much smaller proportion of the community (see e.g., Zoetendal et al. 2002). These communities appear to be more stable and restrictive than those of invertebrates. For example, the intestinal microbiota of vertebrates, although modified, is not lost with starvation (Fuller and Turvey 1971; Tannock and Savage 1974; Tannock 1997). One of the challenges in these analyses has been, and continues to be, the ability to distinguish the "tourist" from the "resident" coevolved microbiota.

2.3.2. The Microbial Consortia of Gnathostome Vertebrates – Their Interface with the Combinatorial Immune System

Correlating with the evidence that vertebrates have an abundant and diverse microbiota is the presence in gnathostome vertebrates of a different type of immune system than that of the invertebrates, that is, one that includes a combinatorial immune system (Fig. 2.2). What immunologists refer to as the "big bang" in the evolution of the immune system occurred at the agnathan–gnathostome transition (Marchalonis et al. 2002). At this point, a series of protein families appeared that are rarely, if ever, found in the invertebrates. These include the immunoglobulins (Igs), T-cell receptors (TCRs), the proteins of the major histocompatibility complex (MHC), and the recombination activating (RAG) proteins. The basic mechanism by which somatic, that is, V(D)J, recombination occurs in the vertebrates was elucidated in the 1990s with the discovery of the recombination activating genes, *Rag1* and *Rag2* (Agrawal et al. 1998). These genes appear in the vertebrate lineage at the agnathan–gnathostome transition. *Rag1* shares significant similarity to microbial integrases, a finding that has led to the suggestion that its presence in the gnathostomes is a result of a lateral gene transfer that occurred near the time of the divergence of this clade, about 450 million years ago (Bernstein et al. 1996). Thus, an interacting suite of characters appeared with these vertebrates that is not operative in the ancestral invertebrates or agnathans. One should be careful to note that homologues for most or all of these genes

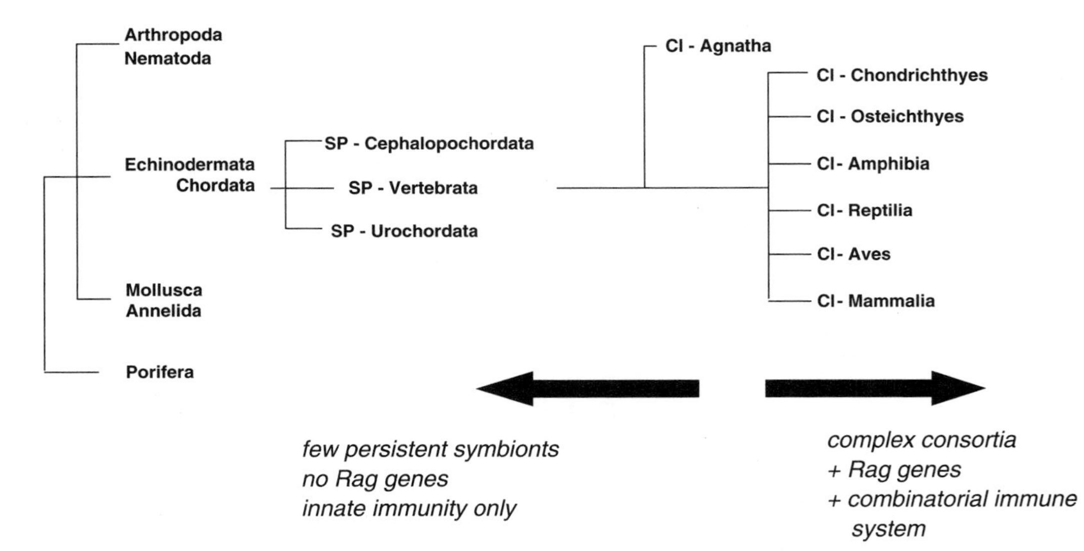

Figure 2.2. Divergence of the gnathostome vertebrates and the evolution of the combinatorial immune system.

might be found as precursors among invertebrate species (e.g., one might expect lateral gene transfer of microbial integrases to have occurred more than once), but all evidence suggests that the complex interactions of all elements associated with the combinatorial immune response are the exclusive property of the gnathostome vertebrates.

The mucosal surfaces, which are the principal sites of host–microbiota interaction, represent a large proportion of the immune system (Gaskins 1997). It is estimated that more than seventy percent of all of the immune cells of the body are located within the intestine (Kagnoff 1993; Kraehenbuhl and Neutra 1992). Further, the number of immunoglobulin(Ig)-secreting cells present in the mouse and human gut exceeds by severalfold the total number of Ig-secreting cells from all other lymphoid organs (Mestecky 1987; van der Heijden et al. 1987; Brandtzaeg et al. 1999). Thus, the mucosal surface, at least of the mammalian body, is the major "immune organ." From an evolutionary point of view, the commitment that mammals, for example, make to the development, function, and maintenance of the immune system is staggering.

By what mechanisms do vertebrates maintain their complex consortial relationships with microbes along these mucosal surfaces throughout the life of an individual and over evolutionary time? One possibility that has not yet been suggested or explored is that:

■ *the presence of a complex, co-evolved consortium of microorganisms is a shared-derived character of the vertebrates, and that selection pressure on the combinatorial immune system is driven by the requirement that the host maintain these communities in balance.*

The mathematical models that are independently presented by Seymour in this volume support this theory (see Chapter 9). They predict that diversity of the microbiota associated with an animal will be increased when one imposes a more complex immune system. Biologists have a long way to go in exploring this issue, but some testable predictions can be made. For the resident microbial consortium to drive selection on a character as complex as the combinatorial immune system, harboring such a community would have to confer some advantage to the host, providing a strategy not available to invertebrates. For example, the presence of permanent co-evolved gut microbiota in the vertebrates may confer increased efficiency in the assimilation of food, that is, for a given caloric input, vertebrates may be better at converting the material into usable products. Very few data are currently available on this subject. However, it is known that, for the same growth rate, germ-free mice must eat about a third more food than mice with a conventional microbiota. Thus, instead of the approximately 10^{14} intestinal bacteria exacting a net cost for carriage, they may actually increase host efficiency. A great deal of comparative research within the vertebrates, as well as between vertebrates and invertebrates, will have to been done before these types of issues can be resolved.

Some data are available, however, wherein characterizations of the consortia that associate with vertebrates are demonstrating that they are not *sensu strictu* "commensal" as previously thought; that is, neither the host cells nor the microbial consortia are "eating at the same table" with little or no effect on one another. Many benefits of the microbial consortium to the biology of vertebrate hosts have been recognized for years, for-example, the provision of vitamins and essential amino acids and protection against pathogens. More recently, Gordon and coworkers have demonstrated that the microbiotia are critical for aspects of the normal development of the host tissues (reviewed in Xu and Gordon 2003; see Hooper, Chapter 12). For example, they have compared intestinal tissues of germ-free (sterile), gnotobiotic (with known microbial partners), and conventionalized (with the naturally occurring consortium) mice using histology, laser microdissection, and microarray analyses (Hooper et al. 2001). These elegant studies have demonstrated that the microbiota directs critical portions of the development of normal gut microanatomy, cell kinetics, biochemistry, and gene expression. In addition, studies of Gordon and others have indicated that the microbiota is essential for the maturation of the combinatorial immune system (for reviews, see Cebra 1999; Lanning et al. 2000; Umesaki and Setoyama 2000). The fully developed immune system involves a complex tripartite dialogue among the cells of the immune system, epithelia, and its associated microbial community (Gordon et al. 1997). For normal function, elements such as the gut-associated lymphoid tissues (GALT) (Gordon et al. 1997; Cebra et al. 1998) and mucosal-associated invariant T (MAIT) cells (Treiner et al. 2003) require exposure to normal microbiota during development.

Some data are also available demonstrating that the microbiota not only educate the immune system, but the immune system has some influence on the composition of the microbial community. A recent study has shown that the composition of the mouse intestinal microbiota is genetically regulated by variations in the MHC (Toivanen et al. 2001). This genetic determination provides evidence for the co-evolution of the host with its specific microbiota.

Finally, very recent studies of certain immune cell types are demonstrating that they interact with the mucosa and microbiota in unexpected ways. For example, dendritic cells (DCs), previously believed to be associated principally with immunity to infection and antigens, are now being recognized as critical elements in mediating "tolerance" to antigens, including the recognition of the gut microbiota (Mahnke et al. 2003; Steinman et al. 2003; Macpherson and Uhr 2004). The largest population of DCs is along the mucosa, where they can represent one to two percent of the total host cell number (Guermonprez et al. 2002). It was once thought that M cells in Peyer's patches were the only site of bacteria sampling. However, DCs are now known to disrupt tight junctions of mucosal epithelium and send pseudopodia into luminal spaces to sample the environment. Evidence is accumulating that the balance of Th1, Th2, and Th3 cells is modulated through the DCs interacting with the gut microbiota. In a recent review, Grannucci and Ricciardi-Castagnoli (2003) remarked that, "A dual role for lamina propria DCs in mucosal immunity is now emerging: the resident DC population, which has regulatory functions, maintains homeostasis and suppresses immune responses against commensal bacteria; second, during inflammation DCs expressing the chemokine receptor CCR6 are recruited from the blood in response to the macrophage inflammatory protein MIP-3alpha chemotactic gradient formed during inflammatory responses."

2.3.3. The Significance to Disease: Selection for Increased *Vulnerability*, Not Increased Resistance, in the Vertebrates; or, Why Are We So Weak?

It has always been assumed that the combinatorial immune system confers greater resistance to bacterially induced pathogenesis than would occur with reliance solely on the innate immune system. If the innate immune system of invertebrates is capable of ridding the animal of any encroaching pathogen through the recognition of PAMPs, it would seem that nothing additional would be needed to control microbial pathogenesis. If on the other hand, selection has occurred in such a way as to foster association with a complex microbial consortium, one might expect an immune system that is *more permissive,* and one that must be more finely tuned to minor variations in microorganisms so as to distinguish pathogens from nonpathogens (see Chapter 16 for the view that moonlighting proteins may be involved in this fine-tuning). Some bits of intriguing evidence for this possibility already exist, coming from recent analysis of immune system function. For example, zebrafish mutants defective in *Rag1*, and thus incapable of V(D)J recombination, appear to be no less disease resistant than animals carrying the wild-type gene. Furthermore, they grow to adulthood and reproduce normally (Wienholds et al. 2002). These data suggest that the combinatorial immune system is not an essential defense system for these animals. Within the context of the present discussion, of interest would be whether these mutant fish have a normal microbiota in their intestines and whether their growth rates were comparable to those of wild-type animals. As suggested by Matzinger and others, the combinatorial immune system of vertebrates may not function as a non-self recognition system, but rather a system that senses danger, or the damage associated with the activity of a true pathogen (Matzinger 2002). To expand this idea, the combinatorial immune system may be responsible not only for identifying dangerous self (e.g., tumors) and non-self, but also for maintaining balance of the host–microbe ecosystem. However the combinatorial immune system works, its presence has resulted in pathologies that do not occur in the invertebrates, such as autoimmune diseases, which may be the price that vertebrates pay for promoting a complex group of bacterial partners. Some such autoimmune diseases, such as colitis and dental diseases, are now thought to be the result of the inability of the immune system to properly respond to or manage the native microbiota, that is, they are "ecological catastrophes" in the human body.

2.3.4. The Immunological Frontier

These above-described findings and others are suggesting that a new synthesis might be in order: one that integrates our current knowledge of the immune system, incorporating into our view the interplay between the microbiota, the mucosal surfaces, and the internal portions of the animal's body. When the sophisticated techniques that have been applied to mammals are used to study other vertebrates, we will be able to determine whether the presence of essential co-evolved microbial communities is a shared, derived character, and perhaps a key evolutionary innovation, of vertebrates. We have compelling evidence for such communities in mammals; many more data for other vertebrate classes must be gathered before this picture becomes clear. In addition, a wide variety of invertebrates need to be rigorously analyzed before we could conclude that the presence of a consortium is not a major theme in these animals.

With this type of information in hand, we should also be able to determine what role the immune system has in controlling these communities in the vertebrates. A spectrum of possibilities exists. At one end, the immune system may in some way merely be permissive for the microbial communities (tolerance); at the other, a complex co-evolved relationship may exist between

	INVERTEBRATES	VERTEBRATES
Associated Beneficial Microbial Partners		
Intracellular	uncommon, diversity low	common, diversity high
Binary	common	rare
Consortia	common	nonexistent(?)
Host Immune System		
Association with microorganisms	restrictive	permissive
Principal selection pressure on/function	non-self recognition	control of an ecosystem

Figure 2.3. Summary of the hypotheses. The concept presented holds that critical differences exist in the way invertebrates and gnathostome vertebrates interact with microorganisms and that these differences influence, and perhaps drive, the differences in the form and function of their immune systems.

the vertebrate microbiota and the combinatorial immune system. It remains to be determined which scenario characterizes the vertebrates or whether different vertebrate classes are at different places along this spectrum.

From a comparative viewpoint, if invertebrates generally do not and vertebrates generally do have persistent, co-evolved associations with complex microbial communities, one might expect that the innate immune systems of these two groups would reflect these features. For example, invertebrates may not show activities associated with vertebrate-type "tolerance" (e.g., LPS "tolerance" mediated through TLRs). Although no cytokines have yet been unequivocally identified in the invertebrates, the homologues of vertebrate cytokine receptors have been identified in the invertebrates, and "telltale" signs of cytokines have been found in the studies of the activity of invertebrate immune systems. In the context of this discussion, invertebrates may only have proinflammatory cytokines, but not the anti-inflammatory cytokines that allow vertebrates to live with diverse microbial partners. As the data accumulate, we will be better able to assess how the immune system interfaces with the microbiota, and our understanding of the similarities and differences between invertebrates and vertebrates will become increasingly sophisticated.

2.4. CONCLUSIONS

The author recognizes that the ideas presented herein represent a kind of horizon analysis (Fig. 2.3). Future data may reveal that the selection on the evolution of the combinatorial immune system was largely, or completely, independent of the microbiota. At this point, however, the ideas fit the data as well as any available, and they are testable. If they do turn out to have some merit, biologists may see the development of a new synthesis that breaks the nineteenth century dichotomy between pathogenic and environmental microbiology, and incorporates immunology into the ecology of microbial communities that associate persistently and beneficially with animals.

ACKNOWLEDGMENTS

I thank my postdoctoral and graduate students, as well as my colleagues, for thoughtful discussions on the subject of this chapter. I thank the Rockefeller Foundation for the opportunity to present these ideas at the meeting in Bellagio, Italy, in the fall of 2003, and the National Science Foundation for their support of that meeting.

REFERENCES

Agrawal, A., Eastman, Q. M., and Schatz, D. G. (1998). Transposition mediated by RAG1 and RAG2 and its implications for the evolution of the immune system. *Nature* **394**, 744–751.

Barbieri, E., Pasteur, B. J., Hughes, D., Zurek, L., Moser, D. P., Teske, A., and Sogin, M. L. (2001). Phylogenetic characterization of epibiotic bacteria in the accessory nidamental gland and egg capsules of the squid *Loligo pealei* (Cephalopoda: Loliginidae). *Environmental Microbiology* **3**, 151–167.

Baumann, P., Baumann, L., Lai, C. Y., Rouhbakhsh, D., Moran, N. A., and Clarke, M. A. (1995). Genetics, physiology, and evolutionary relationships of the genus *Buchnera*: intracellular symbionts of aphids. *Annual Review of Microbiology* **49**, 55–94.

Bell, D., Young, J. W., and Banchereau, J. (1999). Dendritic cells. *Advances in Immunology* **72**, 255–324.

Bernstein, R. M., Schluter, S. F., Bernstein, H., and Marchalonis, J. J. (1996). Primordial emergence of the recombination activating gene 1 (RAG1): sequence of the complete shark gene indicates homology to microbial integrases. *Proceedings of the National Academy of Sciences USA* **93**, 9454–9459.

Boyle, P. J., and Mitchell, R. (1978). Absence of microorganisms in crustacean digestive tracts. *Science* **200**, 1157–1159.

Brandtzaeg, P., Farstad, I. N., Johansen, F. E., Morton, H. C., Norderhaug, I. N., and Yamanaka, T. (1999). The B-cell system of human mucosae and exocrine glands. *Immunology Reviews* **171**, 45–87.

Broderick, N. A., Raffa, K. F., Goodman, R. M., and Handelsman, J. (2004). Census of the bacterial community of the gypsymoth larval midgut by using culturing and culture-independent methods. *Applied and Environmental Microbiology* **70**, 293–300.

Buchner, P. (1965). *Endosymbiosis of Animals with Plant Microorganisms.* New York: Interscience.

Cebra, J. J. (1999). Influences of microbiota on intestinal immune system development. *American Journal of Clinical Nutrition* **69**, 1046S–1051S.

Cebra, J. J., Periwal, S. B., Lee, G., Lee, F., and Shroff, K. E. (1998). Development and maintenance of the gut-associated lymphoid tissue (GALT): the roles of enteric bacteria and viruses. *Developmental Immunology* **6**, 13–18.

Cruden, D. L. and Markovetz, A. J. (1987). Microbial ecology of the cockroach gut. *Annual Review of Microbiology* **41**, 617–643.

Deplancke, B. and Gaskins, H. R. (2001). Microbial modulation of innate defense: goblet cells and the intestinal mucus layer. *American Journal Clinical Nutrition* **73**, 1131S–1141S.

Diaz-Torres, M. L., McNab, R., Spratt, D. A., Villedieu, A., Hunt, N., Wilson, M., and Mullany, P. (2003). Novel tetracycline resistance determinant from the oral metagenome. *Antimicrobial Agents and Chemotherapy* **47**, 1430–1432.

Dillon, R. J. and Dillon, V. M. (2004). The gut bacteria of insects: nonpathogenic interactions. *Annual Reviews of Entomology* **49**, 71–92.

Douglas, A. E. (1989). Mycetocyte symbiosis in insects. *Biological Reviews* **64**, 409–434.

Douglas, A. E. (1994). *Symbiotic Interactions*. Oxford: Oxford University Press.

Finch, C. (1990). *Longevity, Senescence, and the Genome*. Chicago: The University of Chicago Press.

Fuller, R. and Turvey, A. (1971). Bacteria associated with the intestinal wall of the fowl (*Gallus domesticus*). *Journal of Applied Bacteriology* **34**, 617–622.

Garland, C. D., Nash, G. V., and McMeekin, T. A. (1982). Absence of surface-associated microorganisms in adult oysters (*Crassostrea gigas*). *Applied and Environmental Microbiology* **44**, 1205–1211.

Gaskins, H. R. (1997). Immunological aspects of host/microbiota interactions at the intestinal epithelium. In *Gastrointestinal Microbiology*, ed. R. L. Mackie, B. A. White, and R. E. Isaacson, pp. 537–587. New York: Chapman & Hall.

Gordon, J. I., Hooper, L. V., McNevin, M. S., Wong, M., and Bry, L. (1997). Epithelial cell growth and differentiation. III. Promoting diversity in the intestine: conversations between the microflora, epithelium, and diffuse GALT. *American Journal of Physiology* **273**, G565–G570.

Granucci, F. and Ricciardi-Castagnoli, P. (2003). Interactions of bacterial pathogens with dendritic cells during invasion of mucosal surfaces. *Current Opinion in Microbiology* **6**, 72–76.

Guerinot, M. L. and Patriquin, D. G. (1981). The association of N_2-fixing bacteria with sea urchins. *Marine Biology* **62**, 197–207.

Guermonprez, P., Valladeau, J., Zitvogel, L., Thery, C., and Amigorena, S. (2002). Antigen presentation and T cell stimulation by dendritic cells. *Annual Review of Immunology* **20**, 621–67.

Hedges, J. I. (1987). Organic matter in seawater. *Nature* **330**, 205–206.

Hentschel, U., Hopke, J., Horn, M., Fredrich, A. B., Wagner, M., Hacker, J., and Moore B. S. (2002). Molecular evidence for a uniform microbial community in sponges from different oceans. *Applied and Environmental Microbiology* **68**, 4431–4440.

Hooper, L. V., Wong, M. H., Thelin, A., Hansson, L., Falk, P. G., and Gordon, J. I. (2001). Molecular analysis of commensal host–microbial relationships in the intestine. *Science* **291**, 881–884.

Janeway, C. A. and Medzhitov, R. (2002). Innate immune recognition. *Annual Review of Immunology* **20**, 197–216.

Janeway, C. A., Travers, P., Walport, M., and Shlomchik, M. (2001). *Immunology.* New York: Garland Press.

Kagnoff, M. F. (1993) Immunology of the intestinal tract. *Gastroenterology* **105**, 1275–1280.

Kaufman, M. R., Ikeda, Y., Patton, C., van Dykhuisen, G., and Epel, D. (1998). Bacterial symbionts colonize the accessory nidamental gland of the squid *Loligo opalescens* by horizontal transmission. *Biological Bulletin* **194**, 36–43.

Kazor, C. E., Mitchell, P. M., Lee, A. M., Stokes, L. N., Loesche, W. J., Dewhirst, F. E., and Paster, B. J. (2003). Diversity of bacterial populations on the tongue dorsa of patients with halitosis and healthy patients. *Journal of Clinical Microbiology* **41**, 558–563.

Knoll, A. H. (2003). *Life on a Young Planet.* Princeton: Princeton University Press.

Kolenbrander, P. E. (2000). Oral microbial communities: biofilms, interactions, and genetic systems. *Annual Review of Microbiology* **54**, 413–437.

Kraehenbuhl, J. P., and Neutra, M. R. (1992) Molecular and cellular basis of immune protection of mucosal surfaces. *Physiological Review* **72**, 853–879.

Lanning, D., Zhu, X., Zhai, S. K., and Knight, K. L. (2000). Development of the antibody repertoire in rabbit: gut-associated lymphoid tissue, microbes, and selection. *Immunology Reviews* **175**, 214–228.

Lilburn, T. G., Schmidt, T. M., and Breznak, J. A. (1999). Phylogenetic diversity of termite gut spirochaetes. *Environmental Microbiology* **1**, 331–345.

Macpherson, A. J. and Uhr, T. (2004) Induction of protective IgA by intestinal dendritic cells carrying commensal bacteria. *Science* **303**, 1662–1665.

Mahnke, K., Knop, J., and Enk, A. H. (2003). Induction of tolerogenic DCs: 'you are what you eat.' *Trends in Immunology* **24**, 646–651.

Marchalonis, J. J., Kaveri, S., Lacroix-Desmazes, S., and Kazatchkine, M. D. (2002). Natural recognition repertoire and the evolutionary emergence of the combinatorial immune system. *The Federation of American Societies for Experimental Biology Journal* **16**, 842–848.

Matzinger, P. (2002). The danger model: a renewed sense of self. *Science* **296**, 301–305.

McFall-Ngai, M. J. (2002). Unseen forces: the influence of bacteria on animal development. *Developmental Biology* **242**, 1–14.

McFall-Ngai, M. J. and Toller, W. W. (1991). Frontiers in the study of the biochemistry and molecular biology of vision and luminescence in fishes. In *The Molecular Biology and Biochemistry of Fishes*, ed. P. Hochachka and T. Mommsen, pp. 77–107. New York: Elsevier.

McMullin, E. R., Bergquist, D. C., and Fisher, C. R. (2000). Metazoans in extreme environments: adaptations of hydrothermal vent and hydrocarbon seep fauna. *Gravity Space Biological Bulletin* **13**, 13–23.

Medzhitov, R. and Janeway, C. A. (1997). Innate immunity: the virtues of a non-clonal system of recognition. *Cell* **91**, 295–298.

Mestecky, J. (1987). The common mucosal immune system and current strategies for induction of immune responses in external secretions. *Journal of Clinical Immunology* **7**, 265–276.

O'Neill, S. L., Hoffmann, A. A., and Werren, J. H. (1997). *Influential Passengers: Inherited Microorganisms and Arthropod Reproduction.* New York: Oxford University Press.

O'Riordan, M. and Portnoy, D. A. (2002). The host cytosol: Front-line or home front? *Trends in Microbiology* **10**, 361–364.

Paster, B. J., Boches, S. K., Galvin, J. L., Ericson, R. E., Lau, C. N., Levanos, V. A., Sahasrabudhe, A., and Dewhirst, F. E. (2001). Bacterial diversity in human subgingival plaque. *Journal of Bacteriology* **183**, 3770–3783.

Pianka, E. R. (1970). On r and K-selection. *American Naturalist* 104, 592–597.

Prim, P. and Lawrence, J. M. (1975). Utilization of marine plants and their constituents by bacteria isolated from the gut of echinoids (Echinodermata). *Marine Biology* **33**, 167–173.

Rinkevich, B. (1999). Invertebrate versus vertebrate innate immunity: in the light of evolution. *Journal of Immunology* **50**, 456–460.

Roberts, F. A. and Darveau, R. P. (2002). Beneficial bacteria of the periodontium. *Periodontology 2000* **30**, 40–50.

Rohwer, F., Seguritan, V., Azam, F., and Knowlton, N. (2002). Diversity and distribution of coral-associated bacteria. *Marine Ecology Progress Series* **243**, 1–10.

Saffo, M. B. (1992). Invertebrates in endosymbiotic associations. *American Zoologist* **32**, 557–565.

Schloss, P. D. and Handelsman, J. (2003). Biotechnological prospects from metagenomics. *Current Opinion in Biotechnology* **14**, 303–310.

Schmitt-Wagner, D., Friedrich, M. W., Wagner, B., and Brune, A. (2003). Axial dynamics stability, and interspecies similarity of bacterial community structure in the highly compartmentalized gut of soil-feeding termites (*Cubitermes* spp.). *Applied and Environmental Microbiology* **69**, 6018–6024.

Steinman, R. M., Hawigerm D., and Nussenzweig, M. C. (2003). Tolerogenic dendritic cells. *Annual Review of Immunology* **21**, 685–711.

Tannock, G. W. (1997). Modification of the normal microbiota by diet, stress, antimicrobial agents and probiotics. In *Gastrointestinal Microbiology*, ed. R. L. Mackie, B. A. White, and R. E. Isaacson, pp. 434–465. New York: Chapman & Hall.

Tannock, G. W. and Savage, D. C. (1974). Influences of dietary and environmental stress on microbial populations in the murine gastrointestinal tract. *Infection and Immunity* **9**, 591–598.

Thorsen, M. S. (1999). Abundance and biomass of the gut-living microorganisms (bacteria, protozoa, fungi) in the irregular sea urchin *Echinocardium cordatum*. (Spatangoida: Echinodermata). *Marine Biology* **133**, 353–360.

Toivanen, P., Vaahtovuo, J., and Eerola, E. (2001). Influence of major histocompatibility complex on bacterial composition of fecal flora. *Infection and Immunity* **69**, 2372–2377.

Treiner, E., Duban, L., Bahram, S., Radosavljevic, M., Wanner, V., Tilloy, F., Affaticati, P., Gilfillan, S., and Lantz, O. (2003). Selection of evolutionarily conserved mucosal-associated invariant T cells by MR1. *Nature* **422**, 164–169.

Trier, J. S. (1991). Structure and function of intestinal M cells. *Gastroenterology Clinics of North America* **20**, 531–547.

Umesaki, Y. and Setoyama, H. (2000). Structure of the intestinal flora responsible for development of the gut immune system in a rodent model. *Microbes and Infection* **2**, 1343–1351.

van der Heijden, P. J., Stock, W., and Bianchi, A. T. J. (1987). Contribution of immunoglobulin-secreting cells in the murine small intestine to the total 'background' immunoglobulin production. *Immunology* **62**, 551–555.

Wienholds, E., Schulte-Merker, S., Walderich, B., and Plasterk, R. H. A. (2002). Target-selected inactivation of the zebrafish rag1 gene. *Science* **297**, 99–102.

Wilson, K. H. (1997). Biota of the human gastrointestinal tract. In *Gastrointestinal Microbiology*, ed. R. L. Mackie, B. A. White, and R. E. Isaacson, pp. 39–58. New York: Chapman & Hall.

Wright, S. H. and Manahan, D. T. (1989). Integumental nutrient uptake by aquatic organisms. *Annual Review of Physiology* **51**, 585–600.

Xu, J. and Gordon, J. L. (2003). Inaugural article: honor thy symbionts. *Proceedings of the National Academy of Sciences USA* **100**, 10452–10459.

Young, D., Hussel, T., and Dougan, G. (2002). Chronic bacterial infections: living with unwanted guests. *Nature Immunology* **3**, 1026–1032.

Zoetendal, E. G., von Wright, A., Vilpponen-Salmela, T., Ben-Amor, K., Akkermans, A. D. L., and de Vos, W. M. (2002). Mucosa-associated bacteria in the human gastrointestinal tract are uniformly distributed along the colon and differ from the community recovered from feces. *Applied and Environmental Microbiology* **68**, 3401–3407.

CHAPTER 3

Co-evolution of bacteria and their hosts: A marriage made in heaven or hell?

Jörg Hacker, Ulrich Dobrindt, Michael Steinert, Hilde Merkert, and Ute Hentschel

3.1. INTRODUCTION

Bacteria–host interactions are traditionally characterised as symbiotic, commensal, or pathogenic. In the latter, one partner benefits to the detriment of the other, the result being cell or tissue damage, or even death of the organism. Symbiotic (mutualistic) bacteria–host interactions are balanced relationships with reciprocal benefit. Commensal interactions imply the coexistence of at least two different organisms without detriment but also without obvious benefit. As the definitions of bacteria–host interactions are confluent, many bacteria–host interactions fall in between two categories. The interactions between bacteria and their hosts are not static, rather they represent an arms race in which adaptive changes in one partner must be matched by countermeasures of the other to maintain equilibrium. Figuratively speaking, a pathogenic interaction may be viewed as the initial stages of the arms race whereas symbiosis represents a battle that has been settled towards mutual benefit. Because bacteria have higher replication rates, rendering them more susceptible to evolutionary adaptations, they may be considered as catalysts that drive the evolution of their respective hosts, leading to speciation and diversification. In this chapter, the evolutionary, mechanistic, and molecular aspects of bacteria–host interactions are reviewed using the commensal microbiota of sponges, the pathogenic *Legionella*–host interactions, and the different types of enterobacteria–host interactions as model systems.

3.2. MARINE SPONGES AND MICROORGANISMS: AN ANCIENT TYPE OF INTERACTION

The phylum sponges (Porifera) forms one of the deepest radiations of the Metazoa, with a fossil record dating back more than 580 million years (see Chapter 1 for more information on sponge–bacteria interactions). The earliest confirmed sponge fossils were found in South China in Precambrian rock deposits (Li et al. 1998). These fossils are noteworthy for the extraordinarily well-preserved soft tissues containing amoebocytes and even embryos. Well over 1,000 sponge fossils have been described within fifteen genera and thirty species in Cambrian rock deposits, suggesting an early radiation of this phylum. The phylum Porifera is divided into three classes, the Calcarea (calcareous sponges), the Hexactinellida (glass sponges), and the Demospongiae, which contain more than ninety percent of the sponges living today. An estimated 9,000 living sponge species are found on tropical reefs, but they are also found with increasing latitudes and even in freshwater lakes and streams.

Sponges are diploblast metazoans that lack true tissues or organs. In spite of their simple organisation, genome sequencing has revealed highly homologous genes to vertebrate analogues (Müller et al. 2001). As sessile filter-feeders, they pump large volumes of water through a specialised canal system, termed the aquiferous system. The filtration capacities of sponges are remarkably efficient, amounting to 24,000 L kg^{-1} day^{-1}, an accomplishment that is unsurpassed in the animal kingdom. Although typical seawater contains 10^5 to 10^6 bacteria per milliliter, the seawater expelled from the sponge is essentially sterile (e.g., Turon et al. 1997). Food particles such as unicellular algae and bacteria are removed from the seawater and translocated into the mesohyl interior, where they are rapidly digested. The mesohyl serves as a scaffold that is made up of extracellular matrix and constitutes much of the sponge body. Single, amoeboid sponge cells, termed archaeocytes, move freely through the mesohyl matrix and digest food particles by phagocytosis.

Many species of the Demospongiae are known to contain large numbers of bacteria within their tissues, which may contribute up to fifty-seven percent of the tissue volume (for review, see Hentschel et al. 2003) (Fig. 3.1). Bacterial numbers amount to 6×10^8 to 8×10^9 ml^{-1} sponge extract, exceeding the bacterial concentrations of seawater by two to four orders of magnitude (Friedrich et al. 2001). The vast majority of bacteria are located extracellularly within the mesohyl matrix, where they are separated from the surrounding seawater by contiguous host cell barriers. Bacteria are also found within vacuoles of archaeocytes, where they appear in various stages of digestion. In some cases, bacteria are located within the nuclei of certain sponge cells (Friedrich et al. 1999). The sponge surfaces, the canal system, and choanocyte chambers are noticeably free of bacteria.

Several recent studies have addressed the phylogenetic diversity of microbial communities associated with marine sponges using 16S rDNA library

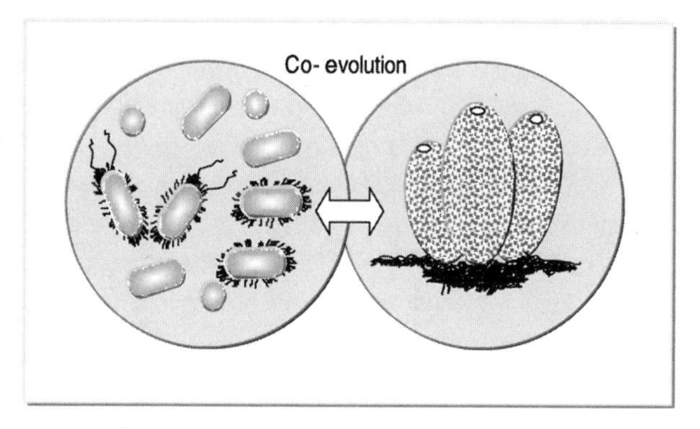

Figure 3.1. Marine sponges as experimental model systems for commensal and possibly ancient bacteria–host interactions.

construction and fluorescence in situ hybridisation (Webster et al. 2001; Hentschel et al. 2002). The sponges assessed in these studies are taxonomically only distantly related, have geographically non-overlapping distribution patterns, and contain host-specific secondary metabolite profiles. In spite of these phenotypic differences, these sponges contain a surprisingly uniform microbial signature (Hentschel et al. 2002, 2003). Particularly, representatives of the poorly characterised divisions Chloroflexi (formerly green non-sulphur bacteria), Acidobacteria, and Actinobacteria, as well as Alpha-, Gamma-, and Deltaproteobacteria, are abundant phylotypes in the gene libraries. Moreover, a novel candidate division, Poribacteria, which has recently been discovered in various marine sponges, is characterised by the presence of a nucleoid-containing organelle (Fieseler et al., 2004). In addition to the planctomycetes, this is only the second report of cell compartmentalisation in prokaryotes, a feature that was so far considered exclusive to the eukaryotic domain. Archaea have so far been identified in the sponge *Axinella mexicana* (Preston et al. 1996). It is noteworthy that none of these sponge-specific bacteria have so far been cultured or are related to culturable bacteria.

We have established marine sponges (Porifera) as model systems for commensal and possibly ancient bacteria–host interactions. Even though they represent the lowest extant metazoan phylum, they have a strikingly complex histocompatibility system, elements of the innate and probably also the adaptive immune system (Müller et al. 1999), and the ability to differentiate between self and non-self (Fernandez-Busquets and Burger 1999). The current hypothesis holds that the microorganisms supplement the nutrition of the sponge. The sponge-associated microorganisms ensure their existence within host tissues by protecting themselves from digestion by phagocytic sponge cells (Wilkinson et al. 1984). Mechanisms by which evasion of phagocytosis may be accomplished include (i) the presence of surface structures such as slimes and extra membranes or (ii) the production of metabolites that deter host archaeocytes.

In comparison to the many pathogenic and symbiotic types of interactions, the commensal ones have been traditionally the most difficult to study. However, understanding commensal microbial consortia is of major importance because bacteria rarely exist as monocultures in nature. It is well known that complex microbial consortia play important roles in various ecological contexts (Hooper and Gordon 2001; Russell and Rychlik 2001; Rickard et al. 2003 – and various chapters in this book). Moreover, it is becoming increasingly clear that they also have an immediate effect on nutrition, immune system control, and the development of the invertebrate and vertebrate hosts (Stappenbeck et al. 2002; Hentschel et al. 2003). More information on these aspects can be found in Chapters 2, 9, and 13. The commensal microbiota also represents a reservoir for the evolution of microbial pathogens (e.g., Dobrindt et al. 2003). With the implementation of molecular tools for microbial community analysis, it is now possible to define the members of the microbial communities (culturable and non-culturable), their spatial and temporal distribution as a function of the environment, of the host development, and following experimentally induced disturbance. This has led to a more comprehensive understanding of commensal bacteria–host interactions that goes well beyond the one-bacterium-one-host concept. For more information on bacterial communities and their analysis by functional metagenomics, see Chapter 8.

3.3. THE *LEGIONELLA*-HOST SYSTEM: PATHOGENIC INTERACTIONS

Legionella pneumophila is naturally found in freshwater habitats like rivers and lakes, where the bacterium parasitises protozoa. Upon aerosol formation from man-made water systems, *L. pneumophila* can enter the human lung and cause a severe form of pneumonia called Legionnaires' Disease. The pathogenesis of Legionnaires' Disease is mainly the result of the ability of *L. pneumophila* to grow intracellularly within professional phagocytic cells. The striking similarities of the *Legionella* life cycle in protozoa and human macrophages include the uptake of the pathogen, the formation of a *Legionella*-specific vacuole that does not follow the endosomal pathway, and finally the lysis of the host cell (for recent reviews, see Steinert et al. 2002, 2003).

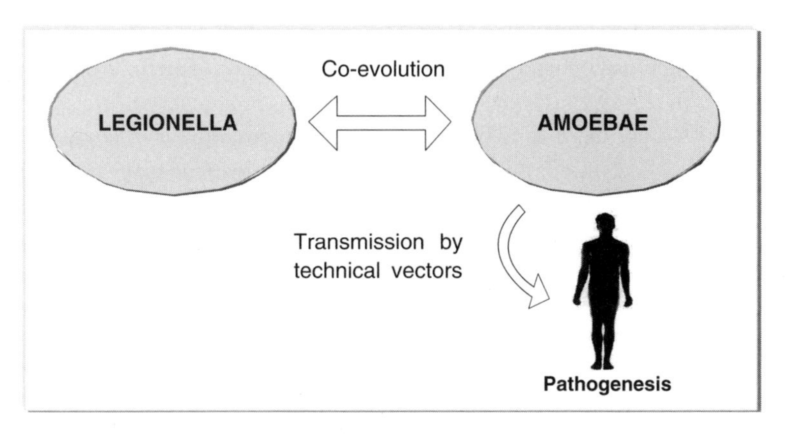

Figure 3.2. The preadaptation of pathogenic *Legionella* to amoebae hosts enhances pathogenesis in humans.

The list of different protozoan species which are known to support *Legionella* growth increases continuously, and *Acanthamoeba, Hartmannella,* and *Naegleria* are most commonly isolated from *Legionella*-contaminated warm-water systems. These hosts do not only provide nutrients for intracellularly growing legionellae, they also shelter the bacteria when environmental conditions become unfavourable (Steinert et al. 1997; Steinert et al. 1998; Grimm et al. 2001). The infection of humans occurs exclusively by inhalation of contaminated aerosols which can be produced by air conditioning systems, cooling towers, whirlpools, and shower heads. Histological studies with infected lung tissue identify intra- and extra-cellular bacteria in phagocytes, fibroblasts, and epithelial cells. It is estimated that legionellosis contributes to one to thirteen percent of all pneumonias in North America and Western Europe.

The interaction of *L. pneumophila* and professionally phagocytic cells is probably the result of many million years of co-evolution (Fig. 3.2). This is suggested by the complex intracellular life cycle that is strictly regulated. After internalisation, the intracellular bacteria reprogramme the endosomal-lysosomal degradation pathway of the host cell. In a number of studies, it has been shown that the establishment of the intracellular niche of *L. pneumophila* requires a membrane-bound secretion apparatus similar to the type IV conjugational transfer systems. This apparatus, which is encoded by a set of *dot* (**d**efective in **o**rganelle **t**rafficking) and *icm* (**i**ntra**c**ellular **m**ultiplication) genes, exports virulence factors that inhibit the phagolysosome fusion and reprogrammes the *Legionella*-bearing vacuole (Coers et al. 1999). In addition, genes encoding surface structures like Mip, flagella, and others also contribute to the pathogenicity of *Legionella* (Heuner and Steinert 2003; Köhler et al. 2003).

The similarities during the infection of protozoa and macrophages also include bacterial differentiation processes which correspond to the respective growth phases of the bacteria. In both host systems, *L. pneumophila* alters its morphology and physiology upon entry into the host cell. These replicative bacteria are more sodium resistant, do not express flagella, and display reduced cytotoxicity. At mid-log phase, *Legionella* replicates by binary fission, with a doubling time of approximately two hours, which results in a host cell that is filled with bacteria. During the late replicative phase, the *Legionella* phagosome merges with lysosomes without detrimental consequences for the enclosed bacteria (Heuner and Steinert 2003). After the exploitation of the host, *Legionella* enters the post-exponential growth phase in which motility and virulence traits that promote transmission to a new host are expressed. These post-exponential phase bacteria, which are finally released from a depleted host cell, are short, thick, flagellated, and highly motile. Consistent with the need of *Legionella* to find and infect protozoan host cells in natural aquatic habitats, flagellation and certain virulence factors are co-regulated (Heuner and Steinert 2003).

Since infection of humans is a dead-end in the evolution of *Legionella*, the complex interplay of legionellae and their respective host cells must have evolved in a "marriage" with protozoa (Hägele et al. 2000; Steinert et al. 2003). The broad protozoal host spectrum in the environment and the exploitation of very basic cellular mechanisms of eukaryotes obviously pre-adapted *Legionella* to infect human cells (Skriwan et al. 2002; Hentschel et al. 2000; Steinert et al. 2002). Therefore, *Legionella* can be viewed as an aquatic microbe that goes astray upon human infection. This also suggests that protozoa in general could represent a driving force in the evolution of pathogenicity. This has recently been suggested for pathogenic *Chlamydia* (Horn et al. 2004).

3.4. THE ENTEROBACTERIA–HOST SYSTEM: AN INTERACTION OF MANY FACETS

The Enterobacteriaceae comprise a distinct phylogenetic cluster that shares a common ancestor with other Gammaproteobacteria. This prokaryotic family comprises 40 genera with 200 species. Many representatives of this lineage live in intimate association with hosts either as pathogens,

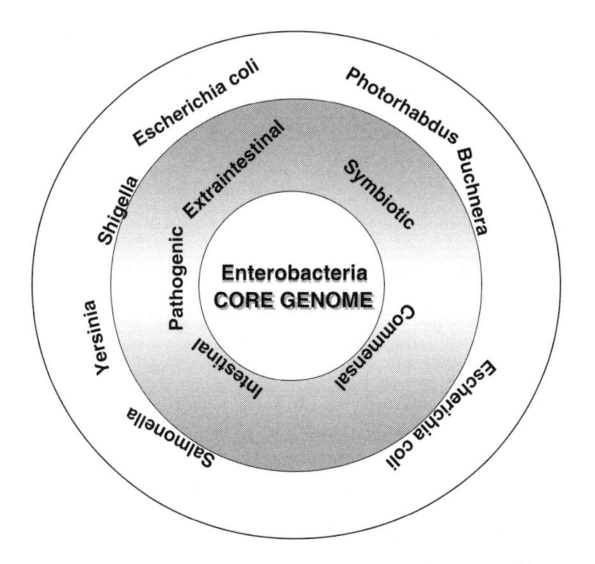

Figure 3.3. Various commensal, pathogenic, and symbiotic lineages of the enterobacteria.

commensals, or symbionts (Steinert et al. 2000) (Fig. 3.3). The enterobacteria, such as *Escherichia coli*, *Enterobacter cloacae*, or *Serratia marcescens*, are members of the commensal gut flora of humans and many animals. Enterobacteria and their close relatives are also found as symbionts of non-mammalian hosts, such as aphids (*Buchnera* spp.), ants (*Blochmannia* spp.), and nematodes (*Photorhabdus luminescens*) as well as in the rhizosphere (*Klebsiella pneumoniae*). Enterobacterial pathogens may cause different intestinal (*Yersinia enterocolitica*, *Salmonella enterica*, *Shigella* spp., *E. coli*) as well as extra-intestinal infections (*Salmonella typhi*, *Yersinia pestis*, *E. coli*) and are therefore of clinical relevance. In the following, the similiarities and differences between different enterobacterial genomes and the implications for the interaction with their respective hosts are discussed using *E. coli* as a model system.

The emergence of enterobacteria is quite recent on an evolutionary scale and coincides with the emergence of mammalian organisms. The original *E. coli* were probably inhabitants of the vertebrate intestinal flora. Selective pressures, such as high bacterial densities (more than 10^{11} bacteria per gram colon content) and species competition (more than 500 bacterial species), were probably the driving forces to promote the expansion of strains into other ecological niches (Fig. 3.4). Accordingly, whereas the commensal *E. coli* populate the large bowel, pathogenic variants are found in the urogenital tract, on intestinal mucosal surfaces, and in the blood. Interestingly,

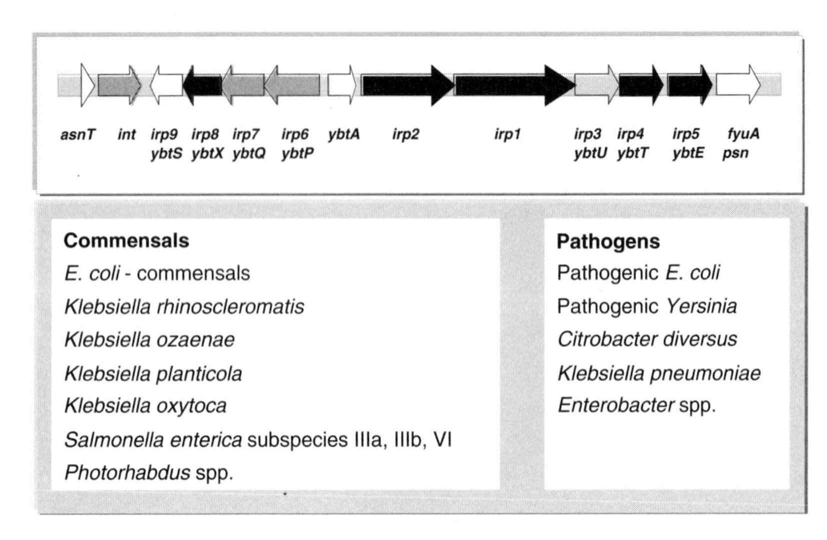

Figure 3.4. Host niches of pathogenic enterobacteria.

in pathogenic *E. coli*, *Shigella* spp., *S. enterica*, and *Yersinia* spp., the majority of the essential virulence determinants are localised on mobile genetic elements. The enterobacteria can therefore be considered as a distinct phylogenetic group that contains certain "core properties" which allow their existence in or on mammalian hosts. With the acquisition of "flexible properties," sometimes by single genetic events, they succeeded in conquering several other niches. This event led to the emergence of diseases over a relatively short evolutionary time period, affecting both humans and animals alike.

Whereas the core genome of the various *E. coli* strains includes the genes for essential cellular processes, such as growth and replication, the flexible gene pool consists of an individual assortment of strain-specific genetic information (Hacker and Carniel 2001). The flexible gene pool may contain mobile and accessory genetic elements, such as plasmids, transposons, insertion sequence elements, prophages, and non-functional fragments thereof, as well as DNA elements termed genomic islands and islets. The latter DNA regions represent specific foreign DNA entities and vary in size (Hacker and Kaper 2000; Hacker et al. 2003). Insertions and deletions of chromosomal regions ranging from a few base pairs to more than 100 kilobases (kb) have been observed (Blattner et al. 1997; Perna et al. 2001; Welch et al. 2002). The accessory genetic elements can be selfish (e.g., insertion sequence elements) or they can provide a benefit for the host (e.g., genomic islands), or they can do both (e.g., resistance determinants on integrons or transposons, virulence-associated genes on bacteriophages, or virulence plasmids). Several types of these elements can be transferred laterally and are present in probably all of the major bacterial phylogenetic groups, thus contributing to the inter- and intra-species variability in genome content (Hacker and Kaper 2000; Dobrindt and Hacker 2001).

Horizontal gene transfer plays an important role in the evolution of the enterobacterial species and results in very dynamic and diverse enterobacterial genome structures. It has been estimated that about eighteen percent of the genome of the *E. coli* strain MG1655 represents horizontally acquired sequences (Lawrence and Ochman 1998). The evolution of pathogenic *Shigella* from a non-pathogenic *E. coli* ancestor resulted from the loss of the *ompT* and *cadA* genes in combination with the acquisition of at least two pathogenicity islands and one virulence plasmid (Nakata et al. 1993; Ochman and Groismann 1995; Maurelli et al. 1998). Moreover, the non-pathogenic laboratory strain MG1655 and the Shiga toxin–expressing *E. coli* strain EDL933 shared a common ancestor about 4.5 million years ago. The parallel gain and loss of mobile genetic elements, such as bacteriophages, plasmids, and the LEE pathogenicity island, in different lineages of pathogenic *E. coli* enabled the evolution of the many different clonal lineages of *E. coli* (Reid et al. 2000).

A striking example of the extent of horizontal gene transfer is the so-called high pathogenicity island (HPI) that was initially discovered in pathogenic *Yersinia* spp. (Fig. 3.5). The core element of the HPI has since then been

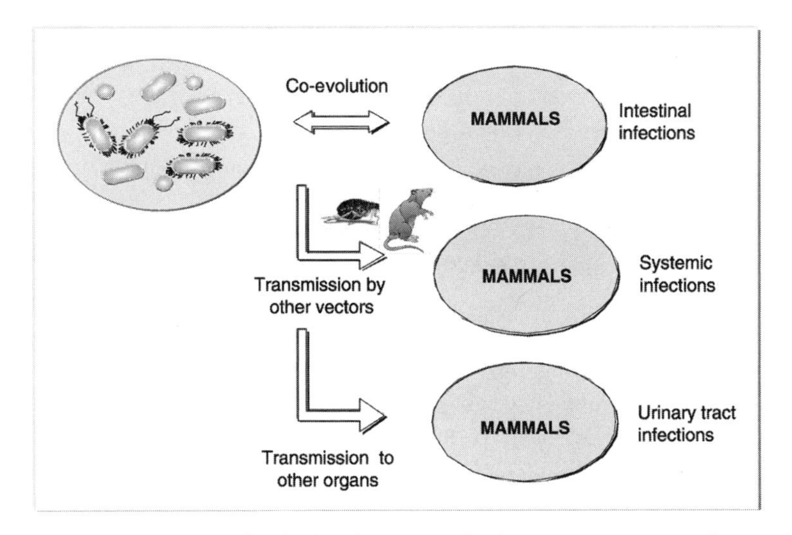

Figure 3.5. The presence of the high pathogenicity island (HPI) in various enterobacteria.

detected in a broad range of other pathogenic enterobacteria, such as *E. coli*, *Enterobacter* spp., *Citrobacter* spp., and *K. pneumoniae* and therefore represents a broad host-range genomic island which contributes to the adaptation to different niches. The presence of this island confers a growth advantage because of an increased capability of iron uptake, which is advantageous in the environment as well as in different hosts (Karch et al. 1999; Schubert et al. 2000; Bach et al. 2000). This genomic island has also been detected in many non-human pathogenic *S. enterica* subspecies III and VI, non-pathogenic *Klebsiella* spp., and commensal *E. coli* isolates (Ölschläger et al. 2003; Dobrindt et al. 2003). Moreover, an HPI homologue exists in *Photorhabdus luminescens*.

Horizontal gene transfer also plays a role in bacteria of the genus *Photorhabdus*. Even though these bacteria are closely related to *E. coli*, they employ very different life strategies. *P. luminescens* bacteria undergo a complex life cycle that involves a symbiotic stage during which the bacteria are carried in the guts of nematodes and a pathogenic stage in which susceptible insects are killed by the combined action of nematode and bacteria. Complete sequencing of the *P. luminescens* genome has revealed many virulence factors that are similar to those of other enterobacteria (ffrench-Constant et al. 2000). These include insecticidal toxins, RTX-like toxins, proteases, lipases, and various antibiotics. Moreover, type III secretion mechanisms have been identified. The fact that virulence factors such as the Yops of *Yersinia* have been identified in *P. luminescens* suggests that horizontal gene transfer may also occur between vertebrate and invertebrate pathogens and even symbionts. Accordingly, many mobile genetic elements, such as insertion elements, transposons, and several large plasmids, have been identified the genome of *P. luminescens* (Duchaud et al. 2003; ffrench-Constant et al. 2003). As more sequence information becomes available, the gap between seemingly different life strategies of bacteria that live in intimate associations with animal hosts will likely become closer.

The process of genome reduction plays a role in several obligate endosymbionts (Ochman and Moran 2001). Accordingly, the genome of the obligate symbionts of aphids, *Buchnera* sp. APS, is considered a subset of the *E. coli* genome (Shigenobu et al. 2000). At about 640 kb, it is extremely small in comparison to free-living enterobacteria (4–5 megabases) and reflects the obligate symbiotic lifestyle of this intracellular organism (Wernegreen et al. 2000). The genome size reduction reflects the stable and nutrient-rich habitat of *Buchnera* in their obligate symbiotic relationship with aphids. The *Buchnera* genome lacks genes involved in the biosynthesis of cell-surface components, such as lipopolysaccharides and phospholipids, genes involved in fermentation, chemotaxis/motility, and osmotic adaptation, as well as those for the entry and exit of eukaryotic cells. Although *Buchnera* contains fewer genes encoding for amino acid biosynthesis than free-living enterobacteria, the pathways for the biosynthesis of essential amino acids are still present. With the exception of two small plasmids, there are no accessory genetic elements such as bacteriophages, IS sequence elements, or other repetitive sequences present in the genome. The fraction of pseudogenes, unique genes, and regulatory genes is much smaller in *Buchnera* sp. SPA than in other free-living bacteria (Shigenobu et al. 2000).

These cumulative data demonstrate that fitness determinants present in genomes of non-pathogenic bacteria, for example, the HPI, may also function as virulence-associated genes of pathogenic variants. This underlines the importance of bacterial fitness and adaptability for successful colonisation/ infection of a given host. The processes and concepts of horizontal gene transfer and genomic islands in their various ecological contexts contribute to an overall fitness, which, in turn, enhances the survival of the respective bacterial species. In conclusion, the picture of the microbial genome as a dynamic entity has far-reaching implications not only for pathogenesis, but also for more general biological processes, including symbiosis and generally, fitness and adaptation.

3.5. CONCLUSIONS

The era of genome sequencing has revealed pathogenesis, symbiosis, and commensalism as different manifestations of the same theme. It is now possible to draw correlations between the structural organisation of a microbial genome and the lifestyle of the respective bacterium. Although it has been known for a long time that genome variability leads to evolutionary changes, the extent and magnitude of these events has only recently become appreciated. Particularly, the above-mentioned variable elements of the bacterial chromosome can lead to what has suitably been termed "evolution in quantum leaps." Because the bacterial chromosome is more susceptible to changes, bacterial genome dynamics can have an immediate impact upon the evolution of their respective hosts (see Chapter 5). For the understanding of bacteria–host interactions, it is of equal importance to consider the host side of an infection. Parameters such as immunity and susceptibility play an important role for the overall outcome. Finally, regardless of whether the marriage has been made in heaven or hell, bacteria are catalysts of co-evolution.

REFERENCES

Bach, S., de Almeida, A., and Carniel, E. (2000). The *Yersinia* high-pathogenicity island is present in different members of the family *Enterobacteriaceae*. *FEMS Microbiology Letters* **183**, 289–294.

Blattner, F. R., Plunkett, G., Bloch, C. A., Perna, N. T., Burland, V., Riley, M., Collado-Villes, J., Glasner, J. D., Rode, C. K., Mayhew, G. F., Gregor, J., Davis, N. W., Kirkpatrick, H. A, Goeden, M. A., Rose, D. J., Mau, B., and Shao, Y. (1997). The complete genome sequence of *Escherichia coli* K-12. *Science* **277**, 1453–1462.

Coers, J., Monahan, C., and Roy, C. R. (1999). Modulation of phagosome biogenesis by *Legionella pneumophila* creates an organelle permissive for intracellular growth. *Nature Cell Biology* **1**, 451–453.

Dobrindt, U., Agerer, F., Michaelis, K., Janka, A., Buchrieser, C., Samuelson, M., Svanborg, C., Gottschalk, G., Karch, H., and Hacker, J. (2003). Analysis of genome plasticity in pathogenic and commensal *Escherichia coli* isolates by use of DNA arrays. *Journal of Bacteriology* **185**, 1831–1840.

Dobrindt, U., and Hacker, J. (2001). Whole genome plasticity in pathogenic bacteria. *Current Opinions in Microbiology* **4**, 550–557.

Duchaud, E., Rusniok, C., Frangeul, L., Buchrieser, C., Givaudan, A., Taourit, S., Bocs, S., Boursaux-Eude, C., Chandler, M., Charles, J. F., Dassa, E., Derose, R., Derzelle, S., Freyssinet, G., Gaudriault, S., Medigue, C., Lanois, A., Powell, K., Siguier, P., Vincent, R., Wingate, V., Zouine, M., Glaser, P., Boemare, N., Danchin, A., and Kunst, F. (2003). The genome sequence of the entomopathogenic bacterium *Photorhabdus luminescens*. *Nature Biotechnology* **21**, 1307–1313.

Fernandez-Busquets, X. and Burger, M. M. (1999). Cell adhesion and histocompatibility in sponges. *Microscopical Research Techniques* **44**, 204–218.

ffrench-Constant, R. H., Waterfield, N., Burland, V., Perna, N. T., Daborn, P. J., Bowen, D., and Blattner, F. R. (2000). A genomic sample sequence of the entomopathogenic bacterium *Photorhabdus luminescens* W14: potential implications for virulence. *Applied and Environmental Microbiology* **66**, 3310–3329.

ffrench-Constant, R., Waterfield, N., Daborn, P., Joyce, S., Bennett, H., Au, C., Dowling, A., Boundy, S., Reynolds, S., and Clarke, D. (2003). *Photorhabdus*: towards a functional genomic analysis of a symbiont and pathogen. *FEMS Microbiology Reviews* **26**, 433–456.

Fieseler, L., Horn, M., Wagner, M., and Hentschel, U. (2004) Discovery of a novel candidate phylum '*Poribacteria*' in marine sponges. *Applied and Environmental Microbiology* **70**, 3724–3732.

Friedrich, A. B., Fischer, I., Proksch, P., Hacker, J., and Hentschel, U. (2001). Temporal variation of the microbial community associated with the mediterranean sponge *Aplysina aerophoba*. *FEMS Microbiology Ecology* **38**, 105–113.

Friedrich, A. B., Merkert, H., Fendert, T., Hacker, J., Proksch, P., and Hentschel, U. (1999). Microbial diversity in the marine sponge *Aplysina cavernicola* (formerly *Verongia cavernicola*) analyzed by fluorescence *in situ* hybridisation (FISH). *Marine Biology* **134**, 461–470.

Grimm, D., Ludwig, W., Brand, B. C., Schleifer, K. H., Michel, R., Hacker, J., and Steinert, M. (2001). Development of 18S rRNA-targeted oligonucleotide probes for specific detection of *Hartmannella* and *Naegleria* in *Legionella*-positive environmental samples. *Systemic and Applied Microbiology* **24**, 76–82.

Hacker, J. and Carniel, E. (2001). Ecological fitness, genomic islands and bacterial pathogenicity. A Darwinian view of the evolution of microbes. *EMBO Reports* **2**, 376–380.

Hacker, J., Hentschel, U., and Dobrindt, U. (2003). Prokaryotic chromosomes and disease. *Science* **301**, 790–793.

Hacker, J. and Kaper, J. B. (2000). Pathogenicity islands and the evolution of microbes. *Annual Review of Microbiology* **54**, 641–679.

Hägele, S., Köhler, R., Merkert, H., Schleicher, M., Hacker, J., and Steinert, M. (2000). *Dictyostelium discoideum*: a new host model system for intracellular pathogens of the genus *Legionella*. *Cellular Microbiology* **2**, 165–171.

Hentschel, U., Dobrindt, U., and Steinert, M. (2003). Commensal bacteria make a difference. *Trends in Microbiology* **11**, 148–150.

Hentschel, U., Fieseler, L., Wehrl, M., Gernert, C., Steinert, M., Hacker, J., and Horn, M. (2003). Microbial diversity of marine sponges. In *Molecular Marine Biology of Sponges*, ed. W. E. G. Müller, pp. 60–88. Heidelberg: Springer Verlag.

Hentschel, U., Hopke, J., Horn, M., Friedrich, A. B., Wagner, M., Hacker, J., and Moore, B. S. (2002). Molecular evidence for a uniform microbial community in sponges from different oceans. *Applied and Environmental Microbiology* **68**, 4431–4440.

Hentschel, U., Steinert, M., and Hacker, J. (2000). Common molecular mechanisms of symbiosis and pathogenesis. *Trends in Microbiology* **8**, 226–231.

Heuner, K. and Steinert, M. (2003). The flagellum of *Legionella pneumophila* and its link to the expression of the virulent phenotype. *International Journal of Medical Microbiology* **293**, 133–143.

Hooper, L. V. and Gordon, J. I. (2001). Commensal host–bacterial relationships in the gut. *Science* **292**, 1115–1118.

Horn, M., Collingro, A., Schmitz-Esser, S., Beier, C. L., Purkhold, U., Fartmann, B., Brandt, P., Nyakatura, G. J., Droege, M., Frishman, D., Rattei, T., Mewes, H. W., and Wagner, M. (2004). Illuminating the evolutionary history of chlamydiae. *Science* **304**, 728–730.

Karch, H., Schubert, S., Zhang, D., Zhang, W., Schmidt, H., Ölschläger, T., and Hacker, J. (1999). A genomic island, termed high-pathogenicity island, is present in certain non-O157 Shiga toxin-producing *Escherichia coli* clonal lineages. *Infection and Immunity* **67**, 5994–6001.

Köhler, R., Fanghänel, J., König, B., Lüneberg, E., Frosch, M., Rahfeld, J. U., Hilgenfeld, R., Fischer, G., Hacker, J., and Steinert, M. (2003). Biochemical and functional analysis of the Mip protein: influence of the N-terminal domain and PPIase activity on the virulence of *Legionella pneumophila*. *Infection and Immunity* **71**, 4389–4397.

Lawrence, J. G. and Ochman, H. (1998). Molecular archaeology of the *Escherichia coli* genome. *Proceedings of the National Academy of Sciences USA* **95**, 9413–9417.

Li, C. W., Chen, J. Y., and Hua, T. E. (1998). Precambrian sponges with cellular structures. *Science* **279**, 879–882.

Maurelli, A. T., Fernández, R. E., Bloch, C. A., Rode, C. K., and Fasano, A. (1998). "Black holes" and bacterial pathogenicity: a large genomic deletion that enhances the virulence of *Shigella* spp. and enteroinvasive *Escherichia coli*. *Proceedings of the National Academy of Sciences USA* **95**, 3943–3948.

Müller, W. E., Blumbach, B., and Muller, I. M. (1999). Evolution of the innate and adaptive immune systems: relationships between potential immune molecules in the lowest metazoan phylum (Porifera) and those in vertebrates. *Transplantation.* **68**, 1215–1227.

Müller, W. E., Schroder, H. C., Skorokhod, A., Bunz, C., Müller, I. M., and Grebenjuk, V. A. (2001). Contribution of sponge genes to unravel the genome of the hypothetical ancestor of Metazoa (Urmetazoa). *Gene* **276**, 161–173.

Nakata, N., Tobe, T., Fukuda, I., Suzuki, T., Komatsu, K., Yoshikawa, M., and Sasakawa, C. (1993). The absence of a surface protease, OmpT, determines the intercellular spreading ability of *Shigella*: the relationship between the *ompT* and *kcpA* loci. *Molecular Microbiology* **9**, 459–468.

Ochman, H. and Groisman, E. A. (1995). The evolution of invasion in enteric bacteria. *Canadian Journal of Microbiology* **41**, 555–561.

Ochman, H. and Moran, N. (2001). Genes lost and genes found: evolution of bacterial pathogenesis and symbiosis. *Science* **292**, 1096–1098.

Ölschläger, T. A., Zhang, D., Schubert, S., Carniel, E., Rabsch, W., Karch, H., and Hacker, J. (2003). The high-pathogenicity island is absent in human pathogens of *Salmonella enterica* subspecies I but present in isolates of subspecies III and VI. *Journal of Bacteriology* **185**, 1107–1111.

Perna N. T., Plunkett III G., Burland V., Mau B., Glasner J. D., Rose D. J., Mayhew G. F., Evans P. S., Gregor J., Kirkpatrick H. A., Pósfai G., Hackett J., Klink S., Boutin A., Shao Y., Miller L., Grotbeck E. J., Davis N. W., Lim A., Dimalanta E. T., Potamousis K. D., Apodaca J., Anantharaman T. S., Lin J., Yen G., Schwartz D. C., Welch R. A., and Blattner F. R. (2001). Genome sequence of enterohaemorrhagic *Escherichia coli* O157: H7. *Nature* **409**, 529–533.

Preston, C. M., Wu, K. Y., Molinski, T. F., DeLong, E. F. (1996) A psychrophilic crenarchaeon inhabits a marine sponge: *Cenarchaeum symbiosum* gen. nov., sp. nov. *Proceedings of the National Academy of Sciences USA* **93**, 6241–6246.

Reid, S. D., Herbelin, C. J., Bumbaugh, A. C., Selander, R. K., and Whittam, T. S. (2000). Parallel evolution of virulence in pathogenic *Escherichia coli. Nature* **406**, 64–67.

Rickard, A. H., Gilbert, P., High, N. J., Kolenbrander, P. E., and Handley, P. S. (2003). Bacterial coaggregation: an integral process in the development of multi-species biofilms. *Trends in Microbiology* **11**, 94–100.

Russell, J. B. and Rychlik, J. L. (2001). Factors that alter rumen microbial ecology. *Science* **292**, 1119–1122.

Schubert, S., Cuenca, S., Fischer, D., and Heesemann, J. (2000). High-pathogenicity island of *Yersinia pestis* in enterobacteriaceae isolated from blood cultures and urine samples: prevalence and functional expression. *Journal of Infectious Diseases* **182**, 1268–1271.

Shigenobu, S., Watanabe, H., Hattori, M., Sakaki, Y., and Ishikawa, H. (2000). Genome sequence of the endocellular bacterial symbiont of aphids *Buchnera* sp. APS. *Nature* **407**, 81–86.

Skriwan, C., Fajardo, M., Hägele, S., Horn, M., Michel, R., Krohne, G., Schleicher, M., Hacker, J., and Steinert, M. (2002). Various bacterial pathogens and symbionts infect the amoeba *Dictyostelium discoideum. International Journal of Medical Microbiology* **291**, 615–624.

Stappenbeck, T. S., Hooper, L. V., and Gordon, J. I. (2002). Developmental regulation of intestinal angiogenesis by indigenous microbes via Paneth cells. *Proceedings of the National Academy of Sciences USA* **99**, 15451–15455.

Steinert, M., Birkness, K., White, E., Fields, B., and Quinn, F. (1998). *Mycobacterium avium* bacilli grow saprozoically in coculture with *Acanthamoeba polyphaga* and survive within cyst walls. *Applied and Environmental Microbiology* **64**, 2256–2261.

Steinert, M., Emödy, L., Amann, R., and Hacker, J. (1997). Resuscitation of viable but nonculturable *Legionella pneumophila* Philadelphia JR32 by *Acanthamoeba castellanii*. *Applied and Environmental Microbiology* **63**, 2047–2053.

Steinert, M., Hentschel, U., and Hacker, J. (2000). Symbiosis and pathogenesis: evolution of the microbe-host interaction. *Naturwissenschaften* **87**, 1–11.

Steinert, M., Hentschel, U., and Hacker, J. (2002). *Legionella pneumophila*: an aquatic microbe goes astray. *FEMS Microbiology Reviews* **26**, 149–162.

Steinert, M., Leippe, M., and Roeder, T. (2003). Surrogate hosts: invertebrates as models for studying pathogen-host interactions. *International Journal of Medical Microbiology* **293**, 321–332.

Turon, X., Galera, J., and Uriz, M. J. (1997). Clearance rates and aquiferous systems in two sponges with contrasting life-history strategies. *Journal of Experimental Zoology* **278**, 22–36.

Webster, N. S., Wilson, K. J., Blackall, L. L., and Hill, R. T. (2001). Phylogenetic diversity of bacteria associated with the marine sponge *Rhopaloeides odorabile*. *Applied and Environmental Microbiology* **67**, 434–444.

Welch R. A., Burland V., Plunkett G. 3rd, Redford P., Roesch P., Rasko D., Buckles E. L., Liou S. R., Boutin A., Hackett J., Stroud D., Mayhew G. F., Rose D. J., Zhou S., Schwartz D. C., Perna N. T., Mobley H. L., Donnenberg M. S., and Blattner F. R. (2002). Extensive mosaic structure revealed by the complete genome sequence of uropathogenic *Escherichia coli*. *Proceedings of the National Academy of Sciences USA* **99**, 17020–17024.

Wernegreen, J. J., Ochman, H., Jones, I. B., and Moran, N. (2000). Decoupling of genome size and sequence divergence in a symbiotic bacterium. *Journal of Bacteriology* **182**, 3867–3869.

Wilkinson, C. R., Garrone, G., and Vacelet, J. (1984). Marine sponges discriminate between food bacteria and bacterial symbionts: electron microscope radioautography and *in situ* evidence. *Proceedings of the Royal Society of London B* **220**, 519–528.

CHAPTER 4

Industrial revolution and microbial evolution

Fernando de la Cruz and Julian Davies

4.1. INTRODUCTION

There is significant controversy over the topic of the molecular mechanisms of evolution between supporters of the role of horizontal gene transfer (HGT) (de la Cruz and Davies 2000; Ochman et al. 2000; Bushman 2002) versus those who support a strict "tree of life" model based on ribosomal RNA sequences of living organisms (Kurland et al. 2003). We would like to propose in this chapter that this dispute is somewhat artificial and the dichotomy is essentially fragile. Both sides are fundamentally correct because both processes are important, but complementary. This is so because the corresponding mechanisms involve different time scales, although they both play indisputable roles in the evolution of the haploid unicellular prokaryotes. The different time scales underscore the two levels at which HGT and classical evolution play independent roles in bacterial genome organization. In the same way that optical and electron microscopy operate on different physical scales to reveal different levels of cell structure and function, so short- and long-term evolutionary processes operate with different genetic processes to reach their end points.

It is difficult to assess the significance of events that occurred billions of years ago. Evolution is largely a series of nonreproducible (historical) events of which we see a final, polished, global result, best represented by the tree of life. On the other hand, we can look directly, and thus dissect with microbiological tools, events of the past 200 years, providing a magnified view of the evolutionary process over a (relatively) short period of time that has been important in terms of human association with microbes in health and in the environment. Within this short time span, we can look at concrete consequences of identifiable selective forces on bacterial populations. As it turns out, both views give contradictory results. The dichotomy is more apparent than real.

Long-term (or classical) evolution is a process of "long wavelength" magnitude (eons or millions of years) spanning a period of some 3.5 billion years, starting with the last common ancestor(s) that proceeded through independent unicellular organisms with major diversions into the formation of protists and eukaryotes. See Chapter 1 for further discussion on this topic. These evolutionary pathways chart the formation of the major bacterial taxa as we identify them now (Woese 2000). Long-term evolution was driven by numerous selection factors (most of which are unknown), although the appearance of oxygen in the biosphere (itself a product of prokaryotes – see Chapter 1) was certainly a major component at a given time. The genetic mechanisms participating in vertical evolutionary pathways can only be inferred in retrospect, but nucleotide sequence analysis of modern microbial genomes indicates that significant gene exchange occurred between both related and unrelated genera (Garcia-Vallve et al. 2000; see Chapter 3). In fact, every new bacterium sequenced up to now contains perhaps thirty percent of its genome formed by DNA unrelated to anything else (obviously acquired DNA). On the other hand, the vertically continuous fraction of the genome more or less loosely follows the tree of life, so its phylogeny seems congruent and nonproblematic. As a result, most of the changes that took place by HGT in long-term evolution are understandably blurred when observed after so many millions of years because, in the long term, they did not contribute sufficient genetic material to obscure the slow, continuous influence of random mutagenesis and selection on genome evolution. Thus in the long term, the effects of this type of genetic creativity (HGT) are likely to be concealed.

Unlike long-term evolution, short-term evolution provides opportunities to examine changes as they happen, by observation of genetic mechanisms that affect the structure and function of populations over years instead of eons. The scientific literature is rich with analyses of the changes in microbial function that have occurred in the recent past. We can identify phenomena that took place over the last two centuries or so, since the beginnings of the industrial revolution. During this time, the human population of the earth increased by almost an order of magnitude and concomitantly the industrial revolution bestowed diverse, intensive, and novel selective pressures on the biosphere because of increasing industrial activities and xenobiotic contamination. Most of this began in the early 1800s with the birth of the heavy chemical industry in Germany.

When we look at the effects of this brief and stressful era of microbial evolution (which is still in process) we immediately perceive how bacteria reacted to the challenges brought about by new selective pressures. Responses to xenobiotic compounds or antibiotic resistance clearly bring HGT to center stage (see de la Cruz and Davies 2000; de la Cruz et al. 2002). Looking only from the viewpoint of experimental short-term evolution, HGT seems certainly to be the most important driving force in prokaryotic adaptation. The development of resistance to a multitude of toxins defined directions of microbial evolution. In principle, this intense period of evolution should have been more amenable to close observation and experimental analysis. Unfortunately, this often occurs too long after the event, so science has to rely, once again, on retrospective rather than prospective investigation. There have been many opportunities for the latter, but few have been taken.

4.2. THE MECHANISMS OF SHORT-TERM EVOLUTION

During the intensive industrialization of the world's economies, significant evolutionary change has taken place, affecting many forms of life in the biosphere by the same genetic processes as during the extended period of classical evolution. However, microbes were presumably the most capable of rapid response to a changing environment, which was made possible by the use of HGT, which we shall see is a highly cooperative, community-associated process. The interposition of mutation and HGT determined the nature of the events. There is a good understanding of these events because the end products are the survivors of defined chemical and physical insults. In many instances, the genetic components involved can be identified using the tools of molecular biology. Does this series of events represent a "capsule" of cellular evolution, or is it a different phenomenon? We believe it is the former and thus focus our discussion on the evolutionary role of designed chemotherapeutic agents, although the same principles apply equally to the bacterial evolution of recalcitrance to xenobiotics and industrial pollutants in the environment. A significant component of microbial change may be described in terms of evolved biotransformation and associated transport mechanisms.

Massive industrial activity is not the only causative element in microbial evolution, since in more recent times, domestic activity has played an increasing role (perhaps less well recognized). For example, there is a growing, misperceived, and widely promoted notion that all microbes are dangerous and that human life is constantly threatened by bacteria. Extensive advertising promotes the need to avoid this danger by the use of chemical agents, and it is estimated that there are some 800 products containing biocides on the market in Europe and North America; the amounts exceed those used for therapy. The use of biocides may have significant consequences as a result of inciting survival responses from a microbial population under stress, leading to a new wave of cooperative evolution involving the same genetic mechanisms, as in the case of antibiotic use. This theme is picked up again by Rino Rappuoli in the last chapter in this volume.

Microbial responses to antibiotics are considered the prime example of evolution to chemical pressure and have been studied extensively. Interestingly, most antibiotics are natural products that are normally present at low concentrations in the environment and resistance mechanisms to these inhibitors are widely distributed in nature. The rapid growth of the pharmaceutical industry since the 1940s was associated with the release of unnaturally large quantities of these biologically active compounds into the biosphere. In many environments, they exist at concentrations that are orders of magnitude higher than normal (natural) levels; it can be said that the earth is essentially bathed in a dilute solution of antibiotics!

Antibiotic resistance is, in reality, the product of a highly interactive system involving many different types of microbes. Microbial evolution is a cooperative process based on community structure and dynamics; complex interactions between different genera and species are required to achieve effective HGT. At another level, antibiotic resistance is the result of a systems biology process involving cellular interactions among the host, the pathogen(s), the commensal population, and the antibiotic. HGT is not a simple process because heterologous genes are acquired by new hosts. Gene adaptation usually involves gene tailoring for functional expression, a process that probably requires passage of genetic information through a variety of different hosts resident in the same community. Orthologues of many of the genes encoding antibiotic-inactivating enzymes (based on nucleic acid and protein sequence comparisons) have been identified in a variety of bacteria (commensals). The genes of origin are likely active in (largely unknown) metabolic functions and not as antibiotic resistance determinants in their hosts of origin. Only rarely is an antibiotic resistance phenotype manifest in the primary host, and gene sequence evolution to permit expression is required to establish a resistance phenotype.

Mutation may also be a cooperative process because mutator genes are subject to HGT. In addition, the genetic and biochemical process of mutation (e.g., hypermutability) is influenced strongly by environmental factors. Many antibiotics (and xenobiotics) are themselves mutagens, and others can activate or repress the expression of DNA repair functions that lead to increased frequencies of mutation in different hosts. Several examples of the latter have been described in the recent literature, and some commonly used antibiotics are among the active agents identified. The roles of small molecules in maintaining community function and population is poorly understood and cannot be overestimated. Most bacterial strains have the genetic capacity to produce biologically active peptides, polyketides, and other types of small molecules (<3,000 daltons) and a bewildering array of these molecules are found in nature. The combination of these activities with those of industrial pollutants (e.g., antibiotics) is likely to provoke extensive metabolic and genetic responses in bacterial populations. Many of these responses may lead to permanent alterations (mutation and HGT), leading to enhanced antibiotic resistance (for example). The notion of an enormous and readily accessible bacterial resistance gene pool has been mooted for some time, and mechanisms of gene recruitment have been demonstrated in the laboratory. Again, we must emphasize the critical involvement of interactions in microbial communities; the evolution and establishment of antibiotic resistance phenotypes does not occur in a vacuum!

4.3. THE CONSTRAINTS OF LONG-TERM EVOLUTION

The classic evolutionary tree illustrates impressive conservation of protein sequence, as well as an indubitable congruence in the evolution of many cellular proteins, such as those essential for cell survival. Many reasons have been put forward to explain these important characteristics, which are certainly causally related. Massive congruence in protein sequence emphasizes the importance of well-defined arrangements of metabolic networks in the cell. Given that many biochemical mechanisms exist for gene shuffling, why is there so little evidence of its occurrence (or is it common and we have no way of recognizing it)? The answer has to be because cellular protein networks are so precisely highly articulated. It is difficult to change cell components and not incur loss of fitness or lack of competitiveness. For example, it seems to be difficult to bring about (seemingly) minute sequence changes in most of the central proteins in the cell. More than 500 (?) proteins (the minimal cell genetic backbone) are highly conserved among the entire bacterial kingdom. In an analysis of fifty-seven of these enzyme sets, Doolittle et al. (1996) found an average of thirty-seven percent identity (full range covering a span from twenty to fifty-seven percent) between eubacterial and eukaryotic proteins. This level of conservation is exceedingly high for what we know are the essential residues for enzyme activity. If mutations are rarely allowed, we can easily imagine the barriers to HGT. This is what has been shown, for example, with the gene components responsible for DNA synthesis and the transcriptional and translational machineries, the so-called "informational" genes. They are highly conserved and much more recalcitrant to HGT than are "operational" genes (Jain et al. 1999), probably because they need to interact (and to avoid interacting) with many other protein components of the cell. All this implies that there are many more "lethal" mutations than can be identified in laboratory studies. As a specific example, in *Mycobacterium tuberculosis*, many mutations can be identified when selecting for rifampicin resistance in the laboratory; however, only three of these mutations account for eighty-six percent of those found in clinical isolates (Ramaswamy and Musser 1998). The same is true for streptomycin resistance mutations (in ribosomal protein RpsL). In summary, it would appear that about thirty percent of all amino acid positions are essential in the broad cellular context, and are thus highly conserved across the entire bacterial kingdom. We cannot but think that this fact represents the formidable integration and subsequent rigidity of bacterial protein networks, which are comprised of modules of pathways. An additional consideration, namely that many of these key enzyme sets (e.g., glycolysis and TCA cycle) contain proteins that extensively "moonlight," is described in Chapter 17, in which the consequences for evolution of proteins are also considered.

4.4. RECENT BACTERIAL HISTORY

Here we discuss events that underscore the roles of HGT in short-term bacterial evolution: since 1940 (for antibiotic resistance) or since the late 1800s (for xenobiotic tolerance and transformation). When wide-scale antibiotic use commenced early in the 1950s, microbial geneticists predicted that development of resistance during the clinical use of antibiotics would be unlikely. This conclusion was based on studies of mutation to antibiotic resistance in laboratory experiments and obviously lacked any knowledge of HGT and the extent to which antibiotics would be used. This expectation proved erroneous (de la Cruz and Davies 2000); microbes do not listen to geneticists! Similarly, when the fluoroquinolone (FQ) antibiotics, an entirely synthetic class of DNA gyrase inhibitors with a complex mode of action, were introduced into clinical practice in the 1970s, it was predicted that resistance would require multiple mutations and thus be a very low-frequency event. Bacterial populations failed to listen to reason in this case also; resistance to the FQs developed rapidly as a result of mutations in DNA gyrase gene combined with increased efflux from the cell. FQ-resistant strains are now common among a variety of human and animal bacterial pathogens. Induced hypermutability appears to have been a factor in this case, and although the FQs are not natural products (and not structurally related to any known bacterial product, although a quinolone molecule has been identified as a quorum-sensing autoinducer), plasmid-determined FQ resistance has been identified recently. It is worth noting that plasmid-determined (HGT) resistance to the sulfonamides and trimethoprim, both synthetic antibacterials, is common in clinical situations. This would not have been predicted. Parenthetically, the origins of these resistance determinants remain unknown.

Erythromycin (a macrolide antibiotic) was introduced into the therapeutic armamentarium for the treatment of Gram-positive infections, especially those resistant to penicillin (and its analogues); the antibiotic was also favored for the treatment of patients allergic to beta-lactams. In spite of the fact that erythromycin has some unpleasant side effects (it causes gastric disturbances because of a highly specific interaction with the motilin receptor), it has been used extensively, and a variety of mechanisms of erythromycin resistance have been characterized. A large number of derivatives have been synthesized, and several compounds with improved pharmacologic characteristics have been introduced. The introduction of each new compound has been followed by the appearance of resistant strains with mutationally altered efflux systems or changes in the ribosome; in addition, plasmid-mediated resistance due to methylation of critical sites in 23S rRNA is widespread in a number of important pathogens. Other resistant isolates inactivate macrolides by enzymatic modification of the drug molecule (the resistance genes have been identified on multidrug resistance integrons of Gram-negative bacteria). This plenitude of resistance functions illustrates the incredible ability of microbes to mount resistance responses against toxic agents. One of the (rare) spin-offs of the use of antibiotics such as the macrolides and the aminoglycosides (which target 16S rRNA) is that high-resolution three-dimensional analysis of the binding of these drugs to the ribosome has provided amazingly detailed information on ribosome structure and function; this may yet lead to the discovery of new types of translation inhibitors.

There are many other examples of the evolution of multidrug-resistant bacterial pathogens; any biochemically plausible mechanism of resistance is possible, and combined with HGT, this illustrates the extraordinary resiliency provided by bacterial communities. As described in the essay of Hacker and Kaper (2000), the same holds true for the evolution of pathogens (see Chapter 3). Microbial communities operate on every scale and in any environment. A proper understanding of the genetics and physiology of this type of cooperation is essential to the understanding of antibiotic resistance.

Humans are considered to be the world's greatest evolutionary force, and the industrial activities of the human population have provoked many significant evolutionary changes, largely the result of pollution of the biosphere. During the industrial revolution, many toxic molecules (heavy-metal derivatives and organic molecules) were released into the environment . . . and still continue to be released. Many are naturally occurring compounds that are used in unnatural ways and amounts (e.g., antibiotics). The potent biocide triclosan is an interesting modern example; this compound is used extensively as an industrial and household cleaner. 100,000 tons are produced and used every year in Europe; however, the largest producers are in Russia, China, and India. The world total production may be five times as much. This is significantly more than any antibiotic, and it is all released into the environment as mouthwash; in clothing, bedding, and garbage bags; and in other consumer products. Triclosan has a very specific biochemical action and blocks microbial cell growth by preventing lipid biosynthesis; resistance occurs both by enhanced efflux and by mutation of a specific enoyl reductase gene (*fabI*). Interestingly, a relative of the latter is the target of the important antituberculosis drug isoniazid (Inh). The possibility has been raised that triclosan resistance may contribute to the development of multidrug resistance in bacteria by providing selection for resistance gene clusters such as integrons. However, the fact that use of this biocide may preselect resistance in mycobacterial infections has not received much attention. Could the use of compounds such as triclosan generate a reservoir of resistance genes that may be acquired by pathogens in the future? The full consequences of extreme biocide use remain to be seen.

4.5. CONCLUSIONS

Long-term evolution (mega-years) and short-term evolution (tens of years) are clearly different genetic processes, which occur as the result of different evolutionary pressures and mechanisms. In recent times, we have had the opportunity to observe what happens when microbes adapt to a novel environmental challenge (the use of antibiotics and biocides). In many cases, new genes and metabolic pathways are rapidly recruited from unknown gene pools or adapted from preexisting ones, to give a first response to the challenge. Evolution, in the long run, does not respond to single challenges; the effects of many different individual selection processes are averaged, and the genome as a whole is preserved (with a few notable very ancient exceptions of massive HGT, such as endosymbiosis in the formation of mitochondria and chloroplasts). Evolutionary trees are depictions of the ancient evolution of bacteria, during which metabolic networks and the gene expression machinery crystalized through the combination of spontaneous mutation and occasional acquisition of vital new characteristics by HGT. Subsequent events produced "scars" but did not change the general structure of a bacterial cell. Every newly sequenced bacterial genome contains a significant number of genes with no obvious relationship to previously sequenced ones (the "species-specific" genome). We cannot ignore the possibility that the roots of the speciation process lay in the acquisition of some of this idiosyncratic genetic material. The acquisition of the symbiotic plasmid (pSym) converts an *Agrobacterium* strain into a bona fide *Rhizobium*, capable of nodulation and nitrogen fixation. Similarly, large virulence plasmids and other mobile genetic elements convert the commensal *Escherichia coli* into important pathogens like *Shigella* (discussed in more detail in Chapter 3).

Contemporary evolution demands the accommodation of adaptive changes that occurred in the past two centuries. When looking at present trends (specific selection processes), we observe HGT more than anything else; the genome is changed. Primary HGT is frequently followed by genetic adaptation, as has been adequately demonstrated in the formation of families of -lactamases as antibiotic derivatives are introduced in efforts to counter "new" resistant strains. At any given point in time, HGT is the response to acute selective pressures. Mobile genetic elements are expert devices to promote evolutionary change, and their efficacy is dependent on gene capture within microbial communities in the environment. The storm of genetic change settles in long-term evolution. Most of the genetic change in bacteria that we have seen during the period of human-based industrial revolution will be lost when we disappear as a species from Earth (during the next 10 million years). But this is just a blink of an eye in bacterial evolution. Most of the R plasmids and conjugative transposons, xenobiotic degradation pathways, pathogenicity islands, and the plethora of mobile genetic elements that go with them will slowly decay in the bacterial genomes. The central protein backbone will remain essentially untouched, perhaps including one or two interesting additions or alternatives. This will be the only legacy of HGT after the next 10 million years. But in the meantime, it would have allowed bacteria to cope with the invasion of the human race, and thrive among the continuous new challenges that this one species imposes on the planet every year. One may not see HGT when looking at phylogenetic trees, but its impact is inescapable when dealing with present-day challenges to microbial life.

REFERENCES

Bushman, F. (2002). *Lateral DNA Transfer: Mechanisms and Consequences.* Cold Spring Harbor, NY: Cold Spring Harbor Laboratory Press.

de la Cruz, F. and Davies, J. (2000). Horizontal gene transfer and the origin of species: Lessons from bacteria. *Trends in Microbiology* 8, 128–133.

de la Cruz, F., García-Lobo, J. M., and Davies, J. (2002). Antibiotic resistance: How bacterial populations respond to a simple evolutionary force. In *Bacterial Resistance to Antimicrobials.* ed. K. Lewis, pp. 19–36. New York: Marcel Dekker, Inc.

Doolittle, R. F., Feng, D. F., Tsang, S., Cho, G., and Little, E. (1996). Determining divergence times of the major kingdoms of living organisms with a protein clock. *Science* **271**, 470–477.

Garcia-Vallve, S., Romeu, A., and Palau, J. (2000). Horizontal gene transfer in bacterial and archaeal complete genomes. *Genome Research* **10**, 1719–1725.

Hacker, J. and Kaper, J. B. (2000). Pathogenicity islands and the evolution of microbes. *Annual Review of Microbiology* **54**, 641–679.

Jain, R., Rivera, M. C., and Lake, J. A. (1999). Horizontal gene transfer among genomes: The complexity hypothesis. *Proceedings of the National Academy of Sciences USA* **96**, 3801–3806.

Kurland, C. G., Canback, B., and Berg, O. G. (2003). Horizontal gene transfer: A critical view. *Proceedings of the National Academy of Sciences USA* **100**, 9658–9662.

Ochman, H., Lawrence, J. G., and Groisman, E. A. (2000). Lateral gene transfer and the nature of bacterial innovation. *Nature* **405**, 299–304.

Ramaswamy, S. and Musser, J. M. (1998). Molecular genetic basis of antimicrobial agent resistance in *Mycobacterium tuberculosis*: 1998 update. *Tuberculosis Lung Disease* **79**, 3–29.

Woese, C. R. (2000). Interpreting the universal phylogenetic tree. *Proceedings of the National Academy of Sciences USA* **97**, 8392–8396.

CHAPTER 5

Bacteria evolve and function within communities: Observations from experimental *Pseudomonas* populations

Paul B. Rainey

5.1. INTRODUCTION

A central goal of evolutionary and ecosystem ecologists is to understand how different ecological processes (particularly the interactions among individuals and populations that comprise communities) influence the evolution of diversity (richness and abundance of species or genotypes) and the function of ecosystems. At least three interrelated conceptual frameworks are relevant. The first is adaptive radiation, that is, the genetic and ecological process by which a single lineage rapidly diversifies to generate multiple niche-specialist genotypes (Schluter 2000). The second concerns the ecological parameters that determine community structure and assembly (Weiher and Keddy 1999). The third deals with the connection between communities and ecosystem function (Loreau et al. 2002).

The ecological causes of adaptive radiation are embodied in theory that stems largely from Darwin's insights into the workings of evolutionary change (Darwin 1890), but owes much to developments in the 1940s and 1950s attributable to Lack (1947), Dobzhansky (1951), and Simpson (1953). Recent work has seen a reformulation of the primary concepts (Schluter 2000) and growing awareness of the need to develop a genetical theory of adaptive radiation (Raff [1996] discusses ideas arising from the interplay between developmental mechanisms and evolutionary patterns and provides a good introduction).

Ecological and evolutionary processes that shape communities and determine patterns of assemblage have been, and continue to be, the subject of much research and debate. Competition – or more generally, interactions – among types is of primary importance, but equally significant are ecological processes such as predation and parasitism, disturbance and productivity (energy flow), which together affect the outcome of competition and thus determine community composition (for an introduction see Begon et al. 1996). Studies of food web linkages and thus interactions among different trophic levels are of central importance to understanding and establishing rules of community assemblage. Much current research is directed toward developing knowledge of the components of food webs and the direct and indirect affects associated with disruptions to linkages.

The notion that there might be a link between diversity of a community and its function also stems back to Darwin (1890), but owes much to Elton (1958). Community reconstruction experiments have often shown that diverse communities are both more productive and more resistant to invasion by foreign types. However, recent advances have questioned the validity of a strictly causal connection between diversity and function. This area is currently one of the most actively debated in ecology. Loreau et al. (2002) provides a good introduction to current issues.

This briefest of introductions to the central issues concerning the origin and functional significance of diversity may seem a great distance from the sphere of most microbiological interest; however, a greater degree of community and ecosystem-level thinking is needed to comprehend bacteria–host interactions. Indeed, in the context of the theme of this book, I would go as far as to argue that the study of interactions between bacteria (be they beneficial, benign, or detrimental) and eukaryotic hosts is unlikely to generate meaningful insights unless considered within the context of evolving microbial communities. The effects of bacteria on host biology are rarely ever the result of the activities of single cells, or even single clones. Effects typically stem from the activities of collections of cells (populations) and collections of populations (communities). The specific effects of these communities on host biology will, in most cases, be dependent upon the composition of the community in terms of the number of types (species or genotypes), the relative abundance (commonness or rarity) of these types, the nature of the interactions, and the evolutionary processes that operate continuously at all levels. In addition, the particular types and interactions among types will be subject to evolutionary change. By way of example, imagine a pathogenic organism that can only cause harm once it achieves a minimum population threshold. Its ability to achieve this threshold is likely to depend less on the pathogen's specific arsenal of virulence determinants (about which we typically know a great deal) and more on interactions with co-evolving community members (about which we know very little).

Here I wish to emphasise the importance of understanding the activities of genotypes, strains, or species within the context of the community within which they evolve and function. I intend to do this by drawing upon previously unpublished data from simple model populations, which, although poor proxies for the complexity of real communities, allow insight into the basic ecological and evolutionary mechanisms that drive diversification and determine community composition and function. Space restrictions mean that a comprehensive introduction is not possible: The references given above are only intended to provide access to key elements of the literature.

5.2. EVOLUTIONARY DIVERSIFICATION OF EXPERIMENTAL *PSEUDOMONAS FLUORESCENS* POPULATIONS

The study described here centers upon the evolutionary diversification of *Pseudomonas fluorescens* in soda glass vials; microcosms that contain a small amount of nutrient medium. For the purposes of this chapter, I ask that the reader consider the vial a simple host – an overly simplistic concept admittedly, but one that nevertheless suffices for the points I wish to make. An especially important feature of the host is its complexity: glass vials containing a complex medium and maintained without shaking afford, to an evolving bacterial population, a complex physically structured environment.

When placed in a static glass vial the ancestral genotype of *P. fluorescens* (known as the smooth morph [SM]) rapidly diversifies, generating, by mutation, a range of niche specialist genotypes that are maintained by negative-frequency dependent selection (Rainey and Travisano 1998). Among the niche specialists are two common types: fuzzy spreader [FS] and wrinkly spreader [WS]. Each functional type (of which just a single representative genotype is considered here) has distinctive colony morphology in agar plate culture and occupies a well-defined niche in the broth microcosms (Fig. 5.1). It is important to note that diversification is dependent on spatial heterogeneity (ecological opportunity) – no such diversification occurs in the absence of spatial structure (see Rainey and Travisano, 1998).

Much has been written on the ecological and evolutionary processes driving diversification of *P. fluorescens* (for a recent review, see Rainey et al. 2004 and Kassen and Rainey 2004; for initial insight into the genetics of diversification, see Spiers et al. 2002). Here I focus attention on ecological interactions among the newly evolved community. I also consider some of the functional consequences of diversity. Experimental details are as provided in Rainey and Travisano (1998), but microcosms were incubated at 25°C, that is 3°C lower than used previously. Doubling time (plus standard error from three replicates) in unshaken culture at 25°C for the specific SM, WS, and

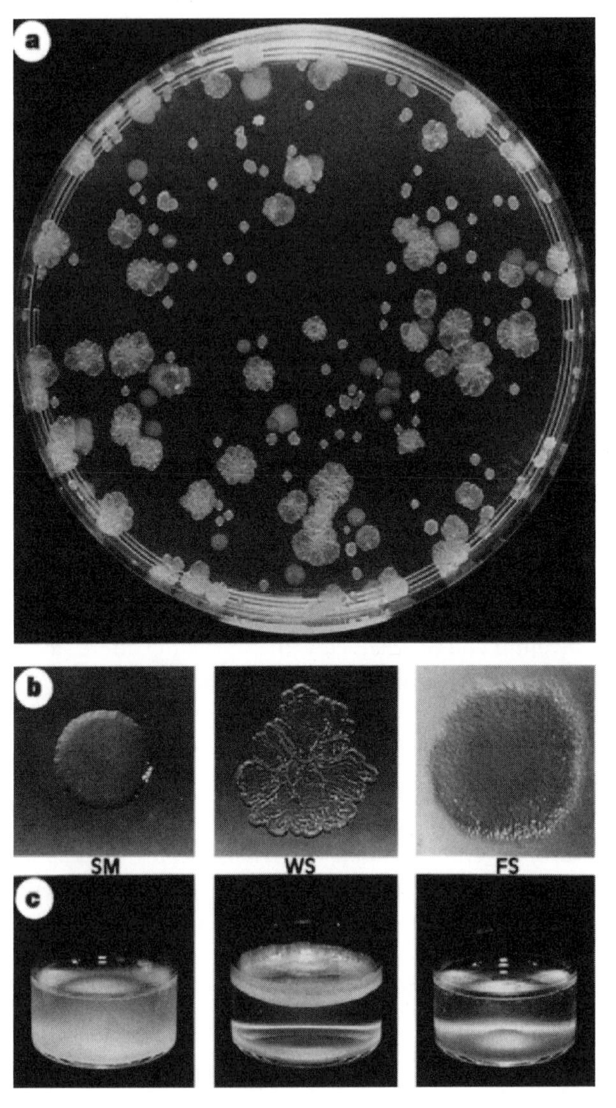

Figure 5.1. Phenotypic diversity and niche specificity among *Pseudomonas fluorescens* SBW25 colonies propagated in spatially heterogeneous microcosms. Microcosms were incubated without shaking to produce a spatially heterogeneous environment. **a**. After seven days, populations show substantial phenotypic diversity, which is seen after plating. **b**. Most phenotypic variants can be assigned to one of three principle morph classes: smooth morph (SM), wrinkly spreader (WS), and fuzzy spreader (FS). **c**. Evolved morphs showed marked niche preferences. Reproduced from Rainey and Travisano (1998).

FS genotypes studied here is as follows: SM, 85 ± 2.1 minutes; WS, 92 ± 1.7 minutes; FS, 100 ± 1.2 minutes.

5.3. ADAPTIVE DIVERSIFICATION OF SM, WS, AND FS IN SPATIALLY HETEROGENEOUS MICROCOSMS

The focus of attention here is the community of diverse genotypes that evolves from ancestral SM and the ecological interactions that shape it. For the sake of completeness, I begin by documenting diversification of WS and FS genotypes as well as the ancestral SM genotype (Fig. 5.2).

Detail of the many different genotypes that arise during the course of each adaptive radiation is not important, but of note is the fact that patterns of diversification depend upon the founding genotype. Diversification of ancestral SM (Fig. 5.2A) has been previously documented, and the patterns here show the typical repeatable emergence of niche specialist genotypes. WS (of which there are many different kinds) is the most common type, with FS being less numerous and taking longer to appear. Scoring cells for defects in motility by plating on semi-solid agar revealed the presence of a nonmotile SM-like genotype that reaches high frequency at day 5. Many minor categories, including mucoid, "fried-egg," and "cartwheel" genotypes are also depicted. Comparison with the patterns of diversification documented previously by Rainey and Travisano (1998) reveals a number of differences, particularly with respect to the abundance of WS genotypes. This is because of the fact that the experiments reported here were performed at 3°C lower than the optimum (28°C) growth temperature of *P. fluorescens* that was used previously. This small reduction in temperature causes a marginal decrease in the growth rate of WS, which significantly impacts upon the strength of the WS mat and thus the ability of WS to dominate the community. For example, at 25°C, WS never exceeds ten percent of the population ($\sim 3 \times 10^8$ cells/ml^{-1}), but at 28°C, WS genotypes comprise more than ninety percent of the population ($\sim 5 \times 10^9$ cells/ml^{-1}).

Diversification of WS (Fig. 5.2B) is limited to the emergence of broth colonizing ancestral (SM)-like cells as depicted by Rainey and Rainey (2003), although the derived SM genotypes are capable of generating patterns of diversity similar to those shown in Figure 5.2A. Evolutionary diversification of FS (Fig. 5.2C) differs considerably from diversification of SM and WS. The FS genotype appears unable to generate, by mutation, genotypes approximating either the ancestral SM type or WS. Indeed the different emergent types are difficult to categorize, and the dynamics of diversification are less

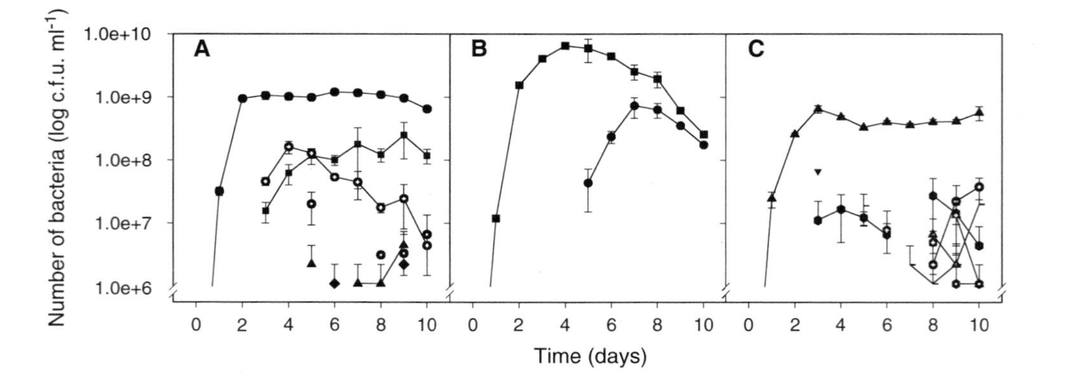

Figure 5.2. Evolutionary diversification of SM (**A**), WS (**B**), and FS (**C**) genotypes of *Pseudomonas fluorescens* SBW25 in spatially heterogeneous microcosms. SM genotypes, circles; WS genotypes, squares; FS genotypes, triangles (upward facing); nonmotile SM-like genotypes are shown in (**A**) as circles containing crosses. Various other symbols depict additional rare genotypes. Data are means and standard errors of three replicate microcosms (microcosms are destructively harvested every 24 hours).

repeatable. However, types ecologically equivalent to SM and WS do arise, although they no longer have the distinctive SM and WS colony morphologies.

5.4. THE ECOLOGY OF DIVERSIFICATION

Previous work established that competitive interactions between the different niche-specialist genotypes were responsible for fueling the process of diversification (Rainey and Travisano 1998; Travisano and Rainey 2000). Evidence of this was obtained by testing for the existence of trade-offs and was achieved by experiments in which the ability of each niche-specialist genotype to invade from rare, a population dominated by another type, was determined. Evidence that different niche-specialists could invade from rare demonstrated the existence of competitive trade-offs which are fully consistent with predictions from theory concerning the role of competition as the engine of divergence (Levene 1953; Abrams 1987). The existence of such trade-offs (arising from the effects of competition – see above) led to experiments which confirmed that the diversity is stable and maintained by negative frequency–dependent selection; the fitness of each genotype being inversely proportional to its frequency with genotypes, being favoured when rare (because resources are most abundant), but not when common (resources are rare and competition intense). The diversity is therefore protected and coexistence of genotypes assured.

In order to gain further insight into the nature of the interactions among co-evolving genotypes, a series of experiments was performed in which the interaction between the main categories of niche-specialist genotypes (SM, FS, WS) was examined. Each genotype was allowed to grow in association with each of the other two types, and population densities were determined on a daily basis. The results of these analyses are presented in Figure 5.3. The dotted lines depict population densities for the founding genotypes (from Fig. 5.2). The solid line in each graph shows fluctuations in the abundance of each genotype when grown with another type. The difference, therefore, between the solid and dotted lines indicates the magnitude of the interaction effect.

The nature and magnitude of the interaction effect is ascertained by comparing the solid and dotted lines. For example, the effect of FS on SM is depicted in the left-hand panel of Figure 5.3A; the effect of SM on FS is shown in the right-hand panel. Figure 5.3B shows the effect of WS on SM, and SM on WS, in the left-hand and right-hand panels, respectively. Figure 5.3C shows the effect of WS on FS, and FS on WS, in the left-hand and right-hand panels, respectively.

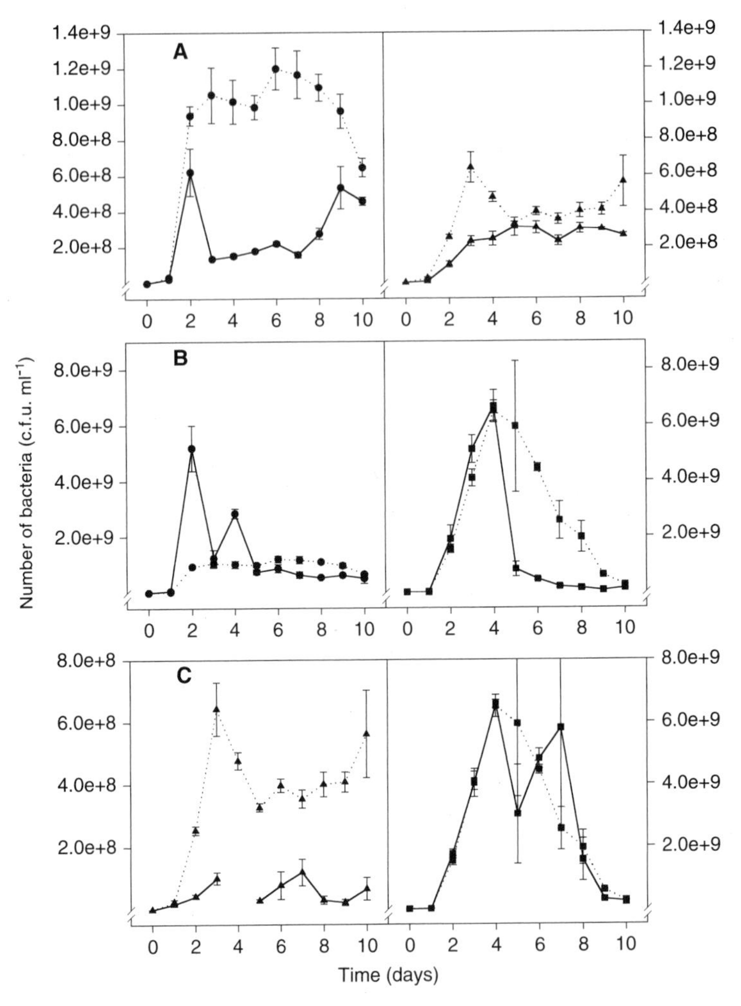

Figure 5.3. Pair-wise interactions among SM, WS, and FS genotypes of *Pseudomonas fluorescens* SBW25 in spatially heterogeneous microcosms. Microcosms were founded by equal numbers of each genotype. The dotted lines depict population densities for the founding genotypes (from Fig. 5.2). The solid line in each graph shows fluctuations in the abundance of each genotype when grown with another type. The difference between the solid and dotted lines indicates the magnitude of the interaction effect. The dotted line in the left-hand panel of (**A**) shows fluctuations in the abundance of the ancestral SM genotype and corresponds to the data for SM in Figure 5.2A. The same data are also included in the left-hand panel of (**B**) (albeit on a differently scaled y-axis). The dotted line in the right-hand panel of Figure (**A**) depicts fluctuations in the abundance of the derived FS genotype over 10 days and is identical to data for FS in Figure 5.2C. Data for FS is also presented in (**C**) (left-hand panel). Fluctuations in the abundance of the derived WS

The interactions reveal a remarkably complex community in which the interactions among genotypes are far more than the sum of the individual components. The full complexity of the community, of which this is just a simplified snapshot (many dozens of different genotypes exist in a 10-day-old microcosm), is hard to envisage. I struggle to comprehend the fact that this community is not the outcome of a long and tortuous period of evolution. That the different types and interactions have arisen during the course of just 10 days of selection (less than 100 generations) is testimony to the extraordinary power of mutation and selection. Equally remarkable is the fact that the derived genotypes (at least of WS) can differ from the ancestral type by as little as a single nucleotide (Spiers et al. 2002; Spiers et al., unpublished) and yet from an ecological perspective, they might as well be a different species – if not a different phyla (Rainey and Rainey 2003).

5.5. INTERFERENCE COMPETITION MARKS THE INTERACTION BETWEEN SM AND FS

A detailed examination of the data in Figure 5.3A reveals that interference competition lies at the heart of the interaction between FS and SM: FS has a strong inhibitory effect on the ancestral SM genotype. This result is surprising and not easily explained given the fact that FS has a doubling time that is approximately 15 minutes longer than the doubling time of SM. SM appears unaffected by FS through the first day of growth; at two days, there is evidence of a negative effect and this becomes clearly apparent at day 3. The most likely explanation for the inhibitory effect of FS on SM is allelopathy, but the precise nature of the factor is unknown. Interestingly, WS is not affected by FS (see below); in fact, evidence of the "immunity" is evident in the left panel of Figure 5.3A, in which the increase in abundance of SM from day 7 onward correlates with the evolutionary emergence of WS (from SM). As WS increases in frequency (data are not presented on the graph to aid clarity), it suppresses the growth of FS and facilitates the growth of SM (see below). The effect of SM on FS (right-hand panel, Fig. 5.3A) also shows evidence of interference competition, but this is

Figure 5.3 (*continued*) genotype are shown in the right-hand panels of (**B**) and (**C**) and are taken from Figure 5.2B. SM genotypes, circles; WS genotypes, squares; FS genotypes, triangles (upward facing). Data are means and standard errors of three replicate microcosms (microcosms are destructively harvested every 24 hours).

less pronounced, with the interaction between FS and SM being strongly asymmetric.

5.6. FACILITATION MARKS THE INTERACTION BETWEEN SM AND WS

The interaction between SM and WS reveals similar surprises that could not have been predicted a priori. A full analysis of this interaction has been presented elsewhere (Rainey and Rainey 2003), but briefly, SM benefits significantly from the presence of WS because of its ability to hitchhike within the WS mat and thereby gain preferential access to oxygen. The rapid fluctuation between days 2 and 5 can be explained by hitchhiking (increase between days 1 and 2), the death of SM cells within the broth phase (decrease between days 2 and 3), continued benefit due to hitchhiking (increase between days 3 and 4), and eventually collapse of the WS mat (day 5). Examination of the right-hand panel of Figure 5.4B shows that WS is unaffected by SM until day 5, when there is a sudden reduction in WS cells due to collapse of the WS mat. The premature collapse of the mat is wrought by the presence of SM, which by invading the mat and contributing nothing to the fabric of the mat, weakens its strength and causes the mat to fall earlier than it otherwise would (see Rainey and Rainey 2003 for a full account). Being maintained by negative frequency–dependent selection, the WS mat is a permanent feature of the evolving communities and reforms following each collapse.

5.7. ASYMMETRICAL INTERFERENCE COMPETITION MARKS THE INTERACTION BETWEEN WS AND FS

The interaction between WS and FS brings no real surprises. Although there is no physical overlap between the two niches, occupancy of the air–liquid interface in the absence of the mat-collapsing SM genotype means that the bottom of the microcosm is likely to become highly anoxic and difficult for FS to colonise. Indeed, FS barely rises to 10^8 cells/ml^{-1} and is undetectable at day 4, the time at which the WS mat population is at its peak. From day 4 onward, even in the absence of SM, the density of WS begins to decline, in part because of the collapse of mats under their own weight. Once the mat has sunk, the FS genotype just manages to increase in number. Although the density of WS fluctuates wildly after day 4 in the presence of FS, FS seems to have little if any impact on WS. The error bars associated with WS after day 4 are typically large, reflecting the fact that mat collapse is in part a stochastic event determined by, for example, random vibrations from equipment.

At this point, it is interesting to contemplate the persistence of FS in the community. Previously we showed that FS was unable to invade, from rare, a population dominated by WS (Rainey and Travisano 1998), whereas all other combinations of the three genotypes could. This indicates that FS is not maintained by a frequency-dependent interaction with WS and if competing on purely ecological grounds with WS, would be driven extinct, a finding which is supported by the data here. This then leads to the question of the emergence of FS in the first place and its maintenance within the community. Its maintenance, surprisingly, is assured, because of the frequency-dependent interactions between SM and WS. Whenever WS becomes common it is at a disadvantage with respect to SM, thus SM prevents WS from completely dominating, which allows the persistence of FS. In other words, SM appears to facilitate the growth of FS, even though FS inhibits the growth of SM! Indeed, the interactions between SM, WS, and FS show evidence of non-transitivity: SM can out-compete WS, WS can out-compete FS, and FS can out-compete SM.

5.8. COMBINED COMMUNITY DYNAMICS OF SM, WS, AND FS

When these experiments were originally conducted, they were done so in part to see whether it was possible to predict, on the basis of pair-wise interaction experiments (Fig. 5.4), the ecological dynamics of a community comprising all three genotypes. The results, shown in Figure 5.5, are partly interpretable on the basis of what was learnt from analysis of the two-way interactions. SM dominates through the first two days, partly because of the fact that it has the fastest doubling time but also because of hitchhiking with WS. At day 3, WS achieves population densities more than ten times that of either SM or FS and coincides with the peak in FS density: the density of SM is at its lowest. In the presence of FS, SM is unable to gain the same benefit from WS that it achieves in the absence of FS. Surprisingly though, FS reaches its highest population density at the time when WS is also at a high point (in the pair-wise interactions, we observed the opposite). From day 4 on, WS density begins to decline, and this is associated with an increase in SM and a decrease in FS. The increase in SM probably reflects increased access to oxygen. The gradual decline from day 6 is typical of these experiments conducted in degrading microcosms.

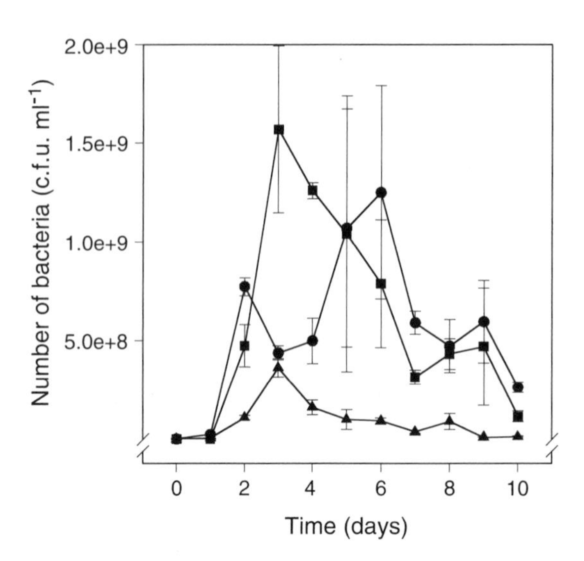

Figure 5.4. Three-way interactions among SM, WS, and FS genotypes of *Pseudomonas fluorescens* SBW25 in spatially heterogeneous microcosms. Microcosms were founded by equal numbers of each genotype. SM genotypes, circles; WS genotypes, squares; FS genotypes, triangles (upward facing). Data are means and standard errors of three replicate microcosms (microcosms are destructively harvested every 24 hours).

5.9. CONNECTING DIVERSITY TO COMMUNITY FUNCTION

The relationship between diversity and productivity is currently the subject of much debate (Loreau et al. 2002). Hodgson et al. (2002) used experimental populations of *P. fluorescens* to test elements of theory and showed a complex relationship between diversity and two measures of ecosystem function: productivity and resistance to invasion by a foreign type. Of some surprise was the range of complex interactions between different functional groups (SM, WS, and FS). Further insights into these interactions have been presented above.

In order to make a link to community function, total cell density (a measure of productivity) from microcosms assembled with different numbers and combinations of functional groups (as depicted in Figs. 5.2, 5.4, and 5.5) is shown in Figure 5.6. The dotted line in each panel shows the productivity of microcosms assembled from three functional groups (SM, WS, and FS) across 10 days. In addition, the left-hand panel depicts the productivity of SM, FS, and SM+FS communities; the central panel depicts the productivity of SM, WS, and SM+WS communities; and the right-hand panel depicts the productivity of WS, FS, and WS+FS communities. In order to summarise

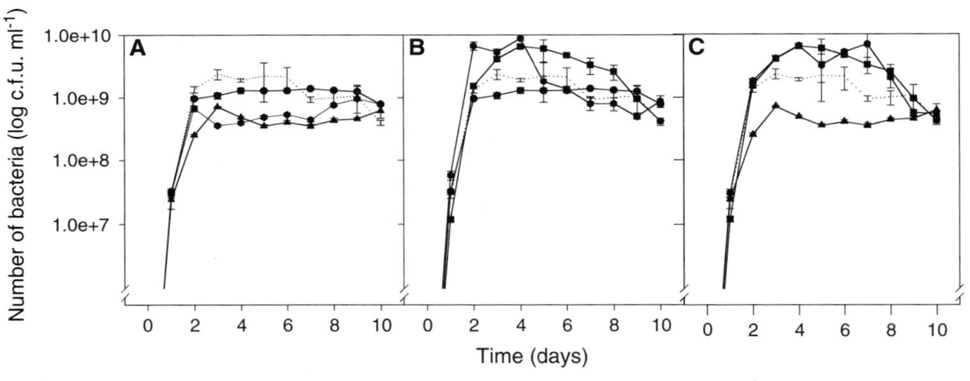

Figure 5.5. Productivity of different community assemblages. The dotted line in each panel shows the productivity of communities comprised of all three genotypes (SM, WS, FS). **A.** the productivity of SM populations (circles), FS populations (triangles) and SM+FS communities (hexagons). **B.** SM populations (circles), WS populations (squares), and SM+WS communities (hexagons). **C.** WS populations (squares), FS populations (triangles), and WS+FS communities (hexagons). Data are means and standard errors of three replicate microcosms (microcosms are destructively harvested every 24 hours).

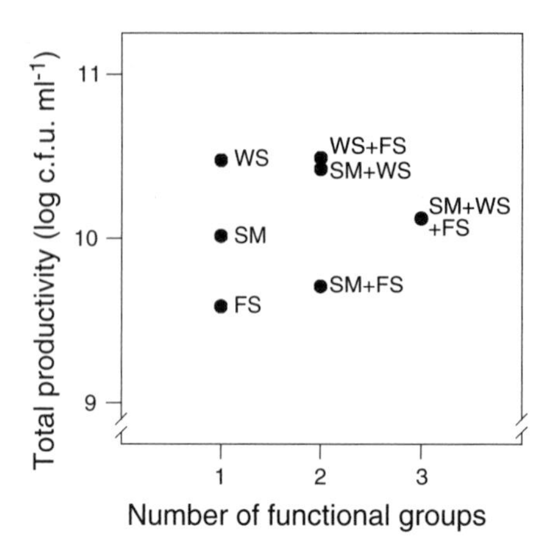

Figure 5.6. Relationship between total productivity and diversity from communities comprising different numbers of functional groups. Labels beside symbols indicate community composition.

these data, the area under each line was integrated to give a total measure of productivity from each microcosm, which was then plotted against the number of functional groups (Fig. 5.6). Within the constraints of this limited data set, here it appears that diversity (number of functional groups) and productivity are not causally connected.

Examination of the composition of the high-productivity communities reveals evidence of a dominance effect (a single type – in this case WS – that contributes disproportionately to productivity). Similarly, the low-productivity microcosms contain FS. Thus, productivity appears to depend largely on the presence or absence of particular functional groups, which are themselves subject to the kinds of interactions discussed above.

These results, like the more detailed analysis by Hodgson et al. (2002), are consistent with a growing stream of work that finds evidence of ecological redundancy and no clear sign that a reduction of diversity adversely affects ecosystem function. However, this interpretation very much depends on the dominance effect – the significance of which in natural communities remains unclear (Huston et al. 2000; Loreau 2001; Wardle 2001;). What is most interesting is not so much the lack of a clear causal connection between diversity and function – indeed it is unclear that one should exist, or that there is functional redundancy, but rather the remarkable scope for interactions to evolve quite by accident, among competing genotypes. That these interactions are as likely to be facilitative as competitive suggests that the interactions themselves, their origin, and particularly the scale at which they operate may be the key to a greater understanding of the functional significance of diversity in bacteria–host interactions.

5.10. CONCLUSIONS

Our current understanding of the interaction between bacteria and their hosts is focused almost exclusively upon the molecular determinants that govern the outcome of specific interactions. Insights stem primarily from the analysis of mutants that display gross defects in terms of their ability to elicit a specific host response. The idea that the effects of bacteria on their hosts may be determined by ecological and evolutionary processes is rarely ever considered.

The experiments outlined here, I hope, have served two main purposes. The first is to demonstrate that, like it or not, recognise it or not, it is impossible to halt evolution. Bacteria, on account of short generation times and large population sizes, are prone to rapid evolutionary change – in the face of ecological opportunity and competition, populations can rapidly diversify. The divergent types that arise may show little resemblance, in terms of core traits, to the bacteria that were the focus of original attention – even over short time frames.

The second purpose is to show that the specific interactions that arise between different members of a microbial community (and between the microbial community and its host) can radically affect the ecological success of individual genotypes. As mentioned in the introduction, the ability of a specific genotype or functional group to elicit a host response is often dependent upon the population reaching a certain threshold level. The probability that this is achieved is likely to depend less on specific virulence determinants and more on the presence or absence of other interacting genotypes within the community. It is interesting to note just how unpredictable such interactions can be.

Returning to the previously suggested analogy of the glass vial as a proxy for a eukaryotic host, consider the possibility that one component of the *P. fluorescens* community, for example, the SM genotype, is capable of eliciting a beneficial effect on its host. In order to do this, it must achieve a population size of more than 10^9 cells/ml^{-1}. Clearly, if FS is a significant component of the community, then the threshold is unlikely to be reached. Conversely, if WS is a significant component, then the threshold has a high probability of being met. Community context matters!

Not surprisingly, these data are not unique in showing that community context matters. A great many studies – some using model bacterial populations – have provided similar insights. Further examples and discussion of relevant issues can be found in Bruno et al. (2003), Horner-Devine et al. (2003) and Day and Young (2004), among others.

The challenges of studying bacterial–host interactions within the context of a more complex and realistic biological framework are considerable. The glass vials used here are a grossly oversimplified proxy for "host," and the vast majority of host-associated microbial communities are many times more diverse than the *P. fluorescens* populations used here. So how do we progress? As always, simple models that allow insight into mechanisms are desirable. It is important that we begin to incorporate ecological and evolutionary dimensions into experimental design. One simple way to do this is to let go of the cherished idea that the effects of one organism on another can be determined solely by examining the effect of mutants inoculated into gnotobiotic hosts (as described in Chapter 12). Indeed, once a broader conceptual framework becomes established, then an entire range of simple but insightful experiments suggest themselves. For example, the effect of a chosen bacterium (and specific mutants) on its host can be studied within the context of the community of microbes typically found associated with that host. Community richness and genotype (species) abundance can be readily manipulated and effects on abundance and host biology monitored. Such experiments stand to provide insight into the ecological mechanisms maintaining diversity within communities and the stability of diversity over time and in the face of changes in ecosystem processes, such as disturbance, productivity, parasitism, and predation. Importantly, clues as to the interactions crucial to the success of the bacterium of interest will emerge: It is even possible that bacteria previously considered benign might suddenly become the focus of attention because of a facilitative effect on the growth of the beneficial organism. Of course, insights such as these also open the door to alternative means of controlling harmful organisms and promoting the growth of beneficial bacteria.

Ultimately we need to develop a greater awareness of basic ecological and evolutionary processes and learn to apply these principles whenever the growth of bacteria is considered – be that in the laboratory environment or the wild, and especially in the context of bacterial–host interactions. The key to a more holistic and comprehensive understanding of biological systems lies at the crossroads between disciplines. The study of bacterial–host interactions, with the significant challenges this field poses, might well be the field that provides just the kind of impetus that brings about a paradigm shift.

ACKNOWLEDGEMENTS

I thank NERC (UK) and the BBSRC (UK) for financial support and members of the laboratory of experimental evolution at Auckland and Oxford for stimulating discussion.

REFERENCES

Abrams, P. A. (1987). Alternative models of character displacement and niche shift. 2. Displacement when there is competition for a single resource. *American Naturalist* **130**, 271–282.

Begon, M., Harper, J. L., and Townsend, C. R. (1996). *Ecology*. Oxford: Blackwell Science.

Bruno, J. F., Stachowicz, J. J., and Bertness, M. D. (2003). Inclusion of facilitation into ecological theory. *Trends in Ecology & Evolution* **18**, 119–125.

Darwin, C. (1890). *The Origin of Species*, 6th ed. London: John Murray.

Day, T. and Young, K. A. (2004). Competitive and facilitative evolutionary diversification. *BioScience*, **54**, 101–109.

Dobzhansky, T. (1951). *Genetics and the Origin of Species*, 3rd ed. New York: Columbia University Press.

Elton, C. S. (1958). *The Ecology of Invasions by Animals and Plants*. London: Methuen.

Hodgson, D. J., Rainey, P. B., and Buckling, A. (2002). Mechanisms linking diversity, productivity and invasibility in experimental bacterial communities. *Proceedings of the Royal Society London B Biology* **269**, 2277–2283.

Horner-Devine, M. C., Carney, K. M., and Bohannan, B. J. M. (2003). An ecological perspective on bacterial biodiversity. *Proceedings of the Royal Society London B Biology* **271**, 113–122.

Huston, M. A., Aarssen, L. W., Austin, M. P., Cade, B. S., Fridley, J. D., Garniey, E., Grime, J. P., Hodgson, J., Lauenroth, W. K., Thompson, K., Vandermeer, J. H., and Wardle, D. A. (2000) No consistent effect of plant diversity on productivity. *Science* **289**, 1255.

Kassen, R. and Rainey, P. B. (2004). The ecology and genetics of microbial diversity. *Annual Review of Microbiology* **58**, 207–231.

Lack, D. (1947). *Darwin's Finches*. Cambridge: Cambridge University Press.

Levene, H. (1953). Genetic equilibrium when more than one ecological niche is available. *American Naturalist* **87**, 331–333.

Loreau, M., Naeem, S., and Inchausti, P. (2002). *Biodiversity and Ecosystem Functioning*. Oxford: Oxford University Press.

Raff, R. A. (1996). *The Shape of Life: Genes, Development, and the Evolution of Animal Form*. Chicago and London: University of Chicago.

Rainey, P. B., Brockhurst, M. A., Buckling, A., Hodgson, D. J., and Kassen, R. (2005). The use of model *Pseudomonas fluorescens* populations to study the causes and consequences of microbial diversity. In *Soil Biodiversity and Ecosystem Function*, ed. R. Bardgett, D. Hopkins, and M. Usher. Cambridge: Cambridge University Press; in press.

Rainey, P. B. and Rainey, K. (2003). Evolution of cooperation and conflict in experimental microcosms. *Nature* **425**, 72–74.

Rainey, P. B. and Travisano, M. (1998). Adaptive radiation in a heterogeneous environment. *Nature* **394**, 69–72.

Schluter, D. (2000). *The Ecology of Adaptive Radiation*. Oxford: Oxford University Press.

Simpson, G. G. (1953). *The Major Features of Evolution*. New York: Columbia University Press.

Spiers, A. J., Kahn, S. G., Bohannon, J., Travisano, M., and Rainey, P. B. (2002). Adaptive divergence in experimental populations of *Pseudomonas fluorescens*. I. Genetic and phenotypic bases of wrinkly spreader fitness. *Genetics* **161**, 33–46.

Travisano, M. and Rainey, P. B. (2000). Studies of adaptive radiation using model microbial systems. *American Naturalist* **156**, S35–S44.

Wardle, D. A. (2001). Experimental demonstration that plant diversity reduces invasibility: evidence of a biological mechanism or a consequence of a sampling effect. *Oikos* **95**, 161–170.

Weiher, E. and Keddy, P. (1999). *Ecological Assembly Rules: Perspectives, Advances, Retreats*. Cambridge: Cambridge University Press.

II Bacterial ecology and the host as an environment

CHAPTER 6

Coral symbiosis: The best and worst of three kingdoms

Eugene Rosenberg

6.1. INTRODUCTION

Coral reefs are not only a joy to behold, they are essential for the health of the oceans and the economies of many countries. Reefs contain an enormous biodiversity, comparable to rain forests. They protect coastlines from erosion caused by the energy of the sea and provide the major source of protein for several underdeveloped countries, as well as the income from the tourist industry. The great biodiversity of coral reefs is the source of new drugs for the pharmaceutical industry. In short, coral reefs are a major asset that should be preserved.

Unfortunately, coral reefs throughout the world are in trouble. During the last few decades, approximately thirty percent of the world's coral reefs have been severely damaged as a result of coral bleaching and other diseases (Hughes et al. 2003). Although the aetiology of coral bleaching and most of the other diseases of corals is not completely understood, what is clear is that disease outbreaks are highly correlated with increased seawater temperatures, resulting from global warming (Rosenberg and Ben-Haim 2002). Based on predicted increases in seawater temperature, Hoegh-Guldberg (1999) has predicted that during the next 20 to 50 years, as seawater temperatures continue to increase, coral reefs will have only remnant populations of reef-building corals.

The fact that corals have succeeded, for thousands of years, in producing the largest and most magnificent biological structures has often been attributed to the special symbiosis between the coral animal and endosymbiotic microalgae, referred to as zooxanthellae. More than fifty percent of the corals' nutrients are derived from the photosynthetic products of zooxanthellae (Muscatine 1990). On a global scale, the most serious disease of corals is the disruption of algae/coral symbioses, resulting in whitening (bleaching) of the coral. The nutritionally depleted bleached corals stop multiplying (Szmant and Gassman 1990), are more susceptible to other diseases (Harvell et al. 2002), and often die. The role of bacteria in the health and disease of corals has, until recently, been largely ignored. After briefly presenting the three most important hypotheses on the relationship between seawater temperature and coral bleaching, I discuss coral-associated bacteria as symbionts and potential opportunistic pathogens.

6.2. THE STRESS HYPOTHESIS OF CORAL BLEACHING

A large number of reports have indicated that coral bleaching occurs shortly after corals are exposed to seawater temperatures 1°C to 2°C higher than their normal maximum (reviewed by Hoegh-Guldberg 1999). Furthermore, laboratory experiments have shown that when temperatures in aquaria are gradually raised to above the maximum that the corals are faced with in their natural habitat, they show bleaching (Jokiel and Coles 1990, 1977). More careful analysis of the field observations and laboratory experiments demonstrated that it was a combination of high temperature and solar radiation that was responsible for the bleaching (Jokiel 2004). These studies have led to the generally accepted hypothesis that stress on the coral results in the disruption of the coral/algae symbiosis (Brown 1997).

During the last few years, there has been a serious effort to explain the stress hypothesis of coral bleaching in terms of photoinhibition (reviewed by Stambler and Dubinsky 2004). According to this model, heat stress interrupts the flow of light energy from the photosystems to the dark reactions, in which carbon dioxide is fixed by the enzyme Rubisco. The light energy, which can not be transferred, builds up and is transferred to oxygen, creating active oxygen (O_2^-). The active oxygen then denatures key photosynthetic proteins of the zooxanthellae, causing an inhibition of photosynthesis. In essence, light, which is necessary for the higher productivity of coral reefs under normal conditions, becomes a liability under condition of higher-than-normal temperatures (Hoegh-Guldberg 1999).

Although there are considerable observational and experimental data supporting the stress hypothesis, the evidence that the mechanism involves photoinhibition is less compelling. One prediction of the photoinhibition model is that any substance that inhibits the dark reactions should induce bleaching. Jones et al. (1998) have reported that cyanide, which blocks dark reactions, leads to symptoms similar to bleaching. Another prediction of the model is that zooxanthellae from bleached corals should be photosynthetically damaged. However, Ralph et al. (2001) demonstrated that zooxanthellae expelled from corals that had been subjected to 33°C were largely unaffected in their photosynthesis and could be heated to 37°C before showing temperature-induced photosynthesis impairment. It is, of course, possible that the zooxanthellae were damaged in the coral, but recovered when expelled. Clearly, more experiments are needed to test the photoinhibition model.

6.3. THE ADAPTIVE HYPOTHESIS OF BLEACHING

Buddemeier and Fautin (1993) formulated the "adaptive bleaching hypothesis" (ABH) of coral bleaching to explain the following three lines of evidence:

1. Metabolically different zooxanthellae could form symbioses with the same host (e.g., Kinzie and Chee 1979).
2. DNA analysis of zooxanthellae from corals demonstrated that they constituted an extraordinarily diverse group of algae (Rowan and Powers 1991).
3. Corals showed a large variation in bleaching susceptibility, occurrence, and recovery (e.g., Oliver 1985; Jokiel and Coles 1990).

The basic principle of the ABH is that by changing combinations of coral hosts and symbiotic algae, "new" ecospecies are created with different environmental potentials and tolerances. With specific regard to coral bleaching, the ABH predicts that hosts that recover from bleaching may acquire different symbiont(s). In essence, this could allow corals to adapt to the higher temperatures in a much more rapid manner than through classical Darwinian evolution. In a recent publication, Buddemeier, 2004 have both clarified some of the phrasing of the hypothesis and generalized it to include viral and microbial disease.

6.4. THE BACTERIAL INFECTION HYPOTHESIS OF CORAL BLEACHING

The hard scleractinian coral *Oculina patagonica* is abundant throughout the Mediterranean Sea. Bleaching of *O. patagonica* was first observed along the Mediterranean coast of Israel in the summer of 1993 (Fine and Loya 1995). Subsequently, extensive bleaching has been recorded each summer when the seawater temperature reaches a maximum of 29°C to 31°C. During the winter, most of the corals recover. Bleaching of colonies of *O. patagonica*

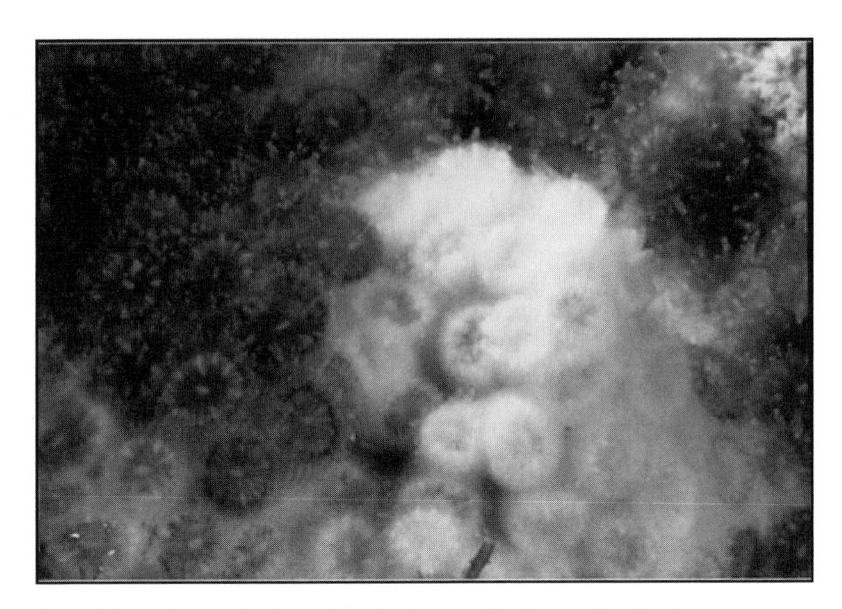

Figure 6.1. A colony of *Oculina patagonica* showing bleached and healthy tissues.

is similar to bleaching of corals in reef systems in: (i) correlation to high seawater temperature, (ii) loss of endosymbiotic zooxanthellae without any apparent tissue damage, (iii) impairment of reproductive capability, and (iv) slow reversibility of the disease when the temperature decreases. The ease of maintaining healthy *O. patagonica* in laboratory aquaria and the predictable cyclic behavior of the disease along the coast of Israel make it an excellent model system to study coral bleaching. A colony of *O. patagonica* containing healthy (dark) and bleached (white) regions is shown in Figure 6.1. Koch's postulates were applied to demonstrate that the etiological agent of the bleaching disease of *O. patagonica* is *Vibrio shiloi* (Kushmaro et al. 1996, 1997). The pathogen has been classified as a new species in the genus *Vibrio*, closely related to *Vibrio mediterranei* (Kushmaro et al. 2001). In controlled aquaria experiments, it was shown that the infection of *O. patagonica* with *V. shiloi* and the resulting bleaching were temperature dependent, occurring only at elevated summer seawater temperatures (Kushmaro et al. 1998). At 28°C, as few as 120 *V. shiloi* cells are sufficient to infect and bleach *O. patagonica*, whereas at 16°C, even one hundred million cells will not cause bleaching. During the last five years, data have been published outlining the sequential steps in the infection of *O. patagonica* by *V. shiloi* and how temperature affects various virulence factors.

6.4.1. Virulence Mechanisms of *V. shiloi*

Chemotaxis

Using the capillary tube assay, it was shown that *V. shiloi* is attracted to the mucus obtained from *O. patagonica* (Banin et al. 2001a). Motility and chemotactic behavior were present when the bacteria were grown either at 16°C or 25°C.

Adhesion

The next step in the infectious process is the adhesion of *V. shiloi* to a β-galactoside–containing receptor on the coral surface (Toren et al. 1998). This was demonstrated by the inhibition of binding to corals by methyl-β-D-galactopyranoside and the specific binding of *V. shiloi* to Sephadex beads containing covalently bound β-galactoside. Adhesion is both coral and bacterium specific. The temperature of bacterial growth was critical for the adhesion of *V. shiloi* to the coral. When the bacteria were grown at the low winter temperature (16°C to 20°C), there was no adhesion to the coral, regardless of what temperature the coral had been maintained. However, bacteria grown at elevated summer temperatures (25°C to 30°C) adhered avidly to corals maintained either at low or high seawater temperatures. The important ecological significance of these findings is that the environmental stress condition (high temperature) is necessary for the coral bleaching pathogen to initiate the infection and become virulent.

The β-galactoside–containing receptor that *V. shiloi* recognizes on the *O. patagonica* surface is present in the coral mucus. This was demonstrated by binding the coral mucus to ELISA plates and showing that the bacteria adhered avidly to the mucus-coated plates (Banin et al. 2001a). Adhesion was inhibited by methyl-β-D-galactopyranoside. Interestingly, the receptor is present only in corals that contain photosynthetically active zooxanthellae. *V. shiloi* does not adhere to bleached corals or to white (azooxanthellate) *O. patagonica* cave corals. When mucus was removed from healthy corals, adhesion of *V. shiloi* was inhibited. Furthermore, addition of 3-(3, 4-dichlorophenyl)-1, 1-dimethyl urea (DCMU), an inhibitor of algal photosynthesis, to mucus-depleted corals prevented the coral from synthesizing the receptor or transporting it to the mucus. Thus, *V. shiloi* is able to distinguish healthy corals from those which are photosynthetically impaired, and it only adheres to healthy zooxanthellate *O. patagonica*.

Penetration and Intracellular Multiplication

Electron micrographs of thin sections of *O. patagonica* following infection with *V. shiloi* demonstrated large numbers of bacteria in the epidermal layer of the coral (Banin et al. 2000). Using monoclonal antibodies specific to *V. shiloi*, it was shown that the observed intracellular bacteria were, in fact, *V. shiloi*. The gentamicin invasion assay was used to measure the kinetics of *V. shiloi* penetration into the epidermal cells. The assay relies on the fact that the antibiotic gentamicin does not penetrate into eukaryotic cells (Isberg and Falkow 1985). Thus, only *V. shiloi* cells that have penetrated into the coral cells escape the killing action of gentamicin. After adhesion was complete in approximately 13 hours, the bacteria began to penetrate into the coral as determined by both total counts and colony-forming units. By 24 hours, forty to fifty percent of the inoculated *V. shiloi* had penetrated into coral cells. From 24 to 72 hours, the intracellular bacteria multiplied (based on total counts), reaching 3×10^8 bacteria per coral fragment. When the infected corals were maintained at the high summer temperatures, the bacteria remained at 10^8–10^9 cm^{-3} for at least two weeks.

Differentiation into the Viable-But-Not-Culturable (VBNC) State

At the same time *V. shiloi* multiplies inside the coral tissue (24 to 48 hours after infection), the number of colony-forming units (cfu) decreases more than a thousandfold. Entry of bacteria into a state described as VBNC has been reported repeatedly with a large number of bacterial species, including several *Vibrio* species (Colwell et al. 1985; Oliver et al. 1991; Lee and Ruby 1995). A bacterium in the VBNC state has been defined as "a cell which can be demonstrated to be metabolically active, while being incapable of undergoing the sustained cellular division required for growth on a medium normally supporting growth of that cell" (Oliver 1993; see also Coates 2003 in this series). Intracellular *V. shiloi* cells fit that definition, but unlike most cases of VBNC that have been studied, this is not brought about by starvation or low temperature. Rather, the entry of *V. shiloi* into the VBNC state occurs inside the coral epidermis, where nutrients are abundant. The fact that intracellular *V. shiloi* is viable, that is, metabolically active, is apparent from the observation that they multiply intracellularly and they score as viable with the Live/Dead Baclight Bacterial Viability Kit. It is possible that the differentiated intracellular VBNC *V. shiloi* becomes dependent on one or more nutrients present in the coral cell. In this regard, it has been shown that VBNC *V. shiloi* is infectious (Israely et al. 2001).

Toxin Production

V. shiloi produces extracellular toxins that block photosynthesis and bleach and lyse zooxanthellae (Ben-Haim et al. 1999). The toxin responsible for inhibition of photosynthesis, referred to as toxin P, is the following proline-rich peptide: PYPVYAPPPVVP (Banin et al. 2001b). In the presence of NH_4Cl, the toxin causes a rapid decrease in the photosynthetic quantum yield of zooxanthellae. Evidence has been presented that the toxin binds irreversibly to algal membranes, forming a channel that allows NH_3, but not NH_4^+, to rapidly pass, thereby destroying the pH gradient across the membrane and blocking photosynthesis (Banin et al. 2001b). This mode of action can help explain the mechanism of coral bleaching. *V. shiloi* also produces heat-sensitive, high-molecular-weight toxins that bleach and lyse isolated zooxanthellae. These toxins have not yet been isolated and characterized. At 29°C, toxin P is produced at ten times the rate that it is produced at 16°C.

Effect of Temperature on *V. shiloi* Virulence Factors

Several of the virulence factors essential for a successful infection of *O. patagonica* by *V. shiloi* are synthesized only at elevated summer seawater temperatures, including:

1. An adhesin on the bacterial cell surface that recognizes a β-galactoside-containing receptor in the coral mucus (Toren et al. 1998; Banin et al. 2001a),
2. Superoxide dismutase (SOD), which allows the bacteria to survive in the oxygen-rich coral tissue (Banin et al. 2003).
3. Toxin P, which binds to zooxanthellae membranes and inhibits photosynthesis (Ben-Haim et al. 1999; Banin et al. 2001b).
4. Enzymes that lyse zooxanthellae (Ben-Haim et al. 1999).

The demonstration that the bacterial adhesin is not produced at temperatures below 20°C provides an explanation for the failure of *V. shiloi* to infect *O. patagonica* in the field during the winter and in the laboratory at low temperatures. However, it does not explain why corals recover during the winter. When corals are infected with *V. shiloi* at the permissive temperature of 28°C, the bacteria adhere, penetrate, and begin to multiply intracellularly. If the infected corals are then shifted slowly (0.5°C per day) to lower temperatures, the bacteria die and the infection is aborted (Israely et al. 2001). The failure of *V. shiloi* to survive inside coral tissue at temperatures below 20°C is due to the fact that it does not produce superoxide dismutase (SOD) at these low temperatures (Banin et al. 2003). This hypothesis was supported by constructing an SOD minus mutant. At 28°C, the mutant adhered to the coral, penetrated into the tissue, and then died. Death occurred only when the coral was exposed to light. The most reasonable explanation for these data is that the high concentration of oxygen and resulting oxygen radicals produced by the zooxanthellae during photosynthesis (Kuhl et al. 1995) are highly toxic to bacteria, and this is proposed to be one of the mechanisms by which corals resist infection. At high temperatures, *V. shiloi* produces a potent SOD that helps it to survive in the coral tissue.

In a more general sense, the data on the importance of SOD for the survival of *V. shiloi* inside coral tissue provides strong evidence for the role of zooxanthellae in coral resistance to bacterial infection. Thus, zooxanthellae not only provide nutrients for coral growth, but also protect the coral against infection by pathogens.

Transmission of the Disease

The observation that *V. shiloi* could not be found inside *O. patagonica* during the winter (Israely et al. 2001) and that the bacterium could not survive in the coral below 20°C (Banin et al. 2003) indicate that bleaching of *O. patagonica* requires a fresh infection each spring, rather than the activation of dormant intracellular bacteria. This posed a question: Where does *V. shiloi* reside during the winter?

Using fluorescence in situ hybridization (FISH) with a *V. shiloi*–specific deoxyoligonucleotide probe, it was found that the marine fireworm *Hermodice carunculata* is a winter reservoir for *V. shiloi* (Sussman et al. 2003). Worms taken directly from the sea during the winter contained 0.6 to 2.9×10^8 *V. shiloi* per worm by FISH analysis. When worms were infected with *V. shiloi* in the laboratory, most of the bacteria adhered to the worm within 24 hours and then penetrated into epidermal cells. By 48 hours, less than 10^{-4} of the intact *V. shiloi* in the worm gave rise to colonies, suggesting that they differentiated inside the worm into the VBNC state.

To test if worms carrying *V. shiloi* could serve as vectors for transmitting the pathogen to *O. patagonica*, worms infected with *V. shiloi* were placed in aquaria containing *O. patagonica*. Corals that came into contact with the infected worms showed small patches of bleached tissue in 7 to 10 days and total bleaching in 17 days. Uninfected worms did not cause bleaching. Thus, *H. carunculata* is not only a winter reservoir for *V. shiloi*, but also a potential vector for transmitting the bleaching disease to *O. patagonica* (Fig. 6.2). It should be emphasized that knowledge about reservoirs and modes of transmission has proven useful in the past for developing technologies for controlling the spread of disease.

6.5. GENERALITY OF THE BACTERIAL HYPOTHESIS OF CORAL BLEACHING

Although bleaching of *O. patagonica* by *V. shiloi* in the Mediterranean Sea is probably the best-studied coral disease, there are several features about it

Figure 6.2. Photograph of the fireworm *H. carunculata* feeding on a coral colony.

that differ from coral bleaching in other parts of the world. To begin with, *O. patagonica* does not exist on coral reefs, but rather as colonies in temperate waters. Second, *O. patagonica* is a recent immigrant to the Mediterranean Sea (Fine et al. 2001) and thus, equilibrium between the coral and local marine organisms may not yet have been established. Third, there is a much larger seawater temperature difference between summer and winter in the Mediterranean Sea off the coast of Israel (ca. 14°C) than is generally found in waters supporting coral reefs. These differences make it difficult to extrapolate from data obtained with the *V. shiloi*/*O. patagonica* model system to coral bleaching that takes place on coral reefs. The recent discovery that bleaching of *Pocillopora damicornis* in the Indian Ocean and Red Sea is the result of an infection by *Vibrio coralliilyticus* greatly strengthens the bacterial infection hypothesis of coral bleaching (Ben-Haim et al. 2003).

6.6. RELATIONSHIPS BETWEEN THE THREE HYPOTHESES OF CORAL BLEACHING

The three hypotheses of coral bleaching reviewed above deal with different aspects of the disruption of the coral/zooxanthellae symbiosis and are in no way exclusive of each other. The stress hypothesis focuses on environmental causes of bleaching. The adaptive hypothesis is an explanation of how corals can adjust to varying environmental conditions, and the bacterial hypothesis is concerned with the causative agent of the disease. Conflicts between these hypotheses become clear only when attempts are made to use the general concepts to explain the "details" of coral bleaching and predict the effect that global warming will have on coral reefs.

In the most general sense, the stress hypothesis can include the bacterial hypothesis and is a prerequisite for the adaptive hypothesis. However, proponents of the stress hypothesis (e.g., Hoegh-Guldberg 1999, 2002) claim that the stress is directed at the zooxanthellae, whereas in the *V. shiloi/O. patagonica* model system of bleaching, it has been shown that the increased temperature exerts its effect by causing the expression of bacterial virulence genes Rosenberg and Falkovitz (2004). The difference between these two points of view is not trivial because it affects the design and interpretation of experiments and bears directly on what measures should be taken to prevent the spread of the disease.

Without consideration of adaptation, coral biologists predict large-scale destruction of coral reefs and extinction of many coral species during the next few decades as a result of global warming (Hughes et al. 2003). The adaptive hypothesis leads to a more optimistic prediction of the future of corals, providing hope (and a mechanism) by which corals can recover from bleaching and "occasionally" become more resistant to the stress by forming a new ecospecies. There is clear evidence that new coral/zooxanthellae ecospecies can be formed (e.g., Kinzie et al. 2001). The major unanswered question is the time scale required for new resistant ecospecies to form in sufficient numbers to avoid the predicted coral reef disaster.

With only minor modifications the adaptive hypothesis can be applied to bleaching resulting from infection by a microbe or virus. In this case, the selection would be not for heat-resistant zooxanthellae but for infection-resistant zooxanthellae. Unfortunately, this does not appear to be happening in the *V. shiloi/O. patagonica* model system. Mass bleaching has occurred every summer for the last 10 years, and there is no evidence that resistant populations are being formed (Fine, personal communication).

6.7. BACTERIAL/CORAL SYMBIOSES

Although largely ignored by coral biologists, the coral surface mucus layer contains a large and diverse microbial population, which will be referred to as coral surface microorganisms (CSM). Using culture-dependent (Ritchie and Smith 2004) and culture-independent (Rohwer and Kelley 2004) methods, it has been shown that each coral species has distinct CSM. The majority of coral-associated bacteria are novel species (Rohwer et al. 2001). Furthermore, the same coral species has similar CSM even when separated by hundreds of kilometers (Rohwer et al. 2002). Even more interesting, the genetic similarity of the bacterial population of one coral species to the population of other coral species fits closely to the corals' evolutionary relationships (Ritchie and Smith 2003). These data suggest that specific bacteria/coral symbioses have existed for millions of years and thus must provide benefits to both partners. What are these benefits?

CSM receive a constant supply of organic nutrients from the coral. The precise composition of the secreted mucus is not known, but it contains lipids, complex polysaccharides, proteins and waxes (Benson and Muscatine 1974; Ducklow and Mitchell 1979a and b; Krupp 1981; Meikle et al. 1988). All these compounds are excellent sources of carbon and energy for bacteria. Paul et al. (1986) determined that coral-associated bacteria grow ten times faster than bacteria in the overlying water column. Bacteria attached to the coral surface may also be shielded from protozoan predators and bacteriophages. The high concentration of bacteria in the coral mucus allows for the kind of efficient cross-feeding that exists in other biofilms.

What benefits do corals derive from their CSM? Although there have been no direct experiments performed to answer this question, there are several plausible possibilities:

1. CSM can provide a protective barrier against opportunistic pathogens by occupying niches and producing antibiotics (Koh 1997; Kelman 2004).
2. CSM can provide the coral with a battery of extracellular enzymes that depolymerize complex polysaccharides and other recalcitrant organic material.
3. CSM may contain nitrogen-fixing bacteria (Frias-Lopez et al. 2002; Rohwer et al. 2002). These bacteria could fix nitrogen during the evening when the coral becomes anaerobic, or aerobically during the day.
4. A portion of the CSM could be ingested by the coral and serve as a rich source of nitrogen, phosphorus, and vitamins. Kushmaro and Kramasky-Winter (2004) have suggested that the coral "farms" bacteria, converts them into large aggregates, and harvests them for food.
5. CSM could scavenge the surrounding water for limiting nutrients, such as phosphorus and iron.

6.8. CONCLUSIONS

As has been described in detail, coral (an animal) has gone into long-term successful tripartite partnership with zooxanthellae (plants), which carry out photosynthesis, and the CSM (prokaryotes), which have the potential to fix nitrogen and degrade complex organic molecules. Evidence is presented that environmental changes (such as elevations in water temperature) that disturb this three-kingdom symbiosis could contribute to coral bleaching and to other coral diseases. Much more research is needed to fully elucidate this most complex of symbiotic relationships.

ACKNOWLEDGMENTS

This work was supported by the Center for the Study of Emerging Diseases and the Pasha Gol Chair for Applied Microbiology.

REFERENCES

Banin, E., Israely, T., Kushmaro, A., Loya, Y., Orr, E., and Rosenberg, E. (2000). Penetration of the coral-bleaching bacterium *Vibrio shiloi* into *Oculina patagonica*. *Applied and Environmental Microbiology* **66**, 3031–3036.

Banin, E., Israely, T., Fine, M., Loya, Y., and Rosenberg, E. (2001a). Role of endosymbiotic zooxanthellae and coral mucus in the adhesion of the coral-bleaching pathogen *Vibrio shiloi* to its host. *FEMS Microbiology Letters* **199**, 33–37.

Banin, E., Sanjay, K. H., Naider, F., and Rosenberg, E. (2001b). A proline-rich peptide from the coral pathogen *Vibrio shiloi* that inhibits photosynthesis of zooxanthellae. *Applied and Environmental Microbiology* **67**, 1536–1541.

Banin, E., Vassilakos, D., Orr, E., Martinez, R. J., and Rosenberg, E. (2003). Superoxide dismutase is a virulence factor produced by the coral bleaching pathogen *Vibrio shiloi*. *Current Microbiology* **46**, 418–422.

Ben-Haim, Y., Banin, E., Kushmaro, A., Loya, Y., and Rosenberg, E. (1999). Inhibition of photosynthesis and bleaching of zooxanthellae by the coral pathogen *Vibrio shiloi*. *Environmental Microbiology* **1**, 223–229.

Ben-Haim, Y., Zicherman-Keren, M., and Rosenberg, E. (2003). Temperature-regulated bleaching and lysis of the coral *Pocillopora damicornis* by the novel pathogen *Vibrio coralliilyticus*. *Applied and Environmental Microbiology* **69**, 4236–4242.

Benson, A. and Muscatine, L. (1974). Wax in coral mucus: Energy transfer from corals to reef fishes. *Limnology & Oceanography* **19**, 810–614.

Brown, B. E. (1997). Coral bleaching: causes and consequences. *Proceedings of the 8th International Coral Reef Symposium* **1**, 65–74.

Buddemeier, R. W. (2004). The adaptive hypothesis of bleaching. In *Coral Health & Disease*, ed. E. Rosenberg and Y. Loya. Berlin and New York: Springer-Verlag, pp. 427–444.

Buddemeier, R. W. and Fautin, D. G. (1993). Coral bleaching as an adaptive mechanism: A testable hypothesis. *BioScience* **43**, 320–326.

Coates. A. R. M. (ed) (2003). *Dormancy and low-growth states in microbial disease.* In *Advances in Molecular and Cellular Microbiology*, vol. 3. Cambridge: Cambridge University Press.

Colwell, R. R., Brayton, P. R., Grimes, D. J., Roszak, D. B., Huq, S. A., and Palmer, L. M. (1985). Viable but non-culturable *Vibrio cholerae* and related pathogens in the environment: implications for release of genetically engineered microorganisms. *Bio/Technology* **3**, 817–820.

Ducklow, H. W. and Mitchell, R. (1979a). Bacterial populations and adaptations in the mucus layers on living corals. *Limnology & Oceanography* **24**, 715–725.

Ducklow, H. W. and Mitchell, R. (1979b). Composition of mucus released by coral reef coelenterates. *Limnology & Oceanography* **24**, 707–714.

Fine, M. and Loya, Y. (1995). The coral *Oculina patagonica*: a new immigrant to the Mediterranean coast of Israel. *Israel Journal of Zoology* **41**, 81.

Fine, M., Zibrowius, H., and Loya, Y. (2001). *Oculina patagonica*, a non-lessepsian scleractinian coral invading the Mediterranean Sea. *Marine Biology* **138**, 1195–1203.

Frias-Lopez, J., Zerkle, A. L., Bonheyo, G. T., and Fouke, B. W. (2002). Partitioning of bacterial communities between seawater and healthy, black band diseased, and dead coral surfaces. *Applied and Environmental Microbiology* **68**, 2214–2228.

Harvell, C. D., Mitchell, C. E., War, J. R., Altizer, S., Dobson, A. P., Ostfeld, R. S., and Samuel, M. D. (2002). Climate warming and disease risks for terrestrial and marine biota. *Science* **296**, 2158–2162.

Hoegh-Guldberg, O. (1999). Climate change, coral bleaching and the future of the world's coral reefs. *Marine & Freshwater Research* **50**, 839–866.

Hoegh-Guldberg, O., Jones, R. J., Ward, S., and Loh, W. K. (2002). Communication arising: Is coral bleaching really adaptive? *Nature* **415**, 601–602.

Hughes, T. P., Baird, A. H., Bellwood, D. R., Cad, M., Connolly, S. R., Folke, C., Grosberg, R., Hoegh-Guldberg, O., Jackson, J. B. C., Kleypas, J., Lough, J. M., Marshall, P., Nystrom, M., Palumbi, S. R., Pandolfi, J. M., Rosen, B., and Roughgarden, J. (2003). Climate change, human impacts, and the resilience of coral reefs. *Science* **301**, 929–933.

Isberg, R. R., and Falkow, S. (1985). A single genetic locus encoded by *Yersinia pseudotuberculosis* permits invasion of cultured animal cells by *Escherichia coli* K-12. *Nature* **317**, 262–364.

Israely, T., Banin, E., and Rosenberg, E. (2001). Growth, differentiation and death of *Vibrio shiloi* in coral tissue as a function of seawater temperature. *Aquatic Microbial Ecology* **24**, 1–8.

Jokiel, P. L. (2004). Temperature stress and coral bleaching. In *Coral Health & Disease*, ed. E. Rosenberg and Y. Loya. Berlin and New York: Springer-Verlag, pp. 401–426.

Jokiel, P. L. and Coles, S. L. (1977). Effects of temperature on the mortality and growth of Hawaiian reef corals. *Marine Biology* **43**, 201–208.

Jokiel, P. L. and Coles, S. L. (1990). Response of Hawaiian and other Indo Pacific reef corals to elevated temperatures. *Coral Reefs* **8**, 155–162.

Jones, R., Hoegh-Guldberg, O., Larkum, A. W. L., and Schreiber, U. (1998). Temperature induced bleaching of corals begins with impairment of dark metabolism in zooxanthellae. *Plant, Cell, and Environment* **21**, 1219–1230.

Kelman, D. (2004). Antimicrobial activity of sponges and corals. In *Coral Health & Disease*, ed. E. Rosenberg and Y. Loya. Berlin and New York: Springer-Verlag, pp 243–254.

Kinzie, R. A. and Chee, G. S. (1979). The effect of different zooxanthellae on the growth of experimentally re-infected hosts. *Biological Bulletin* **156**, 315–327.

Kinzie, R. A., Takayama, M., Santos, S. R., and Coffroth, M. A. (2001). The adaptive bleaching hypothesis: experimental tests of critical assumptions. *Biological Bulletin* **200**, 51–58.

Koh, E. G. L. (1997). Do scleractinian corals engage in chemical warfare against microbes? *Journal of Chemical Ecology* **23**, 379–398.

Kuhl, M., Cohen, Y., Dalsgard, T., Jorgensen, B. B., and Revsbech, N. P. (1995). Microenvironment and photosynthesis of zooxanthellae in scleractinian corals studied with microsensors for O_2, pH and light. *Marine Ecology Progress Series* **117**, 159–172.

Krupp, D. A. (1981). The composition of the mucus from the mushroom coral, *Fungia scutaria*. *Proceedings of Fourth International Coral Reef Symposium* **2**, 69–73.

Kushmaro, A., Banin, E., Stackebrandt, E., and Rosenberg, E. (2001). *Vibrio shiloi* sp. nov: the causative agent of bleaching of the coral *Oculina patagonica*. *International Journal of Systematic Evolutionary Microbiology* **51**, 1383–1388.

Kushmaro, A., and Kramarsky-Winter, E. (2004). Bacteria as a source of coral nutrition. In *Coral Health & Disease*, ed. E. Rosenberg and Y. Loya. Berlin and New York: Springer-Verlag, pp. 231–242.

Kushmaro, Loya, Y., Fine, M., and Rosenberg, E. (1996). Bacterial infection and coral bleaching. *Nature* **380**, 396.

Kushmaro, A., Rosenberg, E., Fine, M., Ben-Haim, Y., and Loya, Y. (1998). Effect of temperature on bleaching of the coral *Oculina patagonica* by *Vibrio shiloi* AK-1. *Marine Ecology Progress Series* **171**, 131–137.

Kushmaro, A., Rosenberg, E., Fine, M., and Loya Y. (1997). Bleaching of the coral *Oculina patagonica* by *Vibrio* AK-1. *Marine Ecology Progress Series* **147**, 159–165.

Lee, K. H. and Ruby, E. G. (1995). Symbiotic role of the viable but nonculturable state of *Vibrio fischeri* in Hawaiian coastal seawater. *Applied & Environmental Microbiology* **61**, 278–283.

Meikle, P., Richards, G. N., and Yellowlees, D. (1988). Structural investigations on the mucus from six species of coral. *Marine Biology* **99**, 187–193.

Muscatine, L. (1990). The role of symbiotic algae in carbon and energy flux in reef corals. *Coral Reefs* **25**, 1–29.

Oliver, J. (1985). Recurrent seasonal bleaching and mortality of corals on the Great Barrier Reef. *Proceedings of the 5th International Coral Reef Symposium* **4**, 201–206.

Oliver, J. D. (1993). Formation of viable but nonculturable cells. In *Starvation in Bacteria*, ed. S. Kjelleberg, pp. 239–272. New York: Plenum Press.

Oliver, J. D., Nilsson, L., and Kjelleberg, S. (1991). Formation of nonculturable *Vibrio vulnificus* cells and its relationship to the starvation state. *Applied & Environmental Microbiology* **57**, 2640–2644.

Paul, J., DeFlaun, M., and Jeffery, W. (1986). Elevated levels of microbial activity in the coral surface monolayer. *Marine Ecology Progress Series* **33**, 29–40.

Ralph, P. J., Gademann, R., and Larkum, A. W. D. (2001). Zooxanthellae expelled from bleached corals at 33°C are photosynthetically competent. *Marine Ecology Progress Series* **220**, 163–168.

Ritchie, K. and Smith, G. W. (2004). Microbial communities of coral surface mucopolysaccharide layers. In *Coral Health & Disease*, ed. E. Rosenberg and Y. Loya. Berlin and New York: Springer-Verlag, pp. 259–264.

Rohwer, F., Breitbart, M., Jara, J., Azam, F., and Knowlton, N. (2001). Diversity of bacteria associated with the Caribbean coral *Montastraea franksi*. *Coral Reefs* **20**, 85–95.

Rohwer, F. and Kelly, S. (2004). Culture-dependent analyses of coral-associated microbes. In *Coral Health & Disease*, ed. E. Rosenberg and Y. Loya. New York and Berlin: Springer-Verlag, pp. 265–278.

Rohwer, F., Serigutan, V., Azam, F., and Knowlton, N. (2002). Diversity and distribution of coral-associated bacteria. *Marine Ecology Progress Series* **243**, 1–10.

Rosenberg, E. and Ben-Haim, Y. (2002). Microbial diseases of corals and global warming. *Environmental Microbiology* **4**, 318–326.

Rosenberg, E., and Falkovitz, L. (2004). The Vibrio shiloi/Oculina patagonica model system of coral bleaching. *Annual Review of Microbiology* **58**, 143–159.

Rowan, R. and Powers, D. A. (1991). A molecular genetic classification of zooxanthellae and the evolution of animal-algal symbioses. *Science* **251**, 1348–1351.

Stambler, N. and Dubinsky, Z. (2004). Stress effects on metabolism and photosynthesis of hermatypic corals. In *Coral Health & Disease*, ed. E. Rosenberg and Y. Loya. Berlin and New York: Springer-Verlag, pp. 195–216.

Sussman, M., Loya, Y., Fine, M., and Rosenberg, E. (2003). The marine fireworm *Hermodice carunculata* is a winter reservoir and spring-summer vector for the coral-bleaching pathogen *Vibrio shiloi*. *Environmental Microbiology* **5**, 250–255.

Szmant, A. and Gassman, N. J. (1990). The effects of prolonged bleaching on the tissue biomass and reproduction of the reef coral *Monastrea annularis*. *Coral Reefs* **8**, 217–224.

Toren, A., Landau, L., Kushmaro, A., Loya, Y., and Rosenberg, E. (1998). Effect of temperature on adhesion of *Vibrio* strain AK-1 to *Oculina patagonica* and on coral bleaching. *Applied and Environmental Microbiology* **64**, 1379–1384.

CHAPTER 7

Interactions between inherited bacteria and their hosts: The *Wolbachia* paradigm

Zoe Veneti, Max Reuter, Horacio Montenegro, Emily A. Hornett, Sylvain Charlat, and Gregory D. Hurst

7.1. INTRODUCTION

The rise to prominence of the bacterium *Wolbachia* has been quite remarkable. Whilst it was first described as an intracellular bacterium of mosquito hosts in the 1930s (Hertig and Wolbach 1924; Hertig 1936), its fastidious nature meant *Wolbachia* remained little studied until the late 1980s and early 1990s. At this time, the development of polymerase chain reaction assay (PCR) allowed easy screening for, and identification of, unculturable species. A small screen of insects by O'Neill et al. (1992) revealed *Wolbachia* to be present in a wide array of arthropods. A later, more complete survey indicated that sixteen to twenty percent of species were infected (Werren et al. 1995). Alongside the recognition that the bacterium was common came an appreciation that it was responsible for a diverse array of manipulations of host reproduction. Yen and Barr (1971) first recognised it as the cause of some incompatible crosses in insects. *Wolbachia* was then identified as a cause of parthenogenesis, feminisation of male hosts, and male killing in different arthropod taxa (Rousset et al. 1992; Stouthamer et al. 1993; Hurst et al. 1999). As a further example of how its interactions with hosts may vary, *Wolbachia* was discovered to be an essential partner of filarial nematodes (Sironi et al. 1995), and some of the symptoms of filariasis are in fact a response to the symbiont rather than the worm (Saint Andre et al. 2002). Whilst this type of interaction was initially thought to be nematode specific, recent studies have indicated it may also occur in insects (Dedeine et al. 2001).

7.2. *WOLBACHIA* AS A BACTERIUM

Wolbachia is a member of the alpha-proteobacteria. This group, from which mitochondria ancestrally derive, also contains a variety of other obligately intracellular bacteria with a range of interactions with their host. The closest relative of *Wolbachia* is the intracellular bacterium *Ehrlichia*, a tick-borne pathogen of horses. Members of the genus *Rickettsia*, which comprises tick-borne pathogens of mammals (e.g., the etiologic agent of scrub typhus), insect-borne pathogens of plants (e.g., papaya top bunchy disease), and obligate insect pathogens, are also close allies. *Anaplasma marginale*, a tick-borne pathogen of cattle, falls within the same group, as do intracellular bacteria associated with plants, such as *Agrobacterium* and *Bradyrhizobium*.

Wolbachia is a typical member of the alpha-proteobacteria. Like many of its relatives, *Wolbachia* has a small genome, varying in size between 1 and 1.6 Mbp in length. It undergoes recombination in natural populations (Jiggins et al. 2001b) and contains insertion elements (Masui et al. 1999) and phage (Masui et al. 2000), but apparently no plasmids (Sun et al. 2001). *Wolbachia* apparently cannot reproduce outside cells and although it can be maintained in cell culture (O'Neill et al. 1997), it has not yet been grown in cell-free media. Within cells, it is found inside a host vacuole, and it possesses a type IV secretion system that is likely used in interplay with the host cell (Masui et al. 2000).

Perhaps the most important aspect of *Wolbachia* to consider is that its mode of transmission within host populations is predominantly vertical. Like mitochondria, *Wolbachia* are passed on in the egg cytoplasm from the mother to the offspring. The bacterium's population biology and strategies to maximise propagation are therefore radically different from the population biology and strategies used by the agents of contagious (i.e., horizontally transmitted) diseases. Because *Wolbachia* are maternally transmitted, their spread relies on infected females producing an above-average number of daughters who again carry the symbiont. The strategies used by *Wolbachia* to achieve this fall into two broad categories. In some cases, the relationship is parasitic in the sense that the bacteria use selfish strategies to promote infection at the expense of optimal host reproduction. In this category falls *Wolbachia*-induced sex ratio distortion, which aims at promoting infection by increasing the proportion of daughters among the offspring of infected females. Another parasitic strategy is cytoplasmic incompatibility that occurs in matings between infected males and uninfected females as a means to impede their reproduction. In other cases, *Wolbachia* have a more mutualistic relationship with their host, playing a positive physiological role to increase the overall productivity of infected individuals.

The dichotomy between mutualistic and parasitic interactions is reflected in the phylogenies of both the symbionts and the hosts. Mutualism is almost completely restricted to *Wolbachia* of filarial nematodes. Very typically for relationships of mutual dependence, nematode–*Wolbachia* associations are old and have been preserved over long periods of time, as evidenced by strict co-cladogenesis (nematode speciation events are mirrored by nodes in the *Wolbachia* phylogeny) (Bandi et al. 1998). In arthropods, parasitism is the rule. As expected in this case, host–symbiont associations are unstable over evolutionary time spans. Very few pairs of sibling species share *Wolbachia* strains (Werren et al. 1995). In most cases, they harbour distantly related strains, acquired by independent horizontal transmission events from other host species. Whilst negligible for infection dynamics within host populations, extensive horizontal transmission has occurred over evolutionary time. It appears to require fairly intimate contact between hosts. Possible routes of transfer are through exchange of haemolymph, movement from a parasitoid to its host (and vice versa), from ectoparasites to hosts, and via predation. To date, there is good evidence for the first pair of mechanisms (Rigaud and Juchault 1995; Heath et al. 1999; Huigens et al. 2004). Exceptional with respect to the frequency of horizontal transmission are some parasitoid wasps. In *Trichogramma kaykai*, horizontal transfer of symbionts occurs commonly even within host populations (Huigens et al. 2000). In this species, females oviposit inside moth and butterfly eggs, and transfer of *Wolbachia* can occur by uninfected larvae cannibalising infected larvae developing in the same butterfly egg.

In this chapter, we examine the mechanism and incidence of each form of manipulation in turn, along with the population biology and evolutionary impact of the symbiosis (defined here in *sensu lato*). We finally discuss whether there really is a *Wolbachia* paradigm, and argue that whilst *Wolbachia* clearly is a very important factor in the ecology and evolution of invertebrates, recent findings indicate that it is one of many, rather than the unique symbiont we once thought.

7.3. SEX RATIO DISTORTION MANIPULATIONS

7.3.1. Parthenogenesis Induction

Parthenogenesis-inducing *Wolbachia* have only been recorded within the haplodiploid taxa of insect and mites, and it is not known whether this exclusivity is the result of functional constraints. Haplodiploidy is a sex determination system in which unfertilised haploid eggs develop into males whereas fertilised (diploid) eggs develop into females. Stouthamer et al. (1990) observed that antibiotic treatment of purely parthenogenetic populations of the parasitoid wasp *Trichogramma* made males reappear. This strongly suggested the involvement of bacteria in the induction of parthenogenesis, which molecular analyses later identified as *Wolbachia* (Stouthamer et al. 1993). Since that time, *Wolbachia*-induced parthenogenesis has been detected in more than forty species of Hymenoptera (Stouthamer 1997), one species of thrip (Arakaki et al. 2001), and a genus of phytophagous mites, *Bryobia* (Weeks and Breeuwer 2001).

The cytogenetic mechanism of *Wolbachia*-induced parthenogenesis is known to vary between host taxa. In the haplodiploid parasitoid wasp *Trichogramma*, it occurs via gamete duplication (Stouthamer and Kazmer 1994). In unfertilised eggs, the two nuclei created after the first mitotic division fuse and restore diploidy. As a result, unfertilised eggs that would have been haploid and thus male, became diploid and therefore developed into females. However, in mites, the mechanism appears to be quite different. In this system, meiotic modifications and not gamete duplication seem to be responsible for the diploidisation process (Weeks and Breeuwer 2001). Therefore, *Wolbachia* appears to have evolved more than one mechanism to achieve parthenogenetic development in arthropods. The *Bryobia* case also suggests parthenogenesis induction could occur in diploid species.

The induction of parthenogenesis is a strong force driving up the prevalence of infection. Infected females can produce up to twice as many daughters as uninfected females, increasing prevalence to a large degree. To this must be added the observation that in some hosts, parthenogenesis-inducing *Wolbachia* may also be transmitted horizontally, as previously discussed (Huigens et al. 2000). Parthenogenesis-inducing *Wolbachia* therefore commonly spread to high prevalence and regularly convert their host species to complete asexual reproduction.

When host species are fully parthenogenetic by virtue of *Wolbachia*, males are no longer produced. It has been conjectured that mutation and selection in this situation act to destroy male and female traits associated with sexual reproduction, thus preventing a return to sexuality. It is certainly notable that many species made parthenogenetic by the presence of *Wolbachia* may produce males following antibiotic treatment, but sexual reproduction is not successful. Failure to court successfully and to transfer sperm are commonly observed, as may female receptivity to courtship and ability to store and process sperm (Gottlieb and Zchori-Fein 2001).

Given the population biology of parthenogenesis induction, it is rather surprising to find some populations of *Trichogramma* in which infection frequencies are relatively low (five to twenty percent of individuals). Having ruled out low rates of vertical transmission or host resistance to symbiont manipulation as possible causes, Stouthamer et al. (2001) demonstrated that the factor impeding the spread of *Wolbachia* is a B chromosome, a second selfish genetic element present in the wasps. The B chromosome is a so-called paternal sex ratio (PSR) factor known from other haplodiploid species. PSRs are paternally transmitted and upon fertilisation of the egg, destroy all paternal chromosomes but itself. In haplodiploids, the haplodisation of the egg results in the zygote devloping into a male and hence a new transmitter of PSR (Werren 1991). PSR counteracts *Wolbachia*-induced parthenogenesis by turning fertilised infected eggs (that normally would have developed into *Wolbachia*-transmitting females) into PSR-transmitting males.

7.3.2. Feminisation

Feminising *Wolbachia* were first recognised in the pill woodlouse, *Armadillidium vulgare* (Rousset et al. 1992; Stouthamer et al. 1993). The authors found that the intracellular microorganisms previously known to feminise genetically male woodlice were *Wolbachia*. Following this, Bouchon et al. (1998) performed an extensive PCR screen and found feminising *Wolbachia* to be quite widespread in terrestrial isopods. Since that time, feminisation has also been recorded in two Lepidoptera, *Ostrinia furnacalis* and *Eurema hecabe* (Kageyama et al. 2002; Hiroki et al. 2002), although the former record turned out to be an example of male killing (Kageyama and Traut 2003).

The process of converting males into females is phenotypically quite well characterised in *A. vulgare*, although details of the molecular mechanisms are still lacking (Rigaud 1997). In *A. vulgare*, the male is the homogametic sex (ZZ) and the female heterogametic (ZW). The Z chromosome carries the gene(s) controlling the development of the androgenic gland, the organ responsible for male hormonal synthesis and male sex differentiation. The W "female determining" chromosome carries suppressors of these gene(s) and hence induces a female differentiation. *Wolbachia* appears to exploit the simplicity of this mechanism, by interfering with androgenic gland differentiation in ZZ individuals infected with *Wolbachia*, resulting in female development (Rigaud 1997). It is tempting to suggest that it emulates the W chromosome to have this effect.

Wolbachia-induced feminisation can have profound effects on the evolution of the sex-determining system of infected populations. As the symbiont spreads through the population, an increasingly large proportion of host females are ZZ "neo-females." The female-determining W chromosome is eventually lost from the population. At this stage, all individuals are ZZ and sex determination is completely taken over by the symbiont. Hosts are female if infected and successfully feminised, and male otherwise. The population sex ratio is determined merely by the transmission efficiency of the bacterium and the efficiency with which it feminises. Populations in which this situation prevails are well known in *A. vulgare* (Rigaud 1997). In this species, evolution can even go further as in some populations the host has been shown to adapt to suppress the transmission or action of the feminiser (Rigaud and Juchault 1992). Here, sex is determined by an interaction between *Wolbachia* and nuclear genes that affect *Wolbachia* transmission.

Ecologically, a weak or moderate female bias may enhance population resilience (i.e., lowering susceptibility of the population to change in size and speeding recovery), as it is female production that influences this characteristic. Only at very high female bias will populations be harmed. Resilience may be further enhanced by adaptation of the host to the biased population sex ratio caused by the feminiser. In a survey of seven species, Moreau and Rigaud (2003) observed that male mating capacity (the capacity of males to inseminate multiple females) was enhanced in five species infected with a feminiser (where there was a great availability of mates to each female) compared with two species that were not infected.

7.3.3. Male Killing

Male-killing *Wolbachia* have been found in the two-spot ladybird, *Adalia bipunctata* (Hurst et al. 1999); the flour beetle, *Tribolium madens* (Fialho and Stevens 2000); one species of *Drosophila*, *D. bifasciata* (Hurst et al. 2000); and the butterflies *Acraea encedon* (Hurst et al. 1999) and *Hypolimnas bolina* (Dyson et al. 2002). Screening of acraeine butterflies suggested fifteen percent of the species harboured a *Wolbachia* male-killer, although this study could not distinguish between a male-killing and a feminisation phenotype (Jiggins et al. 2001a). Little is known about the mechanism of male killing, as it is the latest addition to the known *Wolbachia*-associated phenotypes. Neither the cue used to detect sex nor the mechanism by which death is brought about is known in any detail, apart from the fact that infected males die during embryogenesis. What is interesting is that male-killing *Wolbachia* are found in species with diverse sex determination systems. Lepidoptera, notably, are female heterogametic.

Unlike parthenogenesis and feminisation induction, which can spread solely because of the manipulation of the brood sex ratio of its host, the spread of male-killing strains requires certain ecological or behavioural characteristics in their host. If transmission of the bacteria from a female to the offspring is perfect, and *Wolbachia* has no negative direct effect on female fitness, male killing is a neutral character which will neither favour nor hamper the spread of the symbiont. However, male killing will aid spread if the death of males benefits the surviving sibling female hosts in some way. This can occur if male death reduces antagonistic sibling interactions (like competition or sibling cannibalism), or reduces the rate of host inbreeding (Hurst and Majerus 1993). For ladybirds, females eat their dead brothers, and females from an egg clutch in which the males have died have a higher survival time in the absence of their aphid prey (Hurst et al. 1997). *Acraea* and *Hypolimnas* butterflies lay eggs in clutches, so it is likely that there is competition between siblings.

If death of males benefits the surviving females and vertical transmission is perfect, a male-killer could spread to fixation, consequently leading the host population to extinction because of a shortage of males. The potential for this is seen in studies of two species in which transmission rate is high. First, in *Acraea encedon*, some populations have an extremely high prevalence of male-killing *Wolbachia* and high incidence of unmated females (Jiggins et al. 2002). Second, in some populations of *Hypolimnas bolina* the male-killer infects more than ninety-nine percent of females, with a concomitant 100:1 population sex ratio bias (Dyson and Hurst 2004). However, transmission rate is usually found to be lower outside these extreme cases, and a stable intermediate equilibrium of infected and uninfected individuals is achieved within populations. This is the case in *D. bifasciata*, in which high temperature causes imperfect transmission (Hurst et al. 2001).

The presence of male-killing *Wolbachia* could have a great impact upon host ecology and evolution. Male-killing *Wolbachia* are particularly detrimental to the host as half of the progeny of an infected female dies. As a consequence, there is strong selection on the host to evolve resistance to the action or transmission of the bacteria. Although the selection may be very strong, evidence of resistance genes has been found in only a few male-killing systems, such as in *Drosophila prosaltans* (Cavalcanti et al. 1957). No resistance has been found in other cases, for example, in *A. encedon* (Jiggins et al. 2002), in which high prevalence has been recorded, or in *D. bifasciata* (Hurst et al. 2001).

Another possible impact of male-killing *Wolbachia* is on the pattern of sexual selection. The general insect mating system is one of male-male sexual competition and female choosiness. As the sex ratio becomes more female biased, the situation moves to one in which competition for mates no longer occurs among the now rare males but among the frequent females. This sex-role reversal has been observed in *A. encedon* and *A. encedana*, with virgin females forming lekking (display) swarms on hill tops in female-biased populations only and once mated, leaving the lekking site (Jiggins et al. 2000).

It has also been conjectured that in these systems, males might prefer to mate with uninfected females, as these females would produce sons that would have high reproductive success (Randerson et al. 2000). However, no evidence that the males were selecting uninfected females was found in *A. encedon* (Jiggins et al. 2002).

In addition to a direct effect of the male-killer on the host's evolution as detailed above, there is a possibility that the male-killer can alter the pattern of evolution on the host in general. By killing males, the population may become significantly female biased and the effective population size, N_e, is decreased. With a low N_e, allelic frequencies are more likely to change because of chance events than because of their adaptive value, and drift becomes a more important evolutionary force than selection. As a consequence, male-killers are expected to lower the quantity of standing genetic variation present and make the population less able to respond to selection. There will also be a greater likelihood of deleterious mutations becoming fixed in the population, potentially leading to population extinction or "mutational meltdown."

7.4. CYTOPLASMIC INCOMPATIBILITY

7.4.1. Mechanism

Wolbachia was first linked to reproductive alterations in its host in the case of the cytoplasmic incompatibility (CI) phenotype in the mosquito *Culex pipiens* (Yen and Barr 1971). Since then, it has been recorded as producing this form of reproductive isolation in populations from all major classes of arthropods, including terrestrial isopods (Moret et al. 2001), mites (Breeuwer 1997), and many insects, such as wasps, weevils, bugs, planthoppers, butterflies, moths, beetles, and flies (Hoffmann and Turelli 1997).

Cytoplasmic incompatibility is caused by a modification of sperm during spermatogenesis (*Wolbachia* are physically excluded from the sperm). The modification prevents normal development of a fertilised egg unless rescued by the presence of the same bacterial strain in the egg (Werren 1997). As a consequence of this mechanism, infected males are incompatible with either uninfected females (unidirectional CI) or females infected with different *Wolbachia* strains (bi-directional CI) (O'Neill and Karr 1990). Infected females are completely compatible with both infected and uninfected males. Although CI leads to developmental arrest and death in diplodiploid host species, embryos from incompatible crosses in haplodiploid species occasionally develop into males instead of dying (Breeuwer and Werren 1990).

Although the molecular mechanism of *Wolbachia*-induced CI is still unknown, genetic and cytological data suggest that incompatibility is associated with altered behaviour of paternal chromosomes after fertilisation, and the embryo dies because of asynchronous mitoses (Lassy and Karr 1996; Callaini et al. 1997; Tram and Sullivan 2002). Modification-rescue functions are now characterised through a number of properties, and several projects are trying to identify the factors involved in CI, despite the fact that *Wolbachia* cannot be cultured in standard media (Harris and Braig 2001). Among the factors that have been proposed so far to affect the expression of CI are bacterial and host genetic backgrounds, bacterial density, host age, and mating history (Boyle et al. 1993; Hoffmann et al. 1996; Poinsot et al. 1998; Snook et al. 2000; Reynolds and Hoffmann 2002). These factors can interact in complex ways to affect the strength of CI. However, several studies suggest that the ability of a strain to cause CI is an intrinsic *Wolbachia* trait whereas the role of the host is limited to regulating bacterial numbers, especially in target tissues such as testes, in which an abundance of bacteria are required to induce the phenotype (McGraw et al. 2001; Clark et al. 2003; Veneti et al. 2003). Cytological studies also suggest that the host proteins targeted during CI are cell cycle regulators, because paternal centrosomes appear unaffected during the first mitotic division after fertilisation (Tram et al. 2003).

The genome sequence of a CI-inducing *Wolbachia* strain naturally infecting *Drosophila melanogaster* is complete (Wu et al. 2004), and several others are in progress. Comparative genomics and proteomics will undoubtedly accelerate research in the field and hopefully will lead to the identification of molecular pathways underlying CI. Advanced molecular genetic tools are also available for the host *D. melanogaster*, which make the system ideal for studying host–parasite interactions.

7.4.2. Population Biology of CI

Most maternally transmitted symbionts spread because of alterations they cause in the reproduction of the female host and which "aim" at increasing the production of infected daughters (see the earlier section on sex ratio distorters). Cytoplasmic incompatibilty is atypical in that it is an alteration caused by symbionts residing in males. Its selective advantage is therefore seemingly paradoxical, because host males do not transmit symbionts and incompatibility does not increase the direct reproduction of the symbionts that cause it. The phenomenon can be understood in terms of kin selection (Frank 1997; Hurst 1991; Rousset and Raymond 1991). The mating incompatibility caused by a male host's symbiont specifically impedes the reproduction of uninfected females (CI does not affect infected females) and hence gives a relative reproductive advantage to infected females carrying the symbiont's clonal relatives. By causing CI, a male *Wolbachia* thus indirectly increases its frequency in the population, just as a bee worker promotes its genes by raising siblings in his mother's hive.

The selective advantage of CI is frequency dependent (Caspari and Watson 1959; Hoffmann et al. 1990). At low frequencies (i.e., upon initial invasion), most hosts are uninfected and incompatible matings are rare. Under these conditions, infection spreads very slowly or can even be selected against if the tiny selective advantage of infection is outweighed by imperfect vertical transmission of the symbiont and/or fecundity costs of carrying the symbiont. Invasion of the symbiont then requires random drift to drive infection to a frequency at which incompatible matings are sufficiently common to favour infection. Whether infection will go to complete fixation or stabilise at a frequency lower than 10% prevalence depends on the rate of vertical transmission and the level of incompatibility (both of which raise equilibrium infection frequency) as well as the fecundity cost of carrying the symbiont (reducing equilibrium infection frequency) (Hoffmann et al. 1990).

Detailed information on the parameters governing infection dynamics is available for a number of species (Hoffmann and Turelli 1997). Particularly well studied is the CI *Wolbachia* found in a Californian population of *Drosophila simulans*. Data from field samples and laboratory measurements suggest that vertical transmission is almost perfect, fecundity costs of infection are low or non-existent, and the level of incompatibility is intermediate (the fertility of incompatible crosses is reduced by about forty-five percent) (Turelli and Hoffmann 1995). High rates of vertical transmission and low fecundity costs are in agreement with the conditions predicted to be favourable for the initial spread of *Wolbachia* infections. Furthermore, the equilibrium infection frequency predicted from the parameters is in good agreement with the measured prevalence in the field (Turelli and Hoffmann 1995).

7.4.3. Ecological and Evolutionary Impacts of CI

The strategy of CI is selfish, with *Wolbachia* promoting infection at the expense of optimal transmission of the host's genes (Hurst and Werren 2001). The cost of CI is particularly borne by host males, whose gene transmission is compromised by symbiont-induced incompatibility. Further, these males may have reduced sperm production associated with infection (Snook et al. 2000). If infection is costly in these ways, it may lead to co-evolution between the two parties, during which the host tries to escape manipulation by the symbiont, which in turn aims at circumventing host resistance to its manipulation.

One step of this co-evolutionary process, the evolution of host resistance, has been documented in the fly *D. melanogaster* (Boyle et al. 1993; Hoffmann and Turelli 1997). Populations of this species are naturally infected with a *Wolbachia* strain and show low levels of incompatibility. Transinfection experiments showed that the low level of incompatibility is not the result of particularly benign symbionts but is caused by a partial suppression of CI by the host genome. Indeed, equally low incompatibility was observed in *D. melanogaster* lines transinfected with a symbiont strain that causes strong incompatibility in its natural host, *D. simulans*. The evidence for host resistance is further corroborated by phylogenetic data suggesting that *D. melanogaster* has been associated with *Wolbachia* for longer than *D. simulans*, hence giving it more time to adapt to its symbiont (Solignac et al. 1994; Hoffmann and Turelli 1997).

Although adaptation to the symbiont might be one possible evolutionary change that CI *Wolbachia* can induce in their host, the evolutionary consequences of infection might be much more far reaching. It has been proposed that cytoplasmic incompatibility could play a role in host speciation by reducing gene flow between populations. This could occur in two ways. First, a host population carrying an infection could come into contact with an uninfected population. The resulting unidirectional reduction in gene flow is unlikely to lead to strong genetic divergence of the gene pools on its own. However, the low fitness of incompatible matings might select for mechanisms of pre-zygotic isolation in the uninfected population, such as discrimination against uninfected males. These would then complement cytoplasmic incompatibility to create a symmetrical reduction in gene flow and allow genetic divergence of the two populations. A second scenario involves two populations infected with different *Wolbachia* strains coming into contact. If the two symbiont strains are mutually incompatible, gene flow between the two host populations would be effectively interrupted or reduced, allowing them to diverge genetically even in the absence of any behavioural isolating mechanism evolving in the host.

Compelling evidence is still lacking, leaving the issue as a matter of debate (Werren 1998; Hurst and Schilthuizen 1998; Shoemaker et al. 1999; Bordenstein et al. 2001; Weeks et al. 2002; Bordenstein 2003; Charlat et al. 2003). However, tentative support is available for both of the above scenarios. Shoemaker et al. (1999) reported data in agreement with speciation by unidirectional CI. In a study of two closely related *Drosophila* species, *D. recens* (infected with CI *Wolbachia*) and *D. subquinaria* (uninfected), CI was found to effectively reduce hybridisation between *D. subquinaria* females and *D. recens* males, whereas behavioural isolation was observed in the reciprocal cross. Indirect evidence for the role of bi-directional incompatibility in speciation comes from studies of reproductive isolation between species of the genus *Nasonia*, in particular *N. vitripennis* and *N. longicornis* (Bordenstein et al. 2001). Hybridisation between these two sister species is naturally reduced by bi-directionally incompatible *Wolbachia* infections in the two species. However, populations cleared of *Wolbachia* can interbreed successfully. Discrimination against hybrid matings was weak and occurred in only one direction. These results show that bi-directional CI put the brakes on gene flow between the two species before any mechanisms of pre-zygotic isolation evolved (Bordenstein et al. 2001).

Although the above data are in agreement with scenarios of *Wolbachia*-induced speciation, neither study can establish a causal relationship between incompatibility and speciation. The populations used in Shoemaker's study were sampled at opposite sides of the North American continent, making it difficult to link the observed mate discrimination to selection imposed by incompatible hybrid crosses. Similarly, the two *Nasonia* species do not occur in sympatry (Bordenstein 2003). Hence, even if incompatibility occurs in the absence of strong pre-zygotic isolation, it may not have been necessary for speciation if the two species diverged in allopatry.

7.5. BENEFICIAL INTERACTIONS

Many species of insects do bear beneficial symbionts. For instance, antibiotic treatment is often lethal for aphids because it eliminates their symbiont, *Buchnera*. The same can be seen in filarial nematodes infected with *Wolbachia*, in which tetracycline inhibits nematode growth and may cause infertility. The effect of antibiotics is presumably mediated by the death of the *Wolbachia* within the worms because tetracycline treatment of species of filaria that are naturally free of *Wolbachia* infection does not damage the nematodes. Mutual dependence of symbiont and host is also indicated in their long evolutionary history, with co-cladogenesis (Bandi et al. 1998). The precise role of *Wolbachia* in host function is unknown.

The association of filariae with *Wolbachia* is important in the pathology of the interaction between nematode and mammal host. When *Onchocerca volvulus*, the filarial organism that causes river blindness, dies in host tissues (either naturally or following chemotherapy), it releases *Wolbachia*. Clinical studies have shown that it is this release that causes much of the pathology of river blindness, associated with local inflammatory responses to infection (Saint Andre et al. 2002).

Is the interaction between *Wolbachia* and nematodes unique, or are arthropods also sometimes dependent upon their *Wolbachia*? Total dependence of *Asobara tabida* on one of its *Wolbachia* strains was observed by Dedeine et al. (2001). Treatment of the wasp with antibiotics resulted in females that failed to develop ovaries. Their study suggests an integration of *Wolbachia* into host patterns of oogenesis. This integration is reinforced by the observations by Starr and Cline (2002) of *Wolbachia* "rescue" of the phenotype of certain Sxl^f mutations in *Drosophila*. *Sxl* is the major switch gene in *Drosophila* sex determination and is also involved in female germline development. Interaction between *Wolbachia* and *Sxl* indicates a possible mechanism for dependence of insect oogenesis on the presence of *Wolbachia*.

Beyond these studies, there are a few reports indicating that host fitness is improved by *Wolbachia*. In the mosquito *Aedes albopictus*, Dobson et al. (2002) observed that infected females had greater longevity and fecundity than their isogenic uninfected counterparts. In the flour beetle *Tribolium confusum* and stalk eyed flies, infected males had greater sperm competitive ability (Wade and Chang 1995; Hariri et al. 1998). This may be caused by co-adaptation of the insect to the presence of *Wolbachia* in its tissues, producing a reduction in host physiological function when *Wolbachia* is lost.

7.6. IS THERE A *WOLBACHIA* PARADIGM?

For several years, *Wolbachia* appeared to be a uniquely important associate of insects. During the 1990s, it became clear that it was very common across taxa and had a wide variety of phenotypic interactions with its host. *Wolbachia* was regarded as special (e.g., Knight 2001; Zimmer 2001). However, recent study has demonstrated that although *Wolbachia* is a very important associate of arthropods and nematodes, it is not unique, nor does it show any phenotypes that are not displayed by other symbionts (Weeks et al. 2002).

Inherited microorganisms that distort the sex ratio of their hosts have been initially discovered from their phenotype – hosts that give rise to all-female broods – in which the trait is maternally transmitted. Investigation of the source of these all-female broods have found a variety of different microorganisms associated with the distortion. In the case of male killing, there are at least six different independent evolutions of male killing in eubacterial associates of insects. The eubacteria concerned derive from phylogenetically very disparate clades, representing over 2000 Ma of evolutionary separation.

In the case of feminisation, it has long been recognised that microsporidial infections of *Gammarus* have the same capacity for host feminisation as *Wolbachia*, and there appear to have been several evolutions of feminising ability within the Microspora (Ironside et al. 2003). Feminisation has also been recently attributed to a member of the Cytophaga-Flexibacter-Bacteroides group in *Brevipalpus* mites (Weeks et al. 2001).

Microbial induction of parthenogenesis was until lately thought to be associated only with *Wolbachia*. However, there is evidence that members of *Flavobacterium* induce parthenogenesis in *Encarsia* wasps (Zchori-Fein et al. 2001). Further, Koivisto and Braig (2003) reviewed the evidence for microorganism-induced parthenogenesis in invertebrates and concluded that it was likely that a variety of other microorganisms produce this trait. They point to the case of verrucomicrobial symbionts of *Xiphinema americanum*, a nematode (Vandekerckhove et al. 2000). In this parthenogenetic species, reproduction ceased following the administration of antibiotics.

It is not just sex ratio distortion and parthenogenesis induction that is achieved by other inherited symbionts. Recently, the "holy grail" that made *Wolbachia* special was breached, with the observation that cytoplasmic incompatibility was caused by a member of the Cytophaga-Flexibacter-Bacteroides group now designated *Cardinium* (Hunter et al. 2003). Given that the finding of CI has frequently been of the form "find *Wolbachia* in a PCR screen, then look for CI," it is possible that there is a strong study bias towards finding *Wolbachia* as the cause of CI, and that in fact, several bacteria in addition to *Wolbachia* and *Cardinium* have this ability.

Finally, beneficial symbionts have long been known to be a diverse clade. Members of the alpha-proteobacteria, beta-proteobacteria, gamma-proteobacteria, and flavobacteria all have "mutually dependent" interactions with their hosts. Here, *Wolbachia* has always been appreciated as one of many. Some of the interactions, like the interaction between *Wolbachia* and nematodes, are very ancient (Morand and Baumann 1994; Bandi et al. 1995); others (like the interaction of *Wolbachia* with *Asobara tabida*) are relatively young.

In conclusion, *Wolbachia* is no longer so much the special case. That said, current evidence does still give it a very high incidence compared with other known symbionts (the next most common associate, the CFB bacterium, infects roughly half as many species [Weeks and Stouthamer 2003]). In addition, although none of the phenotypes *Wolbachia* possesses is unique to it, *Wolbachia* is the only bacterium presently known to have all the phenotypes. However, it may well be that the full plasticity of other eubacterial associates of arthropods is just not yet recognised.

7.7. CONCLUSIONS

We know an immense amount more about *Wolbachia* now than we did 15 years ago, when PCR first allowed easy study. But an equally immense amount remains unknown. The unknown elements fall into two major classes: evolutionary and ecological impacts, and aspects of the mechanistic basis of interaction with the host.

With respect to the former, a key question is the extent and type of host response to infection. To what extent is the *A. vulgare* observation of host sex-determination system evolution in response to parasitism true for male-killers? Are there aspects of cell and developmental biology that are covertly responses to the presence of intracellular symbionts? The observation of dependence of *A. tabida* on one of its symbionts certainly indicates a degree of co-evolution between germline formation and *Wolbachia*, and the interaction between *Wolbachia* and mutations in *Sxl* in flies indicates a degree of integration with developmental systems that we did not appreciate previously.

With respect to mechanisms of manipulation and dependence, we still only have the barest understanding of what happens to the host. We do not understand the molecular basis of the manipulations, either from a bacterial or a host perspective. The recently annotated genome sequence of the *Wolbachia* from *D. melanogaster* (Wu et al. 2004), together with the sequences of strains causing different manipulations that are to be completed soon, promises new breakthroughs in these areas. It will be interesting to learn the extent to which *Wolbachia* does interfere with cellular systems, and it will be surprising if the interference is not both subtle and beautiful. It will be very interesting also to discover if all the different manipulations within *Wolbachia* share common mechanistic themes, and whether different bacteria achieve the same manipulation by the same or different means.

However, the most fascinating question remains. Arthropods are very diverse for their systems of sex determination. *Wolbachia* is very diverse in its reproductive parasitic phenotypes. Although it seems intuitively sensible to propose that arthropod diversity has driven *Wolbachia* diversity, the reverse proposition is both tempting and powerful. Could *Wolbachia* and the other inherited parasites found in insects be partly responsible for the diversity of arthropod sex detemination systems that exist?

REFERENCES

Arakaki, N., Miyoshi, T., and Noda, H. (2001). *Wolbachia*-mediated parthenogenesis in the predatory thrips *Franklinothrips vespiformis* (Thysanoptera: Insecta). *Proceedings of the Royal Society of London B* **268**, 1011–1016.

Bandi, C., Anderson, T. J. C., Genchi, C., and Blaxter, M. L. (1998). Phylogeny of *Wolbachia* in filarial nematodes. *Proceedings of the Royal Society of London B* **265**, 2407–2413.

Bandi, C., Sironi, M., Damiani, G., Magrassi, L., Nalera, C. A., Laudani, U., and Sacchi, L. (1995). The establishment of intracellular symbiosis in an ancestor of cockroaches and termites. *Proceedings of the Royal Society of London B* **259**, 293–299.

Bordenstein, S. R. (2003). Symbiosis and the origin of species. In *Insect Symbiosis*, ed. K. Bourtzis and T. A. Miller, pp. 283–304. Boca Raton FL: CRC Press.

Bordenstein, S. R., O'Hara, F. P., and Werren, J. H. (2001). *Wolbachia*-induced incompatibility precedes other hybrid incompatibilities in *Nasonia*. *Nature* **409**, 707–710.

Bouchon, D., Rigaud, T., and Juchault, P. (1998). Evidence for widespread *Wolbachia* infection in isopod crustaceans: molecular identification and host feminization. *Proceedings of the Royal Society of London B* **265**, 1081–1090.

Boyle, L., O'Neill, S. L., Robertson, H. M., and Karr, T. L. (1993). Interspecific and intraspecific horizontal transfer of *Wolbachia* in *Drosophila*. *Science* **260**, 1796–1799.

Breeuwer, J. A. J. (1997). *Wolbachia* and cytoplasmic incompatibility in the spider mites *Tetranychus urticae* and *T. turkestani*. *Heredity* **78**, 41–47.

Breeuwer, J. A. J. and Werren, J. H. (1990). Microorganisms associated with chromosome destruction and reproductive isolation between two insect species. *Nature* **346**, 558–560.

Callaini, G., Dallai, R., and Riparbelli, M. G. (1997). *Wolbachia*-induced delay of paternal chromatin condensation does not prevent maternal chromosomes from entering anaphase in incompatible crosses of *Drosophila simulans*. *Journal of Cell Science* **110**, 271–280.

Caspari, E. and Watson, G. S. (1959). On the evolutionary importance of cytoplasmic sterility in mosquitoes. *Evolution* **13**, 568–570.

Cavalcanti, A. G. L., Falcão, D. N., and Castro, L. E. (1957). "Sex-ratio" in *Drosophila prosaltans* – a character due to interaction between nuclear genes and cytoplasmic factors. *American Naturalist* **91**, 327–329.

Charlat, S., Hurst, G. D. D., and Merçot, H. (2003). Evolutionary consequences of *Wolbachia* infections. *Trends in Genetics* **19**, 217–223.

Clark, M. E., Veneti, Z., Bourtzis, K., and Karr, T. L. (2003). *Wolbachia* distribution and cytoplasmic incompatibility during sperm development: the cyst as the basic cellular unit of CI expression. *Mechanisms of Development* **120**, 185–198.

Dedeine, F., Vavre, F., Fleury, F., Loppin, B., Hochberg, M. E., and Boulétreau, M. (2001). Removing symbiotic *Wolbachia* bacteria specifically inhibits oogenesis in a parasitic wasp. *Proceedings of the National Academy of Sciences USA* **98**, 6247–6252.

Dobson, S. L., Marsland, E. J., and Rattandadechakul, W. (2002). Mutualistic *Wolbachia* infection in *Aedes albopictus*: accelerating cytoplasmic drive. *Genetics* **260**, 1087–1094.

Dyson, E. A., Kamath, M. K., and Hurst, G. D. D. (2002). *Wolbachia* infection associated with all-female broods in *Hypolimnas bolina* (Lepidoptera: Nymphalidae): evidence for horizontal transmission of a butterfly male killer. *Heredity* **88**, 166–171.

Dyson, E. M., and Hurst, G. D. D. (2004). Persistence of an extreme sex ratio bias in a natural population. *Proceedings of the National Academy of Sciences USA* **101**, 6520–6523.

Fialho, R. F. and Stevens, L. (2000). Male-killing *Wolbachia* in a flour beetle. *Proceedings of the Royal Society of London B* **267**, 1469–1473.

Frank, S. A. (1997). Cytoplasmic incompatibility and population structure. *Journal of Theoretical Biology*, **184**, 327–330.

Gottlieb, Y. and Zchori-Fein, E. (2001). Irreversible thelytokous reproduction in *Muscidifurax uniraptor*. *Entomologia Experimentalis et Applicata* **100**, 271–278.

Hariri, A. R., Werren, J. H., and Wilkinson, G. S. (1998). Distribution and reproductive effects of *Wolbachia* in stalk eyed flies (Diptera: Diopsidae). *Heredity* **81**, 254–260.

Harris, H. L. and Braig, H. R. (2001). Sperm nuclear basic proteins in *Drosophila simulans* undergoing *Wolbachia* induced cytoplasmic incompatibility. *Developmental Biology*, **235**, 191.

Heath, B. D., Butcher, R. D. J., Whitfield, W. G. F., and Hubbard, S. F. (1999). Horizontal transfer of *Wolbachia* between phylogenetically distant insect species by a naturally occurring mechanism. *Current Biology* **9**, 313–316.

Hertig, M. (1936). The Rickettsia *Wolbachia pipientis* (gen. and sp. nov.) and associated inclusions in the mosquito, *Culex pipiens*. *Parasitology* **28**, 453–486.

Hertig, M. and Wolbach, S. B. (1924). Studies on Rickettsia-like microorganisms in insects. *Journal of Medical Research* **44**, 329–374.

Hiroki, M., Kato, Y., Kamito, T., and Miura, K. (2002). Feminization of genetic males by a symbiotic bacterium in a butterfly, *Eurema hecabe* (Lepidoptera: Pieridae). *Naturwissenschaften* **89**, 167–170.

Hoffmann, A. A. and Turelli, M. (1997). Cytoplasmic incompatibility in insects. In *Influential Passengers*, ed. S. L. O'Neill, A. A. Hoffmann, and J. H. Werren, pp. 42–80. Oxford: Oxford University Press.

Hoffmann, A. A., Turelli, M., and Harshman, L. G. (1990). Factors affecting the distribution of cytoplasmic incompatibility in *Drosophila simulans*. *Genetics* **126**, 933–948.

Hoffmann, A. A., Clancy, D., and Duncan, J. (1996). Naturally-occurring *Wolbachia* infection in *Drosophila simulans* that does not cause cytoplasmic incompatibility. *Heredity* **76**, 1–8.

Huigens, M. E., Luck, R. F., Klaassen, R. H. G., Maas, M., Timmermans, M., and Stouthamer, R. (2000). Infectious parthenogenesis. *Nature* **405**, 178–179.

Huigens, M. E., de Almeida, R. P., Boons, P. A. H., Luck, R. F., and Stouthamer, R. (2004). Natural interspecific and intraspecific horizontal transfer of parthenogenesis-inducing *Wolbachia* in *Trichogramma* wasps. *Proceedings of the Royal Society of London B* **271**, 509–515.

Hunter, M. S., Perlman, S. J., and Kelly, S. E. (2003). A bacterial symbiont in the *Bacteroidetes* induces cytoplasmic incompatibility in the parasitoid wasp *Encarsia pergandiella. Proceedings of the Royal Society of London B* **270**, 2185–2190.

Hurst, G. D. D., Hurst, L. D., and Majerus, M. E. N. (1997). Cytoplasmic sex ratio distorters. In *Influential Passengers,* ed. S. L. O'Neill, J. H. Werren, and A. A. Hoffmann, pp.125–154. Oxford: Oxford University Press.

Hurst, G. D. D., Jiggins, F. M., von der Schulenburg, J. H. G, Bertrand, D., West, S. A., Goriacheva, I. I., Zakharov, I. A., Werren, J. H., Stouthamer, R., and Majerus, M. E. N (1999). Male-killing *Wolbachia* in two species of insect. *Proceedings of the Royal Society of London B* **266**, 735–740.

Hurst, G. D. D. and Schilthuizen, M. (1998). Selfish genetic elements and speciation. *Heredity* **80**, 2–8.

Hurst, G. D. D. and Werren, J. H. (2001). The role of selfish genetic elements in eukaryotic evolution. *Nature Reviews Genetics* **2**, 597–606.

Hurst, G. D. D, Jiggins, F. M., and Robinson, S. J. W. (2001). What causes inefficient transmission of male-killing *Wolbachia* in *Drosophila*? *Heredity* **87**, 220–226.

Hurst, G. D. D. and Majerus, M. E. N. (1993). Why do maternally inherited microorganisms kill males? *Heredity* **71**, 81–95.

Hurst, G. D. D., Johnson, A. P., Schulenburg, J. H., and Fuyama, Y. (2000). Male-killing *Wolbachia* in *Drosophila*: a temperature-sensitive trait with a threshold bacterial density. *Genetics* **156**, 699–709.

Hurst, L. D. (1991). The evolution of cytoplasmic incompatibility or when spite can be successful. *Journal of Theoretical Biology* **148**, 269–277.

Ironside, J. E., Smith, J. E., Hatcher M. J., Sharpe, R. G., Rollinson, D., and Dunn. A. M. (2003). Two species of feminizing microsporidian parasite coexist in populations of *Gammarus duebeni. Journal of Evolutionary Biology* **16**, 467–473.

Jiggins, F. M., Bentley, J. K., Majerus, M. E. N., and Hurst, G. D. D. (2001a). How many species are infected with *Wolbachia*? Cryptic sex ratio distorters revealed to be common by intensive sampling. *Proceedings of the Royal Society of London B* **268**, 1123–1126.

Jiggins, F. M., von Der Schulenburg, J. H. G., Hurst, G. D. D., and Majerus, M. E. N. (2001b). Recombination confounds interpretation of *Wolbachia* phylogeny. *Proceedings of the Royal Society of London B* **268**, 1123–1126.

Jiggins, F. M., Hurst, G. D. D., and Majerus, M. E. N. (2000). Sex ratio distorting *Wolbachia* cause sex role reversal in their butterfly hosts. *Proceedings of the Royal Society of London B* **267**, 69–73.

Jiggins, F. M., Randerson, J. P., Hurst, G. D. D., and Majerus, M. E. N. (2002). How can sex ratio distorters reach extreme prevalences? Male-killing *Wolbachia* are not suppressed and have near-perfect vertical transmission efficiency in *Acraea encedon*. *Evolution* **56**, 2290–2295.

Kageyama, D., Nishimura, G., Hoshizaki, S., and Ishikawa, Y. (2002). Feminizing *Wolbachia* in an insect, *Ostrinia furnacalis* (Lepidoptera: Crambidae). *Heredity* **88**, 444–449.

Kageyama, D. and Traut, W. (2003). Opposite sex-specific effects of *Wolbachia* and interference with the sex determination of its host *Ostrinia scapulalis*. *Proceedings of the Royal Society of London B* **271**, 251–258.

Knight, J. (2001). Meet the Herod bug. *Nature* **412**, 12–14.

Koivisto, R. K. K. and Braig, H. R. (2003). Microorganisms and parthenogenesis. *Biological Journal of the Linnean Society* **79**, 43–58.

Lassy, C. W. and Karr, T. L. (1996). Cytological analysis of fertilization and early embryonic development in incompatible crosses of *Drosophila simulans*. *Mechanisms of Development* **57**, 47–58.

Masui, S., Kamoda, S., Sasaki, T., and Ishikawa, H. (1999). The first detection of the insertion sequence ISW1 in the intracellular reproductive parasite *Wolbachia*. *Plasmid* **42**, 13–19.

Masui, S., Sasaki, T., and Ishikawa, H. (2000). Genes for the type IV secretion system in an intracellular symbiont, *Wolbachia*, a causative agent of various sexual alterations in arthropods. *Journal of Bacteriology* **182**, 6529–6531.

Masui, S., Kamado, S, Sasaki, T., and Ishikawa, H. (2000). Distribution and evolution of bacteriophage WO in *Wolbachia*, the endosymbiont causing sexual alterations in arthropods. *Journal of Molecular Evolution* **51**, 491–497.

Moran, N. A. and Baumann, P. (1994). Phylogenetics of cytoplasmically inherited microorganisms of arthropods. *Trends in Ecology & Evolution* **9**, 15–20.

McGraw, E. A., Merritt D. J., Droller, J. N., and O'Neill, S. L. (2001). *Wolbachia*-mediated sperm modification is dependent on the host genotype in *Drosophila*. *Proceedings of the Royal Society of London B* **268**, 2565–2570.

Moreau, J. and Rigaud, T. (2003). Variable male potential rate of reproduction: high male mating capacity as an adaptation to parasite-induced excess of females? *Proceedings of the Royal Society of London B* **270**, 1535–1540.

Moret, Y., Juchault, V., and T. Rigaud. (2001). *Wolbachia* endosymbiont responsible for cytoplasmic incompatibility in a terrestrial crustacean: effects in natural and foreign hosts. *Heredity* **86**, 325–32.

O'Neill, S. L. and Karr, T. L. (1992). Bidirectional incompatibility between conspecific populations of *Drosophila simulans. Nature* **348**, 178–180.

O'Neill, S. L., Giordano, R., Colbert, A. M. E., Karr, T. L., and Robertson, H. M. (1992). 16S rRNA phylogenetic analysis of the bacterial endosymbionts associated with cytoplasmic incompatibility in insects. *Proceedings of the National Academy of Sciences USA* **89**, 2699–2702.

O'Neill, S. L., Pettigrew, M. M., Sinkins, S. P., Braig, H. R., Andreadis, T. G., and Tesh, R. B. (1997). *In vitro* cultivation of *Wolbachia pipientis* in an *Aedes albopictus* cell line. *Insect Molecular Biology* **6**, 33–39.

Poinsot, D., Bourtzis, K., Markakis, G., Savakis, C., and Mercot, H. (1998). *Wolbachia* transfer from *Drosophila melanogaster* into *D. simulans*: Host effect and cytoplasmic incompatibility relationships. *Genetics* **150**, 227–237.

Randerson, J. P., Jiggins, F. M., and Hurst, L. D. (2000). Male-killing can select for male mate choice: a novel solution to the paradox of the lek. *Proceedings of the Royal Society of London B* **267**, 867–874.

Reynolds, K. T. and Hoffmann, A. A. (2002). Male age, host effects and the weak expression or non-expression of cytoplasmic incompatibility in *Drosophila* strains infected by maternally transmitted *Wolbachia. Genetical Research* **80**, 79–87.

Rigaud, T. and Juchault, P. (1992). Genetic control of the vertical transmission of a cytoplasmic sex factor in *Armadillidium vulgare* Latr. (Crustacea, Oniscidea). *Heredity* **68**, 47–52.

Rigaud, T. and Juchault, P. (1995). Success and failure of transmission of feminizing *Wolbachia* endoysmbionts in woodlice. *Journal of Evolutionary Biology* **8**, 249–255.

Rigaud, T. (1997). Inherited microorganisms and sex determination of arthropod hosts. In *Influential Passengers*, ed. S. L. O'Neill, A. A. Hoffmann and J. H. Werren, pp. 81–98. Oxford: Oxford University Press.

Rousset, F., Bouchon, D. Pintureau, B., Juchault, V., and M. Solignac. (1992). *Wolbachia* endosymbionts responsible for various alterations of sexuality in arthropods. *Proceedings of the Royal Society of London B* **250**, 91–98.

Rousset, F. and Raymond, M. (1991). Cytoplasmic incompatibility in insects: why sterilize females. *Trends in Ecology and Evolution* **6**, 54–57.

Saint Andre, A., Blacwell, N. M., Hall, L. R., Hoerauf, A., Brattig, N. W., Volkamnn, L., Taylor, M. J., Ford, L., Jise, A. G., Lass, J. H., Diaconu, E., and Pearlman, E. (2002). The role of endosymbiotic *Wolbachia* in the pathogenesis of river blindness. *Science* **295**, 1892–1895.

Sasaki, T., Kubo, T., and Ishikawa, H. (2002). Interspecific transfer of *Wolbachia* between two lepidopteran insects expressing cytoplasmic incompatibility: a *Wolbachia* variant naturally infecting *Cadra cautella* causes male killing in *Ephestia kuehniella*. *Genetics* **162**, 1313–1319.

Shoemaker, D. D., Katju, V. J., and Jaenike, J. (1999). *Wolbachia* and the evolution of reproductive isolation between *Drosophila recens* and *Drosophila subquinaria*. *Evolution* **53**, 1157–1164.

Sironi, M., Bandi, C., Sacchi, L., Disacco, B, Damiani, G., and Genchi, C. (1995). Molecular evidence for a close relative of the arthropod endosymbiont *Wolbachia* in a filarial worm. *Molecular and Biochemical Parasitology* **74**, 223–227.

Snook, R. R., Cleland, S. Y., Wolfner, M. F., and Karr, T. L. (2000). Offsetting effects of *Wolbachia* infection and heat shock on sperm production in *Drosophila simulans*: analyses of fecundity, fertility and accessory gland proteins. *Genetics* **155**, 167–178.

Solignac, M., Vautrin, D., and Rousset, F. (1994). Widespread occurrence of the proteobacteria *Wolbachia* and partial cytoplasmic incompatibility in *Drosophila melanogaster*. *Comptes Rendus de l Academie des Sciences Serie III-Sciences de la Vie-Life Sciences* **317**, 461–470.

Starr, D. J. and Cline, T. W. (2002). A host-parasite interaction rescues *Drosophila* oogenesis defects. *Nature* **418**, 76–79.

Stouthamer, R., Breeuwer, J. A. J., Luck, R. F., and Werren, J. H. (1993). Molecular identification of microorganisms associated with parthenogenesis. *Nature* **361**, 66–68.

Stouthamer, R., Luck, R. F., and Hamilton, W. D. (1990). Antibiotics cause parthenogenetic *Trichogramma* (Hymenoptera/Trichogrammatidae) to revert to sex. *Proceedings of the National Academy of Sciences USA* **87**, 2424–2427.

Stouthamer, R. and Kazmer D. J. (1994). Cytogenetics of microbe-associated parthenogenesis and its consequences for gene flow in *Trichogramma* wasps. *Heredity* **73**, 317–327.

Stouthamer, R., van Tilborg, M., de Jong, J. H., Nunney, L., and Luck, R. F. (2001). Selfish element maintains sex in natural populations of a parasitoid wasp. *Proceedings of the Royal Society of London B* **268**, 617–622.

Stouthamer, R. (1997). *Wolbachia*-induced parthenogenesis. In *Influential Passengers*, ed. S. L. O'Neill, A. A. Hoffmann and J. H. Werren, pp. 102–122. Oxford: Oxford University Press.

Sun, L. V., Foster, J. M., Tzertzinis, G., Ono, M., Bandi, C., Slatko, B., and O'Neill, S. L. (2001). Determination of *Wolbachia* genome size by pulsed gel electrophoresis. *Journal of Bacteriology* **183**, 2219–2225.

Tram, U. and Sullivan, W. (2002). Role of delayed nuclear envelope breakdown and mitosis in *Wolbachia*- induced cytoplasmic incompatibility. *Science* **296**, 1124–1126.

Tram, U., Ferree, P. M., and Sullivan, W. (2003). Identification of *Wolbachia*-host interacting factors through cytological analysis. *Microbes and Infection* **5**, 999–1011.

Turelli, M. and Hoffmann, A. A. (1995). Cytoplasmic incompatibility in *Drosophila simulans*: dynamics and parameter estimates from natural populations. *Genetics* **140**, 1319–1338.

Vandekerckhove, T. T., Willems, A., Gillis, M., and Coomans, A. (2000). Occurrence of novel verrucomicrobial species, endosymbiotic and associated with parthenogenesis in *Xiphinema americanum*–group species (Nematoda, Longidoridae). *International Journal of Systematic and Evolutionary Microbiology* **50**, 2197–2205.

Veneti Z., Clark, M. E., Zabalou, S., Savakis, C. Karr, T. L., and Bourtzis, K. (2003). Cytoplasmic incompatibility and sperm cyst infection in different *Drosophila-Wolbachia* associations. *Genetics* **164**, 545–552.

Wade, K. and Chang, N. W. (1995). Increased male fertility in *Tribolium confusum* beetles after infection with the parasite *Wolbachia. Nature* **373**, 72–74.

Weeks, A. R. and Breeuwer, J. A. (2001). *Wolbachia*-induced parthenogenesis in a genus of phytophagous mites. *Proceedings of the Royal Society of London B* **268**, 2245–51.

Weeks, A. R., Marec, F., and Breeuwer, J. A. J. (2001). A mite species that comprises entirely of haploid females. *Science* **292**, 2479–2482.

Weeks, A. R., Reynolds, K. T., Hoffmann, A. A., and Mann, H. (2002). *Wolbachia* dynamics and host effects: what has (and has not) been demonstrated? *Trends in Ecology and Evolution* **17**, 257–262.

Weeks, A. R., Velten, R., and Stouthamer, R. (2003). Incidence of a new sex-ratio-distorting endosymbiotic bacterium among arthropods. *Proceedings of the Royal Society of London B* **270**, 1857–1865.

Werren, J. H. (1991). The Paternal-Sex-Ratio Chromosome of *Nasonia. American Naturalist* **137**, 392–402.

Werren, J. H. (1998). *Wolbachia* and speciation. In *Endless Forms: Species and Speciation*, ed. D. Howard and S. Berlocher, pp. 245–260. Oxford: Oxford University Press.

Werren, J. H., Windsor, D., and Guo, L. R. (1995). Distribution of *Wolbachia* amongst neoptropical arthropods. *Proceedings of the Royal Society of London B* **262**, 197–204.

Werren, J. H. (1997). Biology of *Wolbachia. Annual Review of Entomology* **42**, 587–609.

Wu, M., Sun, L. V., Vamathevan, J. et al. (2004). Phylogenomics of the reproductive parasite *Wolbachia pipientis w* Mel: A streamlined genome overrun by mobile genetic elements. *PLOS Biology* **2**, 327–341.

Yen, J. H. and Barr A. R. (1971). New hypothesis of the cause of cytoplasmic incompatibility in *Culex pipiens. Nature* **232**, 657–658.

Zchori-Fein, E., Gottlieb, Y., Kelly, S. E., Brown, J. K., Wilson, J. M., Karr, T. L., and Hunter, M. S. (2001). A newly discovered bacterium associated with parthenogenesis and a change of host selection behaviour in parasitoid wasps. *Proceedings of the National Academy of Sciences USA*, **98**, 12555–12560.

Zimmer, C. (2001). *Wolbachia.* A tale of sex and survival. *Science*, **292**, 1093–1095.

CHAPTER 8

Microbial communities in lepidopteran guts: From models to metagenomics

Jo Handelsman, Courtney J. Robinson, and Kenneth F. Raffa

8.1. INTRODUCTION

Most microorganisms live in complex microbial communities. So do their hosts. Eukaryotes have co-evolved with diverse microorganisms on their internal and external surfaces. These microorganisms constitute communities, and the host's fitness may be affected differently by the community than by any single species. Therefore, understanding the structure and function of microbial communities must be integral to the study of host–microbe interactions (see Chapter 5 for further discussion of this point).

Communities are dynamic assemblages whose stability and functions are governed by dependencies and antagonisms among the members. All microbial communities are dynamic and continually responding to their changing physical environments, but those associated with animals must contend with the vicissitudes of their hosts as well. Host tissues may present attractive surfaces and a rich source of nutrients, but they may also expose the microorganisms to extreme conditions due to pH, toxins in food, or precipitous changes in anatomy, such as the shedding of the very tissue that the microorganisms call home.

Spatial and temporal variation presents challenges to those who study communities as well as the microorganisms that comprise them. Although complexity and change typify the natural environment and are thus central to understanding microbial ecology, it is these very features that microbiologists have striven to minimise in ecological models. The emphasis on pure culture over the last century of microbiology has removed microorganisms from their communities and focused on their behaviour in the biologically simple environment of the Petri dish or liquid culture. Although simple model systems have driven the explosion of knowledge in host–microbe interactions over the last two decades, the reality of natural communities demands that we direct attention to complex assemblages as well. The recent surge of interest in community ecology, coupled with the development of powerful new methods for its study present an unprecedented opportunity to dissect the interactions among microorganisms in communities and elucidate how the function of the entire community determines outcomes for its host. In this context, the reader should refer to Chapter 5 for a discussion of the complexity that can occur in even the simplest monospecies culture system over short time periods.

8.2. GLOBAL ECOLOGICAL QUESTIONS

There are several overarching themes that need to be studied to develop a comprehensive view of the function of any community. Key among these are the influence of physical and chemical factors on community structure and function, succession and community development, mechanisms of mutualism, antagonism, and communication among community members, and the robustness of the community when disturbed or invaded.

An appropriate microbial community can provide a powerful model for studying these principles. Some of the basic tenets of ecology have been well-developed in macroecological systems, and microbial ecology can draw on that existing knowledge, whereas others have been difficult to study at the landscape level and may be better suited for development in microbial models. For example, the keystone species concept has been tested with macroorganisms quite effectively. This is illustrated by the work of Paine in a rocky intertidal zone where he demonstrated that the removal of the top predator, the starfish *Pisaster ochraceus*, resulted in the reduction of species diversity from fifteen species to eight. He showed that the starfish maintains diversity by controlling the population sizes of species that outcompete other members in its absence (Paine 1966). The key findings from this seminal work are broadly applicable to a diverse range of terrestrial and aquatic systems, and a broad range of taxa (Hanke et al. 1992; Naiman et al. 1994; Navarrete and Menge 1996; Crooks and Soulé 1999; Gonzalez-Megias and Gomez 2003). This type of study can provide a model for studying microbial community structure and function, and a framework on which to determine whether similar principles govern micro- and macro-organisms living in communities (Andrews 1991).

In contrast to keystone species, our study of biological invasions lacks opportunity for the same type of experimental approaches, because invasions are not conducive to replicated experiments. Biological invasion theory attempts to predict invasion patterns, the nature of successful invaders, conditions conducive to invasion, consequences of invasion, and rates and directions of spread (Shigesada and Kawasaki 1997). The nature of invasions makes it challenging to test theory with a replicated experimental design, as most invasions are unexpected, unwanted, and not studied until after the process has begun. Yet the increasing rate of biological invasions necessitates that more general and predictive theories replace the case-by-case, post hoc approach on which we currently rely (Kareiva 1996). The field of biological control, however, provides many examples of replicated studies that document the behaviour of invaders. Unfortunately, many of the planned introductions of insect biocontrol agents rely on the prior establishment of an earlier invader, and so the applicability of biological control experiences to invasion ecology is uncertain.

In an ironic contradiction, invasions by certain insects and microorganisms are responsible for damage to agriculture and forestry, whereas invasions by others, in the form of biological control agents, have been powerful regulators of invasive pests and pathogens. The study of biological control has informed invasion theory. Biocontrol is often dependent on successful invasion by a natural predator or parasitoid of an insect pest. Such invasions have been remarkably effective in some cases. For example, in 1888, the citrus tree pest *Icerya purchasi* was successfully controlled by the introduction of *Rodolia cardinalis* (Shigesada and Kawasaki 1997). More recent successes include control of the cassava mealybug and green mite in Africa using parasitoids from South America (Bellotti et al. 1999). In addition to reducing populations of pests, invasion by biocontrol agents can also affect native communities adversely, as is illustrated by *Compsilura concinnata*, a parasitoid fly, which was repeatedly introduced to North America to control thirteen pests between 1906 and 1986. This biocontrol agent successfully invaded New England and may now be responsible for the decline in the population of the native, nontarget silk moth (Boettner et al. 2000). Because of this and other examples (Pearson et al. 2000), the potential for further biological control has declined as environmental concerns have risen (Simberloff and Stiling 1996; Daehler and Gordon 1997; Schaffner 2001). Once again, the lack of comprehensive theories that can be empirically validated hinders further advances. Similar problems hinder the field of planned introduction of genetically modified organisms. The enormous potential of this approach is well established, as is our ability to test for potentially direct, immediate effects in closed systems. Unfortunately, most adverse effects that have accompanied other technologies have been delayed, indirect, and only manifested under natural conditions (Kareiva 1996).

Microbiology has provided other systems for exploring invasion biology because some microbiological therapies depend on successful invasions. Biological control of plant disease, for example, involves successful establishment and proliferation in a community by a microorganism, resulting in suppression of plant disease (Cook and Baker 1983; Handelsman and Stabb 1996). Kinkel and Lindow (1993) examined the invasion and exclusion abilities of *Pseudomonas syringae* strains on plant leaves. They found that population size alone did not predict a strain's ability to invade a community or to exclude others from it, and the characteristics of successful invaders and excluders differed.

Invaders of root-associated communities contend with the highly complex community in the rhizosphere. In general, the invader's population size is correlated with biological effects, such as disease suppression, but there are some surprises. Gilbert et al. (1993) found that when *Bacillus cereus* was inoculated onto seeds, its population on the root emerging from the seed diminished rapidly, but it continued to have a global impact on the composition of the microbial community long after it was detectable. The complexity of the rhizosphere community makes it difficult to test the hypotheses that are simple enough to be empirically tractable yet robust enough to incorporate community–plant interactions, and thus, simple models would be useful to track bacterial populations and their lasting impacts on communities through which they pass.

A human gut is an ecosystem in which invasion is of particular interest. The success of gut pathogens depends on their invasive ability. Likewise, probiotics, or bacterial inoculants such as *Lactobacillus* and *Bifidobacterium* spp. might be more effective if they survived and colonized the gut (Bengmark 1998; Holzapfel et al. 1998; Goossens et al. 2003; Guarner and Malagelada 2003). Both pathogens and beneficial bacteria must compete with the indigenous community for attachment sites and nutrients. A healthy gut community is highly resistant to invasion, providing the "barrier effect" or "colonisation resistance" that maintains gut function (Bourlioux et al. 2003; Guarner and Malagelada 2003). But little is known about what makes a microbial community robust to or able to recover from invasion.

There is scant knowledge on which to base the design of successful invaders for probiotics or biological control, or to predict how easily a microbial or macrobiological community will be penetrated, and perhaps altered, by an invader. Developing the right model system will advance microbial and macroecology by providing a context for testing invasion theory. One of our goals is to develop a model system in which hypotheses can be tested rigorously with precisely modulated treatments, controlled variables, and replicated experimental designs.

8.3. MODEL SYSTEMS

The right model system is essential to test ecological principles about microbial community structure, function, succession, and robustness. To be serviceable, a model community must be easily maintained and reproduced and have reproducible composition. There should be means to introduce chemicals or organisms into the community by a process that approximates a natural event, and the community should be sufficiently simple in composition to ensure that all of its members can be studied in culture or by genomics and that the communication networks connecting community members can be mapped.

A number of outstanding model systems have been established for microbial community study over the last two decades (Table 8.1). Each system is suited best to particular investigations. In the following section, we review a few examples of these communities, the advances they have afforded, and their limitations. Many of these model systems involve associations of a single species with a host. Although they may not immediately inform our understanding of complex communities, they have established broad principles of host–microbe interactions and microbe–microbe interactions that are applicable to multispecies communities. The rapid progress made in these simple systems has facilitated the next steps involving complex ones.

8.3.1. *Vibrio*–Squid

Vibrio fischeri and its host, the squid *Euprymna scolopes*, have an extraordinary relationship. The bacteria enter and colonise the light organ of the squid in which the bacteria emit light at night, providing counterillumination that enables the host to avoid detection by its prey, who would otherwise see the squid's shadow in moonlit waters (McFall-Ngai 2000). The partners substantially affect each other's biology – the bacteria affect development of the squid's light organ, and the squid provides a non-competitive niche for the bacteria by preventing colonisation by other species (Foster et al. 2000; Visick and McFall-Ngai 2000; Visick et al. 2000). Despite its exotic, perhaps unique, outcome, study of this system has revealed principles in microbial behaviour and host–microbe interactions that appear to be common throughout the microbial world.

Table 8.1. *Examples of model systems for studying host–microbe interactions*

Symbiosis	Approximate number of species	Strengths and contributions of system	References
Squid–*Vibrio*	1	Study of signal exchange and colonisation; effect of symbiont on host development	McFall-Ngai 2000; Nyholm et al. 2000; Visick and McFall-Ngai 2000; Visick et al. 2000
Legume–*Rhizobium*	1	Study of signal exchange with host; effect on host development; quantitative modelling of interstrain competition	Beattie et al. 1989
Nematode–*Pseudomonas* and *Enterococcus*	1	Identification of virulence genes; comparative virulence in other hosts	Choi et al. 2002; Hendrickson et al. 2001; Jander et al. 2000; Miyata et al. 2003
Gypsy moth	~11	Relatively simple, multispecies	Broderick et al. 2004
Cabbage white butterfly	~10	Relatively simple, multispecies	Robinson et al., unpublished
Termite gut	100–200	Study of carbon, nitrogen, and hydrogen cycling within a community; demonstration of functional redundancy in microbial community	Kane and Breznak 1991; Leadbetter and Breznak 1996; Leadbetter, et al. 1998; Leadbetter et al. 1999; Lilburn et al. 1999; Lilburn et al. 2001; Nakashima et al. 2002
Mouse gut	500–1,000	Effect of bacteria on immune system development	Hooper et al. 2003; Stappenbeck et al. 2002
Human oral cavity	~500	Bacterial adhesion and succession	Paster et al. 2001

Study of the squid–*Vibrio* symbiosis led to the discovery of quorum sensing, a mechanism by which many bacteria sense the population density of their species and express genes accordingly. Population density is detected by accumulation of acylated homoserine lactones (AHLs) whose structures vary, conferring species specificity. Quorum sensing determines expression of the genes responsible for light emission by *Vibrio fisheri* and regulates genes in other organisms that encode virulence factors, antibiotic production, and other functions that require a high density of cells to be useful to the population (Rice et al. 1999).

The squid–*Vibrio* system has shaped thinking about microbial ecology by illustrating communication and cooperative behaviours among members of a population of a single species (McFall-Ngai 2000; Nyholm et al. 2000). In fact, some microbiologists have argued that a single species can comprise a community based largely on the principle that there is communication among members of a population (Buckley 2003 – see also the arguments by Rainey in Chapter 5). Community, however, has historically been used in the ecological literature to describe a *multispecies assemblage* (Begon et al. 1990), and therefore in this chapter we adhere to the traditional definition of community.

8.3.2. *Rhizobium*–Legume Symbiosis

The *Rhizobium*–legume interaction is another system involving one microbial species and its host that has dramatically shaped thinking about microbial ecology. Members of the Rhizobiaceae family infect the roots of their leguminous hosts and induce the formation of an organ, known as a nodule, in which the bacteria fix nitrogen. Both the bacteria and the plant host produce signals that are interpreted by the other partner, leading to initiation of a new developmental program (Peters et al. 1986; Long 2001; Cullimore and Denarie 2003). Legumes release an array of flavonoids that induce the early *nod* genes, which results in the synthesis of Nod factors by the bacterial partner. Nod factors are lipooligosaccharides with chitin-like structures that induce morphological changes in the plant. The signals in this symbiosis are highly specific – in general, each species of bacteria infects one or a few species of legumes, and the flavonoids and Nod factors generally define the host range of the rhizobia. This example of signal exchange differs from quorum sensing in that the exchange of signals is between members of different domains of life, the partners are sensitive to minute concentrations (micromolar) of each other's signals (10^{-6} micromolar), and the events leading to nodulation can be initiated by a single cell (Faucher et al. 1988; Begum et al. 2001).

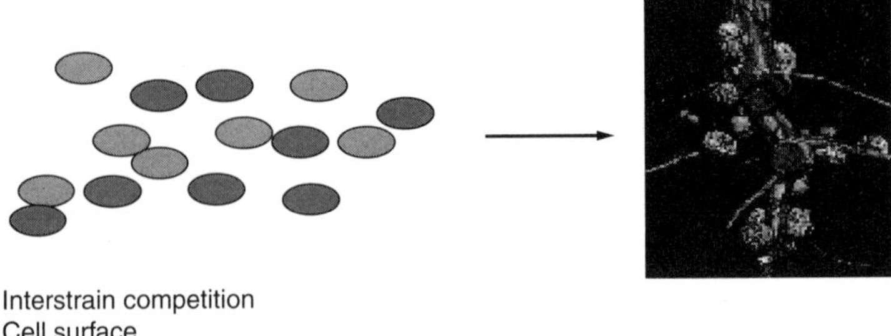

$$\log\left(\frac{P_{\mathrm{KIM5s}}+P_{\mathrm{both}}}{P_{\mathrm{CE3}}+P_{\mathrm{both}}}\right)=C_{\mathrm{KIM5s:CE3}}+k\cdot\log\left(\frac{I_{\mathrm{KIM5s}}}{I_{\mathrm{CE3}}}\right)$$

Figure 8.1. *Rhizobium* competition. When multiple strains compete for infection of leguminous hosts, one strain often dominates in the nodules. The quantitative modelling that has been applied to intrastrain competition may be useful in understanding multispecies communities (Beattie et al. 1989).

The *Rhizobium*–legume system has also provided a simple model for studies of bacterial competition. When strains that are able to nodulate a common host are applied together, the proportion of nodules occupied by each strain is often quite different from their representation in the inoculum. The successful competitor establishes a pure culture inside the nodule, making quantitative studies of competition simpler than in a system that remains in an open environment that is bathed in other bacteria. Quantitative modelling and mutant analysis has revealed an ecological picture of interstrain competition.

These relationships are not linear or simple (Fig. 8.1), and mathematical modelling that describes the relationship has offered insights into the biology (Beattie et al. 1989). Extensive mutant analysis indicates that cell surface features (Handelsman et al. 1984; Milner et al. 1992; Lagares et al. 1992; Bittinger et al. 1997; Thomas-Oates et al. 2003), toxin production (Robleto et al. 1998; Oresnik et al. 1999), nutritional competence, and many other physiological factors determine competitiveness (Triplett and Sadowsky 1992). Plants differ in their ability to admit certain strains into their nodules (Rosas et al. 1998; Laguerre et al. 2003), demonstrating that each partner plays an active role in microbial competition. As in the *Vibrio*–squid interaction, interstrain competition in the *Rhizobium*–legume symbiosis has revealed mathematical and genetic relationships that likely apply to competition in more-complex communities (see Chapter 9 for more details of application of mathematical modelling to such community-based problems).

8.3.3. Insect–Microbe Interactions

The interactions between insects and their microbial symbionts have provided the basis for many principles in symbiosis and have revealed unexpected mechanisms that illustrate how organisms cooperate to perform life functions and gain a competitive edge. The *Buchnera*–aphid symbiosis provides an example of an ancient relationship that has evolved into one of mutual dependence. Analysis of the *Buchnera* genome indicates that the bacterium has lost the capacity to synthesise certain amino acids that it acquires from its host. The *Buchnera* genome has been whittled down to a minimal genome, smaller than genomes of most free-living bacteria. Likewise, the aphid cannot synthesise and depends on *Buchnera* for other essential nutrients (Moran and Mira 2001; Ochman and Moran 2001; Moran 2002, 2003; Moran et al. 2003; Wilcox et al. 2003).

Other insects depend on their bacterial associates for production of mating (Brand et al. 1975, 1976) and cohesion pheromones (Dillon et al. 2002). *Wolbachia* alters reproductive behavior in diverse insects (Plantard et al. 1998, 1999), and an unnamed member of the *Cytophaga-Flexibacter-Bacteroides* group induces parthenogenesis in parasitoid wasps (Zchori-Fein et al. 2001). Insect–bacteria cooperation is discussed in detail in Chapter 7, with emphasis on *Wolbachia*, and Chapter 10, on the the intracellular endosymbionts of the Dryophthoridae.

Most of the arthropod symbioses involve highly specific interactions, often between one bacterial species and one insect. There are a few examples of multispecies communities, such as the associates of the hindgut of termites (discussed in Chapter 2). Degradation of complex carbohydrates and fixation of nitrogen form the metabolic foundation for the community's function (Breznak and Canale-Parola 1972; Breznak et al. 1973; Breznak and Pankratz 1977; Breznak and Kane 1990; Leadbetter and Breznak 1996; Leadbetter et al. 1999; Nakashima et al. 2002; Bakalidou et al. 2002). The intricacies of managing the carbon and nitrogen economy of the hindgut provides evidence for interdependence and co-evolution among the members of the community. A combination of ultrastructural studies, microbiology, and microelectrode measurements indicates that the hindgut is delicately structured to maintain gradients of oxygen and hydrogen, and the microbial community is spatially arranged to take advantage of these and maintain the chemical gradients (Brune 1998). Perhaps the most ecologically intriguing aspect of the termite hindgut community is the functional redundancy. The community contains numerous and diverse spirochetes, for example, that fix nitrogen (Lilburn et al. 1999, 2001). A future challenge in microbial ecology is to determine the role and significance of functional redundancy in communities, and the termite hindgut provides an ideal venue for such studies.

The termite system offers one of the most powerful existing models for studying nutritional interactions among community members. The exchange of energy and elements is well understood and establishes principles that are likely repeated in other microbial communities. However, its relative complexity prevents a complete genomic analysis of the community members and will make it challenging to construct a comprehensive map of its communication networks.

8.3.4. Microbial Communities in Mammals

The human oral cavity has been the site of intense study of microbial succession, which has revealed a delicately orchestrated sequence of colonisation events that lead to reproducible and consistent three-dimensional community architecture. Cleaned teeth or enamel chips in a normal mouth provide the vehicle for tracking the succession of community membership (Kolenbrander et al. 2002). The early colonisers are predominantly streptococci, which attach directly by adhesion to the salivary receptors on the pellicle that coats the tooth surface in mixed-species biofilms (Palmer et al. 2003). A mixture of species then attaches to the streptococci, forming a complex, multispecies biofilm. A member of the mixed biofilm is *Fusobacterium nucleatum*, which provides the platform to which diverse late colonisers attach (Kolenbrander et al. 2002). In addition to the surface contact, the bacteria communicate with diffusible signals of the type 2 quorum sensing class of inducers (Blehert et al. 2003). The spatial and temporal organisation of the oral community both is shaped by and contributes to community function and impact on the host, thereby providing an impressive model system for relating structure and function.

Gnotobiotic animals have provided the basis for some elegant and informative studies of host–microbe interactions. Such animals are raised from offspring that are delivered by caesarean section and raised in sterile conditions. This approach simplifies experiments and definitively isolates the role of one organism. For example, in a stunning piece of work, Hooper et al. (2003) have shown that a member of the mouse gut microbiota (and a major member of the human gut microbiota), *Bacteroides thetaiotaomicron*, induces the normal development of the mouse immune system. They implicated angiogenins, which have been associated previously with tumour-associated angiogenesis, in innate immunity. The angiogenin, Ang4, is secreted into the gut lumen and has bactericidal activity against intestinal microbes. Ang4 expression is induced by *B. thetaiotaomicron*, a predominant member of the gut microflora. Moreover, another angiogenin, Ang1, is found in the circulatory system in mice and humans and exhibits microbicidal activity against systemic bacterial and fungal pathogens, suggesting that angiogenins contribute to systemic responses to infection (Stappenbeck et al. 2002; Hooper et al. 2003) The gnotobiotic mouse system has revealed intricacies of bacterial communication with the mammalian immune system that would have been essentially impossible to pinpoint in a more complex microbial environment. The use of this system for studying gut–bacteria interactions is described in detail in Chapter 12.

To address the next level of complexity in host–microbe interactions, we need model systems that offer analytical power comparable to that provided by gnotobiotic plants and animals but have utility for understanding interactions among community members.

8.4. APPROACHES TO COMMUNITY STUDY

There are few models that approximate the natural events involved in colonisation by a multispecies community as effectively as the tooth biofilm system. In other systems, far more disruptive approaches have been used. Much of microbial community ecology in host–microbe studies is predicated on the "scorched earth" approach, which involves ridding the environment of all its microorganisms and then adding them back singly or in small groups. Although this method has revealed much about interactions of hosts with one microbial species, the resulting community is too different from a natural community to provide a meaningful basis for analysis.

Many powerful approaches in microbiology involve the removal of one element from an otherwise normal biological system. In genetics, for example, we use mutant analysis to isolate the role of a single gene, and in the study of metabolism, we use inhibitors or mutants to isolate the role of an enzyme in a pathway. There are few tools available for microbial ecology that approximate this degree of rigor or yield arguments with the power of those established through genetics and biochemistry. We propose that a method that selectively removes a single species from a community will provide a powerful and precise strategy to understand community interdependencies. As we cannot reach in and remove members of a species in a microbial community by hand as can be done with starfish, for example, we must develop alternative strategies.

8.4.1. The Lepidopteran Gut Community

Our goal is to describe the species diversity in a community and then determine the role of each member of the community in the health of the host and in the stability and function of the community. To be tractable for these studies, the community needs to be readily available and reproducible, portable and contained, readily manipulated by addition of chemicals and organisms, and of relatively simple composition. Communities that meet these criteria reside in the midguts of the lepidopteran insects, such as the gypsy moth.

8.4.2. Overall Approach

The overall approach is to remove each species individually from the midgut community and study the resulting effect on the host and the community. We will measure effects on resistance to disease and toxins, development, and fecundity in the host, and nutritional status and robustness of the microbial community following reduction of the population size of one member. Robustness comprises resistance, stability, and resilience. Resistance is the ability of a community to maintain its structure upon challenge by an invader, stability is the ability to return to its original structure, and resilience is the rate of return to original structure (Begon et al. 1990).

8.4.3. Microbial Diversity in Lepidopteran Midguts

We have characterised the microbiota in the larval midguts of the gypsy moth by culture and culture-independent methods. When fed a sterilised diet, the community is comprised of ten members (Table 8.2), seven of which were culturable. Most of the members of the community belong to the γ-Proteobacteria and Firmicute phyla.

The gypsy moth is a generalist that feeds on 300 to 500 plant species that contain a diverse array of allelochemicals. Therefore, we determined the community composition when the larvae were fed foliage from various tree species. The microbial composition of midguts differed substantially among larvae feeding on a sterilised artificial diet, aspen, larch, white oak, or willow. A culturable *Enterococcus* species, and an *Enterobacter* species that

Table 8.2. *Bacterial phylotypes identified by culturing and culture-independent analysis of third instar gypsy moth midguts based on 16S rRNA gene sequence analysis*

Sequence ID	Bacterial division	Genus	Database matches	Population or presence[a]
		(≥ 95% identity)	(≥ 98% identity)	Artificial diet
Identified by culturing and culture-independent analysis				
NAB1	*γ-Proteobacteria*	*Pseudomonas*	*P. putida*	1.8×10^5
NAB3	*γ-Proteobacteria*	*Enterobacter*	Uncultured soil	4.8×10^6
NAB4	*γ-Proteobacteria*	*Pantoea*	*P. agglomerans*	2.5×10^5
NAB7	Low G+C Gram-positive	*Staphylococcus*	*S. lentus*	8.1×10^4
NAB8	Low G+C Gram-positive	*Staphylococcus*	*S. cohnii*	2.9×10^6
NAB9	Low G+C Gram-positive	*Staphylococcus*	*S. xylosus*	1.3×10^6
NAB11	Low G+C Gram-positive	*Enterococcus*	*E. faecalis*	1.5×10^8
Identified by culture-independent analysis only				
NAB16	*α-Proteobacteria*	*Agrobacterium*		+
NAB17	*γ-Proteobacteria*	*Enterobacter*		+
NAB20	Low G+C Gram-positive			+

[a] Cultured: numbers represent the average population of phylotype based on colony morphology in twenty individual larvae per treatment (cfu/ml gut material). Non-cultured: + indicates presence in guts of insects in diet treatment.

was culturable from some guts but not others were both present in all larvae, regardless of feeding substrate.

8.4.4. Manipulation of Community Composition

Antibiotics

The methods we are using to reduce targeted populations in a community cover a range of specificities. Antibiotics, for example, are not species specific, but by using various drugs of different specificities, the effect on a given species might be isolated. Larvae fed on antibiotics in various combinations

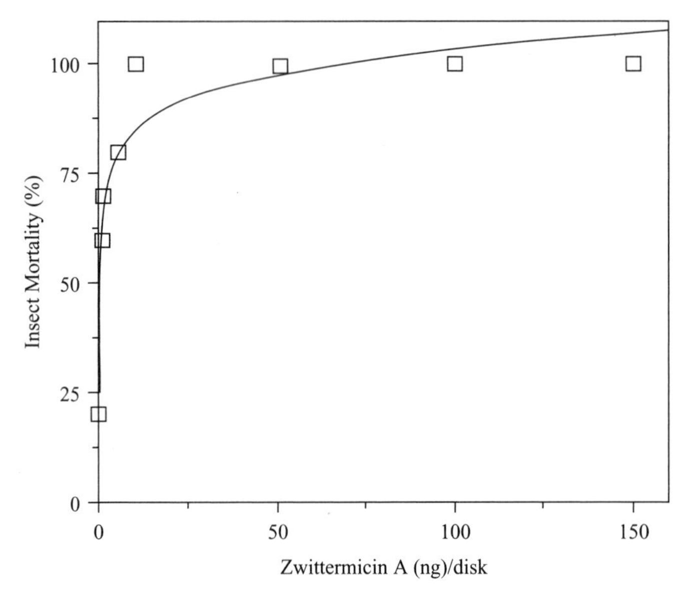

Figure 8.2. Synergy of zwittermicin A and Bt toxin. Zwittermicin A potentiates the activity of the insecticidal toxin produced by *Bacillus thuringiensis* (Broderick et al. 2000).

display numerous signs of reduced health. For example, a cocktail of antibiotics including tetracycline reduces larval survival by fifty percent over the first 20 days after hatching, and the surviving larvae weigh one tenth of the untreated control larvae. Tetracycline alone has a similar effect on development, whereas gentamicin, rifampicin, and penicillin have no effect on survival or growth.

Another antibiotic, zwittermicin A, dramatically increases larval sensitivity to the insecticidal toxin produced by *Bacillus thuringiensis*, although this antibiotic has no measurable effect on the larvae by itself (Fig. 8.2). Zwittermicin A alters the population size of more than one member of the gut community; thus it is not possible to assign the effect on toxin sensitivity to any one member of the community. Gypsy moth larvae fed on aspen leaves have substantially altered gut communities and also show enhanced sensitivity to *B. thuringiensis* and greater susceptibility to virus infection (Hunter and Schultz 1993; Lindroth et al. 1999). In aggregate, these results suggest that a change in the gut microbiota may stunt larval growth, stall development, and make the larvae more sensitive to pathogens and toxin. However, the results do not elucidate the role of any one bacterium or group of bacteria, because all of the treatments affect more than one member of the gut community. To test the hypothesis that a member or members of the gut community contribute to larval development and protect larvae from pathogens requires methods that isolate the effects of the members of the community. One approach to assigning functions to specific organisms is to feed bacteria to the treated insects to determine whether any of the normal residents of the gut rescues the host. However, methods to selectively remove members from the normal community are also essential because they will generate fewer secondary effects and be less disruptive to overall community function.

Phage

Bacteriophage, or phage, are viruses that inject into bacteria a nucleic acid genome that directs synthesis of viral components, leading to the production of more phage particles. Some phage are highly lytic, completing their replication cycle by weakening the bacterial cell wall, which results in cell lysis and release of phage particles. Some phage can release hundreds of infective particles from one infected cell. Their rapid life cycles and high reproductive rates make phage excellent agents to reduce bacterial populations. As agents of bacterial destruction, phage have advantages over antibiotics, such as their high degree of host specificity and amplification over time (Sulakvelidze et al. 2001). The host range of most phage is limited to a single species, and most are selective for certain strains within a species. The host selectivity and explosive killing exhibited by many phage have drawn attention to them as therapeutic agents to cure infectious disease. This idea was introduced by d'Herelle in 1917 (d'Herelle, 1926) and pursued in collaboration with his colleagues in Georgia (formerly USSR), although it did not gain attention until recently in the United States early experiments by d'Herelle and others claimed high survival rates of people treated with phage to control dysentery and cholera, but the experiments were not designed with the rigor of modern experimentation; thus the data, however intriguing, must be interpreted cautiously. A combination of political prejudices and the discovery of antibiotics diverted attention from phage therapeutics in Western medicine until recently (Summers 2001; Sulakvelidze et al. 2001). The last few years have seen renewed interest in phage, as the antibiotic resistance crisis impels microbiologists to find new (or rediscover old, in this case) solutions to infectious disease. Recent reports present rigorous and highly successful in vivo tests of phage to control infectious disease. One study showed that mice infected with *E. faecium* were cured by a single injection of a phage that reproduces in that bacterium. Even mice that were already moribund had a fifty percent survival rate when treated with the phage, compared with zero percent of the untreated ones (Biswas et al. 2002).

Plant pathologists have also used phage to protect plants from infection by bacterial pathogens (Vidaver 1976; Flaherty et al. 2000) and found that virulent phage interfere with the establishment of a bacterial biocontrol agent (Keel et al. 2002), indicating that phage exert influences on population structures outside the controlled conditions of the laboratory. Of particular interest to us is the fact that d'Herelle first observed phage in cultures of a pathogen that caused an epizootic infection of locusts in Mexico in 1909 (d'Herelle 1926). Thus, observations about the role of phage in bacteria associated with insects date back to the very beginning of phage biology.

The history of phage biology and ecology suggests the potential for their use in ecological studies. Their specificity makes them carefully addressed "letter bombs" that will destroy only one component of a community and remain biologically invisible to the rest. The resistance problem encountered in previous studies can be addressed either by using the phage for short-term studies or by combining more than one phage, in which case the frequency of a doubly resistant mutant is lower than the number of bacteria in the population to be addressed.

Bacteriophage exhibit more specificity than antibiotics, infecting only certain members of a species. The advantage of phage is that they can reduce bacterial populations by four to eight orders of magnitude in vitro. A disadvantage is the difficulty in isolating phage for bacteria that we do not yet know how to culture. A concern with phage was that they might not be able to infect in a gut, especially the gypsy moth gut, which has an average pH greater than 12. Variations and combinations of methods will likely lead to the best tools that incorporate phage into the study of community ecology.

8.5. FUNCTIONAL CONNECTIONS AMONG COMMUNITY MEMBERS: CULTURED AND UNCULTURED

A long-ignored aspect of community ecology is the uncultured majority. Most bacteria in environmental samples are not culturable by standard methods. Therefore, to understand the structure and function of microbial communities, we must include the uncultured bacteria in our analyses. Community structure can be analysed by polymerase chain reaction amplification of 16S rRNA genes from DNA directly extracted from the environmental sample. DNA isolated from environmental samples can also be used for functional genomics by cloning into a suitable vector that replicates in a culturable host. This approach, termed metagenomics, has provided insight into uncultured communities in soil, seawater, sponge tissue, and the human oral cavity (Stein et al. 1996; Schleper et al. 1997; Henne et al. 1999; Beja et al. 2000; Rondon et al. 2000; Courtois et al. 2003; Diaz-Torres et al. 2003; Handelsman et al. 2003).

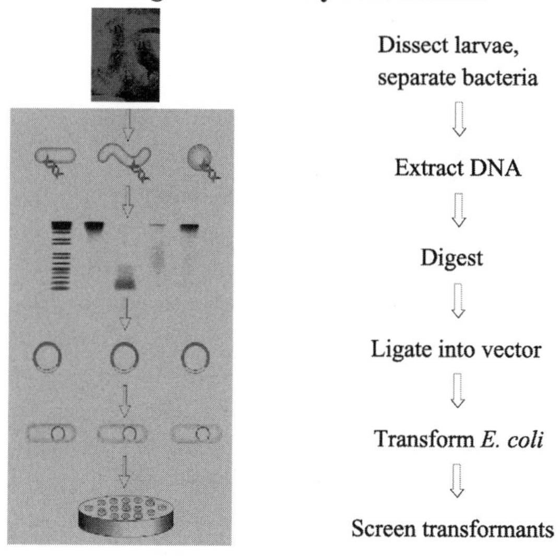

Figure 8.3. Metagenomics provides a means to access the genomes of as-yet unculturable microorganisms by direct extraction of their DNA from mixed communities (Handelsman 2003; Handelsman et al. 1998; Schloss and Handelsman 2003).

We are characterising the cultured and the as-yet unculturable bacteria in the gypsy moth midgut. We have constructed highly redundant libraries from DNA extracted directly from the gut bacteria that have not been subjected to culturing (Fig. 8.3). Preliminary studies indicate that clones in these libraries express novel functions that have not been found among the cultured bacteria. A major focus of this work is to identify molecules that play a role in communication among bacteria – both culturable and unculturable – in the gut environment. This aspect of host–community interactions will add a new dimension to our understanding of the interactions of animals with their associated microorganisms.

8.6. CONCLUSION

The study of the impact of communities on their hosts is at a new intersection. New tools are available for the dissection of communities, and the knowledge of interactions of single species with their hosts lays a strong foundation for the study of multispecies communities and their hosts. Application of molecular methods that address both the uncultured and cultured members of the communities, computational approaches to model quantitative events, and diverse biological and chemical approaches to perturb communities will produce an understanding of the complex networks that maintain the structure of the community and govern its influence on the host.

REFERENCES

Andrews, J. H. (1991). *Comparative Ecology of Microorganisms and Macroorganisms, Brock/Springer Series in Contemporary Bioscience.* New York: Springer-Verlag.

Bakalidou, A., Kampfer, P., Berchtold, M., Kuhnigk, T., Wenzel, M., and Konig, H. (2002). *Cellulosimicrobium variabile* sp. nov., a cellulolytic bacterium from the hindgut of the termite *Mastotermes darwiniensis. International Journal of Systemic Evolutionary Microbiology* **52**, 1185–1192.

Beattie, G. A., Clayton, M. K., and Handelsman, J. (1989). Quantitative comparison of the laboratory and field competitiveness of *Rhizobium leguminosarum* biovar phaseoli. *Applied Environmental Microbiology* **55**, 2755–2761.

Begon, M. Harper, J. L., and Townsend, C. R. (1990). *Ecology: Individuals, Populations and Communities.* 2nd ed. London: Blackwell Scientific Publications.

Begum, A. A., Leibovitch, S., Migner, P., and Zhang, F. (2001). Specific flavonoids induced *nod* gene expression and pre-activated *nod* genes of *Rhizobium leguminosarum* increased pea (*Pisum sativum L.*) and lentil (*Lens culinaris L.*) nodulation in controlled growth chamber environments. *Journal of Experimental Botany* **52**, 1537–1543.

Beja, O., Suzuki, M. T., Koonin, E. V., Aravind, L., Hadd, A., Nguyen, L. P., Villacorta, R., Amjadi, M., Garrigues, C., Jovanovich, S. B., Feldman, R. A., and DeLong. E. F. (2000). Construction and analysis of bacterial artificial chromosome libraries from a marine microbial assemblage. *Environmental Microbiology* **2**, 516–529.

Bellotti, A. C., Smith, L., and Lapointe. S. L. (1999). Recent advances in cassava pest management. *Annual Review of Entomology* **44**, 343–370.

Bengmark, S. (1998). Ecological control of the gastrointestinal tract. The role of probiotic flora. *Gut* **42**, 2–7.

Biswas, B., Adhya, S., Washart, P., Paul, B., Trostel, A. N., Powell, B., Carlton, R., and Merril., C. R. (2002). Bacteriophage therapy rescues mice bacteremic from a clinical isolate of vancomycin-resistant *Enterococcus faecium. Infection and Immunity* **70**, 204–210.

Bittinger, M. A., Milner, J. L., Saville, B. J., and Handelsman. J. (1997). *rosR*, a determinant of nodulation competitiveness in *Rhizobium etli. Molecular Plant Microbe Interactions* **10**, 180–186.

Blehert, D. S., Palmer R. J., Xavier J. B., Almeida J. S., and Kolenbrander, P. E. (2003). Autoinducer 2 Production by *Streptococcus gordonii* DL1 and the biofilm phenotype of a *luxS* mutant are influenced by nutritional conditions. *Journal of Bacteriology* **185**, 4851–4860.

Boettner, G. H., Elkinton, A., and Boettner. C. J. (2000). Effects of a biological control introduction on three nontarget native species of Saturniid moths. *Conservation Biology* **14**, 1798–1806.

Bourlioux, P., Koletzko, B., Guarner, F., and Braesco. V. (2003). The intestine and its microflora are partners for the protection of the host: report on the Danone Symposium "The Intelligent Intestine," held in Paris, June 14, 2002. *American Journal of Clinical Nutrition* **78**, 675–683.

Brand, J. M., Bracke, J. W., Markovetz, A. J., Wood, D. L., and Browne, L. E. (1975). Production of verbenol pheromone by a bacterium isolated from bark beetles. *Nature* **254**, 136–137.

Brand, J. M., Bracke, Britton, L. N., Markovetz, A. J., and Barras, S. J. (1976). Bark beetle pheromones: production of verbenone by a mycangial fungus of *Dendroctonus frontalis*. *Journal of Chemical Ecology* **2**, 195–199.

Breznak, J. A. and Canale-Parola, E. (1972). Metabolism of *Spirochaeta aurantia*. II. Aerobic oxidation oxidation of carbohydrates. *Archiv fur Mikrobiologie* **83**, 278–292.

Breznak, J. A., Brill, W. J., Mertins, J. W., and Coppel, H. C. (1973). Nitrogen fixation in termites. *Nature* **244**, 577–80.

Breznak, J. A. and Pankratz, H. S. (1977). In situ morphology of the gut microbiota of wood-eating termites [*Reticulitermes flavipes* (Kollar) and *Coptotermes formosanus* Shiraki]. *Applied and Environmental Microbiology* **33**, 406–426.

Breznak, J. A. and Kane, M. D. (1990). Microbial H_2/CO_2 acetogenesis in animal guts: nature and nutritional significance. *FEMS Microbiology Reviews* **7**, 309–313.

Broderick, N. A., Goodman, R. M., Raffa, K. F., and Handelsman, J. (2000). Synergy between zwittermicin A and *Bacillus thuringiensis* subsp. kurstaki against gypsy moth (Lepidoptera: Lymantriidae). *Environmenal Entomology* **29**, 101–107.

Broderick, N. A., Raffa, K. F., Goodman, R. M., and Handelsman, J. (2004). Census of the bacterial community of the gypsy moth larval midgut by using culturing and culture-independent methods. *Applied and Environmental Microbiology* **70**, 293–300.

Brune, A. (1998). Termite guts: the world's smallest bioreactors. *Trends in Biotechnology* **16**, 16–21.

Buckley, M. R. (2004). Microbial communities: From life apart to life together: *American Academy of Microbiology*. http:www.asm.org/ASM/Files/CCLibrary Files/FILENAME 000000001262/ASM-sys%20Microbio%20test.PDF.

Choi, J. Y., Sifri, C. D., Goumnerov, R. C., Rahme, L. G., Ausubel, F. M., and Calderwood, S. B. (2002). Identification of virulence genes in a pathogenic strain of *Pseudomonas aeruginosa* by representational difference analysis. *Journal of Bacteriology* **184**, 952–961.

Cook, R. J. and Baker, K. F. (1983). *The Nature and Practice of Biological Control of Plant Pathogens*. St. Paul, Minn.: American Phytopathological Society.

Courtois, S., Cappellano, C. M., Ball, M., Francou, F.-X., Normand, P., Helynck, G., Martinez, A., S Kolvek, S. J., Hopke, J., Marcia S. M. S., August, P. R., Nalin, R. Guerineau, M., Jeannin, P., Simonet, P., and Pernodet, J.-L. (2003). Recombinant environmental libraries provide access to microbial diversity for drug discovery from natural products. *Applied and Enviromental Microbiology*. **69**, 49–55.

Crooks, K. R. and Soulé, M. E. (1999). Mesopredator release and avifaunal extinctions in a fragmented system. *Nature* **400**, 563–566.

Cullimore, J. and Denarie, J. (2003). How legumes select their sweet talking symbionts. *Science* **302**, 575–578.

Daehler, C. C. and Gordon, D. R. (1997). To introduce or not to introduce–trade-offs of non-indigenous organisms. *Trends in Ecology & Evolution* **12**, 424–425.

d'Herelle, F. (1926). *The Bacteriophage and its Behaviour*. Translated by G. H. Smith. Baltimore: The Williams & Wilkins Company.

Diaz-Torres, M. L., McNab, R., Spratt, D. A., Villedieu, A., Hunt, N., Wilson, M., and Mullany, P. (2003). Novel tetracycline resistance determinant from the oral metagenome. *Antimicrobial Agents and Chemotherapy* **47**, 1430–1432.

Dillon, R. J., Vennard, C. T., and Charnley, A. K. (2002). A note: gut bacteria produce components of a locust cohesion pheromone. *Journal of Applied Microbiology* **92**, 759–763.

Faucher, C., Maillet, F., Vasse, J., Rosenberg, C., van Brussel, A. A., Truchet, G., and Denarie, J. (1988). *Rhizobium meliloti* host range *nodH* gene determines production of an alfalfa-specific extracellular signal. *Journal of Bacteriology* **170**, 5489–5499.

Flaherty, J. E., Jones, J. B., Harbaugh, B. K., Somodi, G. C., and Jackson, L. E. (2000). Control of bacterial spot on tomato in the greenhouse and field with H-mutant bacteriophages. *Horticultural Science* **35**, 882–884.

Foster, J. S., Apicella, M. A., and McFall-Ngai, M. (2000). *Vibrio fischeri* lipopolysaccharide induces developmental apoptosis, but not complete morphogenesis, of the *Euprymna scolopes* symbiotic light organ. *Developmental Biology* **226**, 242–254.

Gilbert, G. S., Parke, J. L., Clayton, M. K., and Handelsman. J. (1993). Effects of an introduced bacterium on bacterial communities on roots. *Ecology* **74**, 840–854.

Gonzalez-Megias, A. and Gomez, J. M. (2003). Consequences of removing a keystone herbivore for the abundance and diversity of arthropods associated with a cruciferous shrub. *Ecological Entomology* **28**, 299–308.

Goossens, D., Jonkers, A. Russel, M., Stobberingh, E., Van Den Bogaard, A., and Stockbrügger, R. (2003). The effect of *Lactobacillus plantarum* 299v on the bacterial composition and metabolic activity in faeces of healthy volunteers: a placebo-controlled study on the onset and duration of effects. *Ailmententary Pharmacology and Therapy* **18**, 495–505.

Guarner, F. and Malagelada, J. R. (2003). Gut flora in health and disease. *Lancet* **361**, 512–519.

Handelsman, J., Ugalde, R. A., and Brill, W. J. (1984). *Rhizobium meliloti* competitiveness and the alfalfa agglutinin. *Journal of Bacteriology* **157**, 703–707.

Handelsman, J. and Stabb, E. V. (1996). Biocontrol of soilborne plant pathogens. *Plant Cell* **8**, 1855–1869.

Handelsman, J., Rondon, M. R., Brady, S. F., Clardy, J., and Goodman, R. M. (1998). Molecular biological access to the chemistry of unknown soil microbes: a new frontier for natural products. *Chemical Biology* **5**, R245–249.

Handelsman, J. (2003). Soil – the metagenomics approach. In *Microbial Diversity Bioprospecting*, ed. A. T. Bull Washington, pp. 109–119. American Society for Microbiology Press.

Handelsman, J. (2004). Metagenomics: Application of genomics to uncultured microorganisms. *Microbiology and Molecular Biology Reviews* **68**, 699–678.

Hanke, J. H., Nichols, L. N., and Coon, M. E. (1992). FK506 and rapamycin selectively enhance degradation of IL-2 and GM-CSF mRNA. *Lymphokine and Cytokine Research* **11**, 221–231.

Hendrickson, E. L., Plotnikova, J., Mahajan-Miklos, S., Rahme, L. G., and Ausubel FM. (2001). Differential roles of the *Pseudomonas aeruginosa* PA14 *rpoN* gene in pathogenicity in plants, nematodes, insects, and mice. *Journal of Bacteriology* **183**, 7126–7134.

Henne, A., Daniel, R., Schmitz, R. A., and Gottschalk, G. (1999). Construction of environmental DNA libraries in *Escherichia coli* and screening for the presence of genes conferring utilization of 4-hydroxybutyrate. *Applied and Environmental Microbiology* **65**, 3901–3907.

Holzapfel, W. H., Haberer, P., Snel, J., Schillinger, U., and in't Veld, J. H. J. (1998). Overview of gut flora and probiotics. *International Journal of Food Microbiology* **41**, 85–101.

Hooper, L. V., Stappenbeck, T. S., Hong, C. V., and Gordon. J. I. (2003). Angiogenins: a new class of microbicidal proteins involved in innate immunity. *Nature Immunology* **4**, 269–273.

Hunter, M. D. and Schultz, J. C. (1993). Induced plant defenses breached–phytochemical induction protects an herbivore from disease. *Oecologia* **94**, 195–203.

Jander, G., Rahme, L. G., and Ausubel, F. (2000). Positive correlation between virulence of *Pseudomonas aeruginosa* mutants in mice and insects. *Journal of Bacteriology* **182**, 3843–3845.

Kane, M. D. and Breznak, J. A. (1991). *Acetonema longum* gen. nov. sp. nov., an H_2/CO_2 acetogenic bacterium from the termite, *Pterotermes occidentis*. *Archives of Microbiology* **156**, 91–98.

Kareiva, P. (1996). Developing a predictive ecology for non-indigenous species and ecological invasions. *Ecology* **77**, 1651–1652.

Keel, C., Ucurum, Z., Michaux, P., Adrian, M., and Haas, D. (2002). Deleterious impact of a virulent bacteriophage on survival and biocontrol activity of *Pseudomonas fluorescens* strain CHA0 in natural soil. *Mol. Plant Microbe Interactions* **15**, 567–576.

Kinkel, L. L. and Lindow, S. E. (1993). Invasion and exclusion among coexisting *Pseudomonas syringae* strains on leaves. *Applied and Environmental Microbiology* **59**, 3447–3454.

Kolenbrander, P. E., Andersen, R. N., Blehert, D. S., Egland, P. G., Foster, J. S., and Palmer, R. J. (2002). Communication among oral bacteria. *Microbiology and Molecular Biology Reviews* **66**, 486–505.

Lagares, A., Caetano-Anolles, G., Niehaus, K., Lorenzen, J., Ljunggren, H. D., Puhler, A., and Favelukes, G. (1992). A *Rhizobium meliloti* lipopolysaccharide mutant altered in competitiveness for nodulation of alfalfa. *Journal of Bacteriology* **174**, 5941–5452.

Laguerre, G., Louvrier, P., Allard, M. R., and Amarger, N. (2003). Compatibility of rhizobial genotypes within natural populations of *Rhizobium leguminosarum* biovar viciae for nodulation of host legumes. *Applied and Environmental Microbiology* **69**, 2276–2283.

Leadbetter, J. R. and Breznak, J. (1996). Physiological ecology of *Methanobrevibacter cuticularis* sp. nov. and *Methanobrevibacter curvatus* sp. nov., isolated from the hindgut of the termite *Reticulitermes flavipes*. *Applied and Environmental Microbiology* **62**, 3620–3631.

Leadbetter, J. R., Crosby, L. D., and Breznak, J. A. (1998). *Methanobrevibacter filiformis* sp. nov., A filamentous methanogen from termite hindguts. *Archives of Microbiology* **169**, 287–292.

Leadbetter, J. R., Schmidt, T. M., Graber, J. R., and Breznak, J. A. (1999). Acetogenesis from H_2 plus CO_2 by spirochetes from termite guts. *Science* **283**, 686–689.

Lilburn, T. G., Schmidt, T. M., and Breznak, J. A. (1999). Phylogenetic diversity of termite gut spirochaetes. *Environmental Microbiology* **1**, 331–345.

Lilburn, T. G., Kim, K. S., Ostrom, N. E., Byzek, K. R., Leadbetter, J. R., and Breznak, J. A. (2001). Nitrogen fixation by symbiotic and free-living spirochetes. *Science* **292**, 2495–2498.

Lindroth, R. L., Hwang, S. Y., and Osier, T. L. 1999. Phytochemical variation in quaking aspen: Effects on gypsy moth susceptibility to nuclear polyhedrosis virus. *Journal of Chemical Ecology* **25**, 1331–1341.

Long, S. R. (2001). Genes and signals in the *Rhizobium*-legume symbiosis. *Plant Physiology* **125**, 69–72.

McFall-Ngai, M. (2000). Negotiations between animals and bacteria: the 'diplomacy' of the squid-vibrio symbiosis. *Comparative Biochemistry and Physiology A* **126**, 471–480.

Milner, J. L., Araujo, R. S., and Handelsman, J. (1992). Molecular and symbiotic characterization of exopolysaccharide-deficient mutants of *Rhizobium tropici* strain CIAT899. *Molecular Microbiology* **6**, 3137–3147.

Miyata, S., Casey, M., Frank, D. W., Ausubel, F. M., and Drenkard, E. (2003). Use of the *Galleria mellonella* caterpillar as a model host to study the role of the Type III secretion system in *Pseudomonas aeruginosa* pathogenesis. *Infection and Immunity* **71**, 2404–2413.

Moran, N. A. (2002). Microbial minimalism: genome reduction in bacterial pathogens. *Cell* **108**, 83–6.

Moran, N. A. (2003). Tracing the evolution of gene loss in obligate bacterial symbionts. *Current Opinion Microbiology* **6**, 512–518.

Moran, N. A. and Mira, A. (2001). The process of genome shrinkage in the obligate symbiont *Buchnera aphidicola. Genome Biology* **2** (12): RESEARCH0054.

Moran, N. A., Plague, G. R., Sandstrom, J. P., and Wilcox. J. L. (2003). A genomic perspective on nutrient provisioning by bacterial symbionts of insects. *Proceedings of the National Academy of Sciences USA* **100** Suppl 2, 14543–14548.

Naiman, R. J., Pinay, G., Johnston, C. A., and Pastor, J. (1994). Beaver influences on the long term biogeochemical characteristics of boreal forest drainage networks. *Ecology* **75**, 905–921.

Nakashima, K. I., Watanabe, H., and Azuma, J. I. (2002). Cellulase genes from the parabasalian symbiont *Pseudotrichonympha grassii* in the hindgut of the wood-feeding termite *Coptotermes formosanus. Cellular and Molecular Life Sciences* **59**, 1554–1560.

Navarrete, S. A. and Menge, B. A. (1996). Keystone predation and interaction strength – interactive effects of predators on their main prey. *Ecological Monographs* **66**, 409–429.

Nyholm, S. V., Stabb, E. V., Ruby, E. G., and McFall-Ngai, M. (2000). Establishment of an animal-bacterial association: recruiting symbiotic vibrios from the environment. *Proceedings of the National Academy of Sciences USA* **97**, 10231–10235.

Ochman, H. and Moran, N. A. (2001). Genes lost and genes found: evolution of bacterial pathogenesis and symbiosis. *Science* **292**, 1096–1099.

Oresnik, I. J., Twelker, S., and Hynes, M. F. (1999). Cloning and characterization of a *Rhizobium leguminosarum* gene encoding a bacteriocin with similarities to RTX toxins. *Applied and Environmental Microbiology* **65**, 2833–2840.

Paine, R. T. (1966). Food web complexity and species diversity. *American Naturalist* **100**, 65–75.

Palmer, R. J., Gordon, S. M., Cisar, J. O., and Kolenbrander, P. E. (2003). Coaggregation-mediated interactions of Streptococci and Actinomyces detected in initial human dental plaque. *Journal of Bacteriology* **185**, 3400–3409.

Paster, B. J., Boches, S. K., Galvin, J. L., Ericson, R. E., Lau, C. N., Levanos, V. A., Sahasrabudhe, A., and Dewhirst, F. E. (2001). Bacterial diversity in human subgingival plaque. *Journal of Bacteriology* **183**, 3770–3783.

Pearson, D. E., McKelvey, K. S., and Ruggiero, L. F. (2000). Non-target effects of an introduced biological control agent on deer mouse ecology. *Oecologia* **122**, 121–128.

Peters, N. K., Frost, J. W., and Long, S. R. (1986). A plant flavone, luteolin, induces expression of *Rhizobium meliloti* nodulation genes. *Science* **233**, 977–980.

Plantard, O., Rasplus J.-Y., Mondor, G., LeClainche, I., Solignac, M. (1998). *Wolbachia*-induced thelytoky in the rose gall-wasp *Diplolepis spinosissimae* (Giraud) (Hymenoptera: Cynipidae), and its consequences on the genetic structure of its host. *Proceedings of the Royal Society of London B* **265**, 1075–1080.

Plantard, O., Rasplus J.-Y., Mondor, G., LeClainche, I., Solignac, M. (1999). Distribution and phylogeny of *Wolbachia*-inducing thelytoky in *Rhoditini* 'Aylacini' (Hymenoptera: Cynipidae). *Insect Biochemistry and Molecular Biology* **8**, 185–191.

Rice, S. A., Givskov, M., Steinberg, P., and Kjelleberg, S. (1999). Bacterial signals and antagonists: the interaction between bacteria and higher organisms. *Journal of Molecular Microbiology and Biotechnology* **1**, 23–31.

Robleto, E. A., Kmiecik, K., Oplinger, E. S., Nienhuis, J., and Triplett, E. W. (1998). Trifolitoxin production increases nodulation competitiveness of *Rhizobium etli* CE3 under agricultural conditions. *Applied and Environmental Microbiology* **64**, 2630–2633.

Rondon, M. R., August, P. R., Bettermann, A. D., et al. (2000). Cloning the soil metagenome: a strategy for accessing the genetic and functional diversity of uncultured microorganisms. *Applied and Environmental Microbiology* **66**, 2541–2547.

Rosas, J. C., Castro, J. A., Robleto, E. A., and Handelsman, J. (1998). A method for screening *Phaseolus vulgaris* L. germplasm for preferential nodulation with a selected *Rhizobium etli* strain. *Plant and Soil* **203**, 71–78.

Schaffner, U. (2001). Host range testing of insects for biological weed control: How can it be better interpreted? *Bioscience* **51**, 951–959.

Schleper, C., Swanson, R. V., Mathur, E. J., and DeLong, E. F. (1997). Characterization of a DNA polymerase from the uncultivated psychrophilic archaeon *Cenarchaeum symbiosum*. *Journal of Bacteriology* **179**, 7803–7811.

Schloss, P. D. and Handelsman. J. (2003). Biotechnological prospects from metagenomics. *Current Opinions in Biotechnology* **14**, 303–310.

Shigesada, N., and Kawasaki, K. (1997). *Biological Invasions: Theory and Practice. Oxford Series in Ecology and Evolution*. Oxford and New York: Oxford University Press.

Simberloff, D. and Stiling, P. (1996). How risky is biological control? *Ecology* **77**, 1965–1974.

Stappenbeck, T. S., Hooper, L. V., and Gordon, J. I. (2002). Developmental regulation of intestinal angiogenesis by indigenous microbes via Paneth cells. *Proceedings of the National Academy of Sciences USA* **99**, 15451–15455.

Stein, J. L., Marsh, T. L., Wu, K. Y., Shizuya, H., and DeLong E. F. (1996). Characterization of uncultivated prokaryotes: isolation and analysis of a 40-kilobase-pair genome fragment from a planktonic marine archaeon. *Journal of Bacteriology*. **178**, 591–599.

Sulakvelidze, A., Alavidze, Z., and Morris, J. G. (2001). Bacteriophage therapy. *Antimicrobial Agents and Chemotherapy* **45**, 649–659.

Summers, W. C. (2001). Bacteriophage therapy. *Annual Review of Microbiology* **55**, 437–451.

Thomas-Oates, J., Bereszczak, J., Edwards, E., et al. (2003). A catalogue of molecular, physiological and symbiotic properties of soybean-nodulating rhizobial strains from different soybean cropping areas of China. *Systemic and Applied Microbiology* **26**, 453–465.

Triplett, E. W. and Sadowsky, M. J. (1992). Genetics of competition for nodulation of legumes. *Annual Review of Microbiology* **46**, 399–428.

Vidaver, A. K. (1976). Prospects for control of phytopathogenic bacteria by bacteriophages and bacteriocins. *Annual Review of Phytopathology* **14**, 451–465.

Visick, K. L. and McFall-Ngai, M. (2000). An exclusive contract: Specificity in the *Vibrio fischeri-Euprymna scolopes* partnership. *Journal of Bacteriology* **182**, 1779–1787.

Visick, K. L., Foster, J., Doino, J., McFall-Ngai, M., and Ruby, E. G. (2000). *Vibrio fischeri lux* genes play an important role in colonization and development of the host light organ. *Journal of Bacteriology* **182**, 4578–4586.

Wilcox, J. L., Dunbar, H. E., Wolfinger, R. D., and Moran, N. A. (2003). Consequences of reductive evolution for gene expression in an obligate endosymbiont. *Molecular Microbiology* **48**, 1491–500.

Zchori-Fein, E., Gottlieb, Y., Kelly, S. E., Brown, J. K., Wilson, J. M., Karr, T. L., and Hunter, M. S. (2001). A newly discovered bacterium associated with parthenogenesis and a change in host selection behavior in parasitoid wasps. *Proceedings of the National Academy of Sciences USA* **98**, 12555–12560.

CHAPTER 9

Commensal diversity and the immune system: Modelling the host-as-network

Robert M. Seymour

9.1. INTRODUCTION

As has been highlighted in earlier chapters, many animal hosts sustain large and diverse communities of bacteria and other microorganisms. For example, although there are a few dozen bacterial species that are recognised as being capable of causing pathology, the average human body contains more than 1,000 bacterial species that constitute its commensal microflora (or normal microbiota), living mostly on epithelial (mucosal) surfaces. This is approximately twice the number of species held by London Zoo. How is this possible? More specifically, how do the commensal microflora and the host immune response maintain stable coexistence without the mucosal/epithelial surfaces being in a constant state of inflammation? The fact that this is normally the case has been called the "commensal paradox" (Henderson et al. 1999).

Vertebrates have an acquired as well as an innate immune system, and yet their commensal microflora are generally more diverse than that of invertebrates, which have only an innate immune system. How is this greater diversity in vertebrates maintained in the face of apparently stronger countervailing forces? Could it be that, contrary to received wisdom, the vertebrate combinatorial immune system evolved largely under pressure to maintain a stable and diverse microflora? (See Chapter 2 for a separate perspective on this argument.) In this chapter, I develop a simple mathematical model to investigate within-host commensal diversity and its relation to the strength of the host adaptive immune response.

From the point of view of invading bacteria, the host must contain a large number of micro-niches, a fact that may be expected to lead to diversification through niche specialisation (Pianka 1988; Begon et al. 1990). Looked at in this way, there is no problem in explaining the diversity of the commensal microflora: Where there are opportunities, opportunity takers will evolve. However, there are problems with this simple view associated with the non-isolation of distinct niches within the host. These are the same problems as arise with competition in ecological theory and sympatric speciation in evolutionary theory; namely, without isolation, inter-specific competition for limiting resources leads to competitive exclusion, and sympatry promotes gene flow through random mating, both effects which inhibit diversification (Mayr 1963; Pianka 1988).

Ecologists have been studying community diversity for many decades, and from one perspective, the problem can be conceived as the identification of mechanisms that counteract competitive exclusion. At an evolutionary level, there are the fundamental processes of niche diversification and specialisation, which act to decrease competition (Pianka 1988). At an ecological level, predation can sometimes promote diversity amongst competing prey species if the predator's relative preferences for its prey species follow the same ordinal distribution as the prey species' relative competitive abilities (Begon et al. 1990 – also discussed in Chapter 8). More generally, adaptive foraging can promote complexity in food webs (Kondoh 2003). Temporal stochastic disturbance can also promote diversity by disrupting an underlying competitive dominance hierarchy. For example, different environmental conditions may favour different species at different times, or periodic disease epidemics may have differential effects on species (Krebs 1985; Begon et al. 1990). Spatial effects are known to counteract competitive exclusion in some cases, for example, in communities of three or more species with local non-transitive "rock-scissors-paper" competitive structures (Johnson and Seinen 2002).

In this chapter, I model the host as harbouring a number of distinct sites (niches) where various species of microorganisms can grow and reproduce with various degrees of success. Associated with each site is a distinct specialist species that grows particularly well there, but fares badly at other sites. The model also incorporates a generalist species that thrives equally well at all sites, as well as species that are intermediate between specialists and generalists. A low rate of mutation between closely related species is allowed. If this were all, relatively high (inter-site) community diversity would be generated by niche specialisation, with the long-term result that each site would be completely dominated by its associated specialist (but with low intra-site diversity). However, the model also allows for some migration along routes connecting the sites. The network of sites and possible migration routes defines the "host-as-network" of the title. Migration promotes competitive exclusion and gene flow, and leads to a low-diversity community dominated by a single specialist species. However, the host is not passive in the face of microorganism invasion, but can mount an immune response, which I model as acting adaptively to suppress the microorganism growth rates. The model shows that, under certain trade-off assumptions between the site preferences and growth rates of the microorganism species, a stronger immune response can promote greater commensal community diversity.

The model is outlined in sections 9.3 and 9.4, and more formally in Appendix A. The action of the model immune response is described in section 9.6. The remaining sections, 9.5, 9.7, 9.8, and 9.9, report the results of computer simulations of the model. Further discussion is given in a concluding section, 9.10.

9.2. MEASURING DIVERSITY

There are many possible measures of "community diversity" used by ecologists (Pielou 1975; Begon et al. 1990; Rosenzweig 1999). One of the oldest and simplest is the Shannon entropy:

$$H = -\sum_{s=1}^{S} P_s \ln(P_s), \tag{1}$$

where S is the number of species in the community and P_s is the proportion of species s. This has maximum value $H_{\max} = \ln(S)$ achieved when all species have equal representation. H therefore measures both abundance (number of species) and evenness as represented through the frequency distribution P_s. However, H has the disadvantage that it is rather insensitive to rare species. For example, if a community is dominated by one or a few species, but has many other, comparatively rare and evenly distributed species, then H will give a low value for diversity, even though, intuitively, we would regard such a community as having relatively high diversity. Nevertheless, I shall use H, and some more refined indices derived from it (Appendix B), as a convenient index of diversity in what follows.

9.3. THE 'HOST-AS-NETWORK'

The host animal sustains a community of microorganisms that are distributed at various sites within the host – mostly on epithelial surfaces – at which the microorganisms can grow and reproduce. These sites will in reality be spatially extended, but in this simple model they will be treated as homogeneous local environments. These sites are the *nodes* of the host regarded as network.

Microorganisms can migrate between sites along various pathways within the host. Thus, microorganisms in the small intestine may migrate to the large intestine; some in the mouth may migrate to the tonsils or the lungs; some in the gut may migrate to the urinary tract (see Chapter 13 for discussion of bacteria and the urogenital system). These migration pathways are the *edges* of the host-as-network.

Some paths may be more accessible and frequently used than others. To capture this, I assign a probability c_{ij} with which the first site encountered by a microorganism emigrating from site i is site j, given that it survives. I assume that every migrant either dies or eventually reaches *somewhere*, so that $\sum_j c_{ij} = 1$ for each i. I consider three types of network: *local*, in which all edges with positive connection probabilities join neighbouring sites only, so that more distant sites may only be reached in more than one step; *small worlds*, in which most paths are local, but there are a few long-range paths with positive connection probabilities; and *random*, in which all paths can have positive connection probability. I assume that networks are semi-symmetric in the sense that if $c_{ij} > 0$, then also $c_{ji} > 0$, though these probabilities need not be equal. The three types of network are illustrated for ten sites in Figure 9.1.

I also assume that the most likely migration route is directly back to the site from which a microorganism started; the next most likely are the paths to nearest neighbours, and the least likely are the long-range paths. This is what I call the *conservative migration assumption*. This scheme is represented formally in Appendix A, Equation (A.1). The most conservative site(s) I define to be the site(s) i for which c_{ii} is maximum. Sites can therefore be ranked by *conservatism*.

On reaching a site, a microorganism will attempt to settle or not, according to how hospitable it finds the site. This hospitality will vary according to the species of microorganism, some sites being more favourable – in terms of ease of adhesion and growth potential – to some species than to others. I assume there is a community of possible microorganism species, with individuals of species (or genotype) g having a *preference* for site i represented by a settlement probability $\lambda_{g,i}$. That is, $\lambda_{g,i}$ is the probability that a microorganism of type g will settle (adhere or invade) on reaching site i. Though the molecular mechanisms involved in bacterial adhesion to host cells are complex (Ofek et al. 1994; Svanborg et al. 1996; Whittaker et al. 1996; Henderson et al. 1999), they are very crudely summarised in the model by the settlement probabilities $\lambda_{g,i}$. Then $1 - \lambda_{g,i}$ is the probability that the microorganism does not settle at site i, but instead continues its migration. Each such preference is

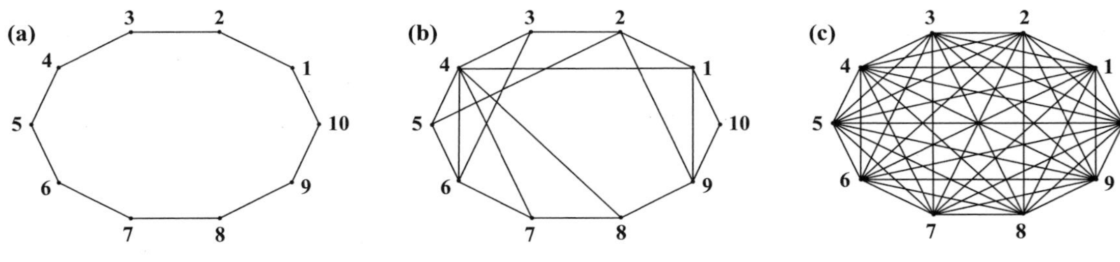

Figure 9.1. Types of network representing the "host." **(a)** A local network with only-nearest-neighbour connections. **(b)** A "small worlds" network with some long-range connections as well as nearest-neighbour connections. **(c)** A random network in which all connections have some non-zero connection probability.

assumed to lie in a fixed range $\lambda_{min} \leq \lambda_{g,i} \leq \lambda_{max}$ with $0 \leq \lambda_{min} < \lambda_{max} \leq 1$. If $\lambda_{min} > 0$, then every species has some probability of settling at any site, and if $\lambda_{max} < 1$, every species has some probability of not settling at any site.

Migration time is measured in search periods. In each search period, there is a fixed probability e that a migrating microorganism will encounter a settlement site. Migrating microorganisms are also subject to additional hazard by being more exposed to the host's immune defenses. This hazard is represented by a fixed survival probability $v < 1$ per search period. As a result, a migrating microorganism will have to settle within a few search periods or else it is likely to be eliminated. The smaller ev is, the faster it will have to settle if it is to survive. This favours species g for which $\lambda_{g,i}$ is large for a large number of sites. On the other hand, if ev is large, then microorganisms can afford to take their time and be more choosey about the site at which they eventually settle. In any case, the conservative migration assumption means that microorganisms return with high probability to the site they have just left, so the net effect of the external hazard is to cull part of the resident population in each search period. This can be thought of as a baseline mortality rate at the site. The model is outlined formally in Appendix A.

9.4. SPECIALISTS AND GENERALISTS: TRADE-OFFS

If a microorganism has many low site preferences, then it may have to search for a long time before it settles at a site on which it can grow well, thereby exposing itself to external hazard for a long time. Conversely, if a microorganism has many high site preferences, then its search time will be short, minimising its exposure to external hazard. However, it pays a price in settling on sites where its growth potential may be modest. This is the fundamental life-history trade-off that a microorganism has to negotiate. I define three distinct life-history strategies reflecting different-trade-offs.

A *generalist* is any species that does not discriminate between sites, and will settle equally readily at any site. A generalist therefore has site preferences that are site independent of the form:

$$\lambda_{gen,i} = \lambda_{max}.$$

At the other extreme are the site-specific specialists. Such a specialist has a maximum preference for a particular site, which it finds especially favourable, but a minimum preference for every other site, which it finds relatively unfavourable. Thus, a *site-k specialist* has site preferences of the form:

$$\lambda_{k-sp,k} = \lambda_{\max} \text{ and } \lambda_{k-sp,i} = \lambda_{\min} \text{ for } i \neq k.$$

There is obviously an extremely large range of possible intermediaries between a generalist and a site-specific specialist. To keep the model tractable, I consider only one intermediate type, which I call a *semi-generalist*. The preference schemes for the three species types of model microorganisms are illustrated in Figure 9.2. The community of model microorganisms therefore consists of one generalist species, and n each of site specialists and semi-generalists (one of each for each site), where n is the number of sites (nodes in the network). For a ten-node network, there are therefore twenty-one distinct species. I use a colour code as in Figure 9.2d to identify these species.

Having defined the model microorganism community by their site preferences, I now consider their reproduction within a site at which they have settled. Assume reproduction (clonal expansion) of species g at site i proceeds at a growth rate $\omega_{g,i}$. Different species at the same site grow at different rates, as does the same species at different sites.

Whether an organism will settle on a site or not will depend on whether it can reproduce at that site. Thus, the growth rates $\omega_{g,i}$ should be positively correlated with the preferences $\lambda_{g,i}$. I assume a relationship of the form

$$\omega_{g,i} = D_g \lambda_{g,i}, \tag{2}$$

where D_g is a parameter representing the degree of *site discrimination* of microorganisms of type g. The idea is that there is a trade-off between site discrimination and growth potential at a given site: the only reason an organism needs to be highly site discriminatory is if it has limited capacity to grow at many sites, and therefore has to be careful where it settles. Conversely, if an organism can grow equally well at many sites, then it need not exhibit strong site-preference discrimination.

The more sites an organism has a high preference to settle on, the *lower* is its discrimination, and conversely, the more sites it has a low preference to settle on, the *higher* is its discrimination. I therefore base a measure of site discrimination on a microorganism's mean site preference:

$$D_g = \frac{1}{1 + \delta \bar{\lambda}_g}, \qquad \bar{\lambda}_g = \frac{1}{n} \sum_{i=1}^{n} \lambda_{g,i}, \tag{3}$$

where δ is a species-independent parameter, which measures the "strength" of the preference-discrimination trade-off. When $\delta = 0$, there is no trade-off,

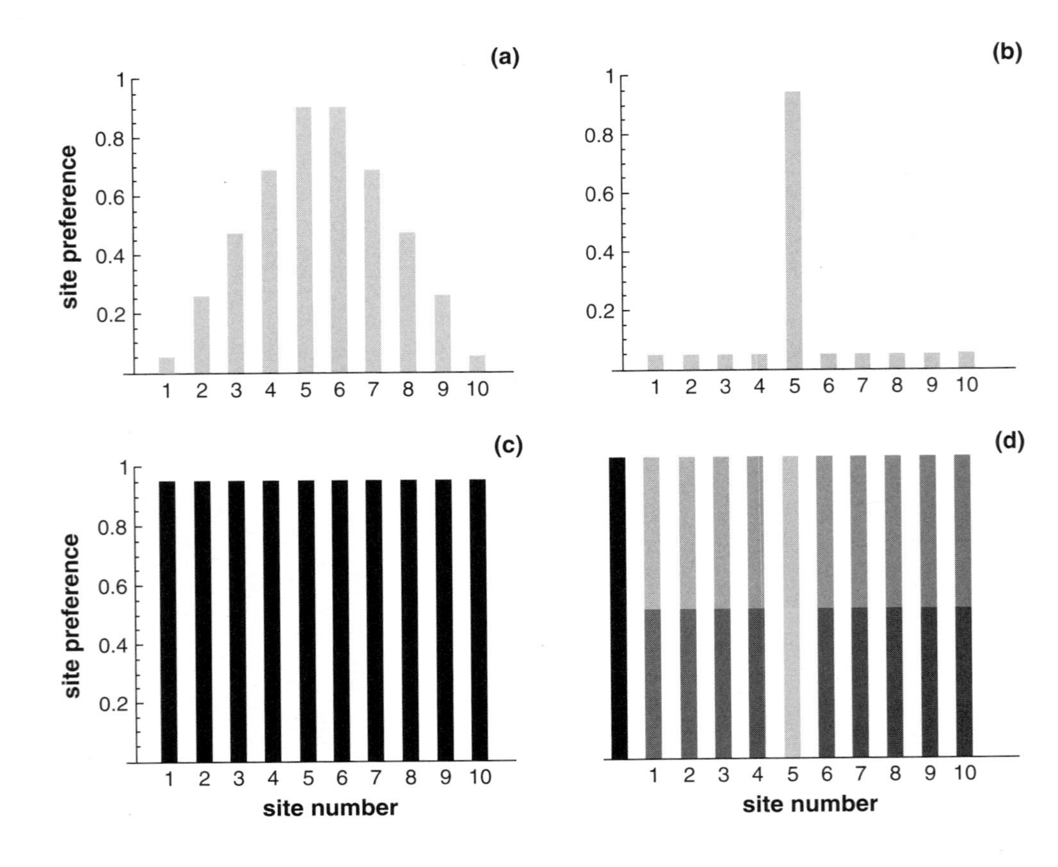
(a)
(b)
(c)
(d)
site preference
site preference
site number
site number
1
0.8
0.6
0.4
0.2
1 2 3 4 5 6 7 8 9 10

$D_g = 1$, and growth rates are equal to preferences. As δ increases, D_g decreases for a given mean preference, and hence acts to suppress growth rates. In particular, if a type has a high mean preference, $\overline{\lambda}_g$, then, for a given $\delta > 0$, its discrimination is low, and this is factored into its site-specific growth rate (2) as a large reduction. Conversely, if a type has a low mean preference, then its discrimination is high, and this is factored into its site-specific growth rate as a small reduction.

For example, a generalist thrives equally well (or badly) at every site, and does not discriminate between them. From (2) and (3), a generalist has growth rate at each site:

$$\omega_{gen,i} = \frac{\lambda_{\max}}{1 + \delta\lambda_{\max}}. \tag{4}$$

Thus, if $\delta = 0$, the generalist pays no growth penalty for its lack of discrimination. However, as δ increases, so does the growth penalty. On the other hand, for a k-specialist, the site-specific growth rates (2) are

$$\omega_{k-sp,i} = \frac{\lambda_{k,i}}{1 + \delta(\Delta\lambda + \lambda_{\min})}, \tag{5}$$

where $\lambda_{k,k} = \lambda_{\max}$ and $\lambda_{k,i} = \lambda_{\min}$ for $i \neq k$, and $\Delta\lambda = (\lambda_{\max} - \lambda_{\min})/n$. Clearly, a generalist grows better than a k-specialist at all sites other than k. However, the k-specialist grows better at site k than the generalist when $\delta > 0$ and $n > 1$.

For a semi-generalist, $\overline{\lambda}_{sgen} = \frac{1}{2}(\lambda_{\max} + \lambda_{\min})$. It follows that a semi-generalist grows better than a generalist on those sites i for which $\lambda_{sgen,i}$ is larger than the threshold value $\lambda^*(\delta) = \lambda_{\max}\{1 + \frac{1}{2}\delta(\lambda_{\max} + \lambda_{\min})\}/(1 + \delta\lambda_{\max})$. When $\delta = 0$, there are no such sites, but the number of possibilities increases as δ increases. Similarly, a semi-generalist grows better than a k-specialist on sites $i \neq k$ provided $\lambda_{sgen,i}$ is larger than the threshold $\lambda_*(\delta) = \lambda_{\min}\{1 + \frac{1}{2}\delta\,(\lambda_{\max} + \lambda_{\min})\}/\{1 + \delta\,(\Delta\lambda + \lambda_{\min})\}$. When $\delta = 0$, this is true for all sites $i \neq k$, but applies to progressively fewer sites as δ increases. However, a k-specialist grows better than any semi-generalist on site k provided $\delta > 0$ and $n > 2$.

←

Figure 9.2. Preference profiles for selected species with corresponding colour coding. **(a)** A site-5 semi-generalist. **(b)** A site-5 specialist. **(c)** A generalist. **(d)** Colour coding for species: black is a generalist; light colours are the specialist at the indicated site; darker shades are the semi-generalists at the indicated sites. The maximum and minimum site-preference values are $\lambda_{\max} = 0.95$, $\lambda_{\min} = 0.05$.

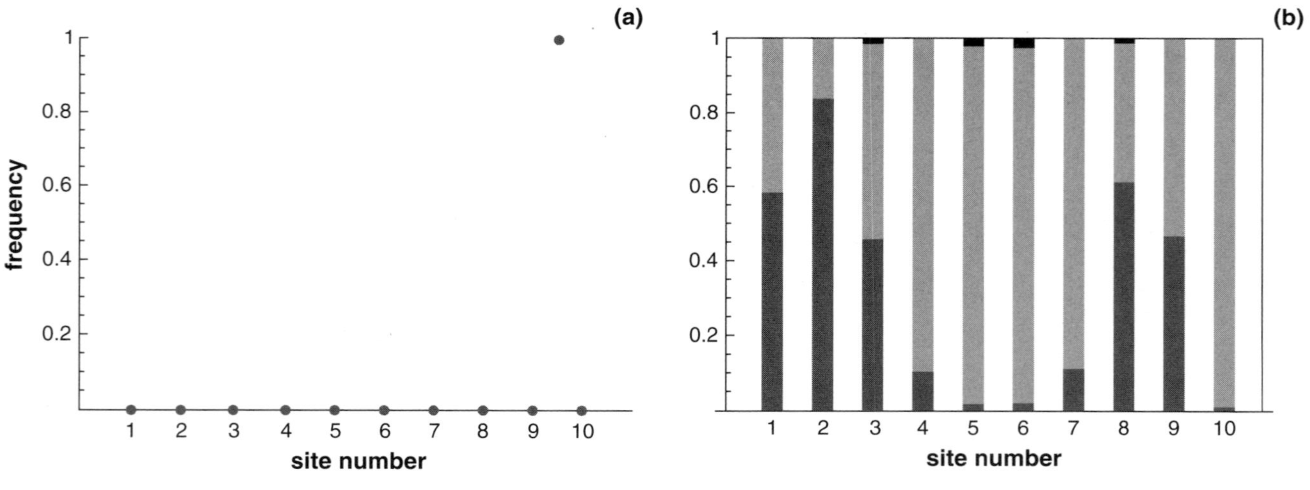

Figure 9.3. **(a)** The stable inter-site distribution $P_{g,i}$ after 250 time steps starting from a uniform distribution of generalists with preference-discrimination trade-off strength $\delta = 3$ and no immune response ($\beta = 0$). The inter-site diversity is $H = 0.039$, which is very low (maximum diversity is $H_{\max} = \ln(21) \approx 3$). **(b)** The intra-site stable distributions at each site. These are dominated by the specialist at site 10, with some representation of the site-8 semi-generalist, and a little of the true generalist. See Appendix A for model and parameters.

I allow a low probability μ of mutation between "neighbouring" species at the beginning of the reproductive phase. With probability $1 - \mu$, a microorganism does not mutate. Neighbours are defined as follows. The generalist species is a neighbour of every semi-generalist species, and can mutate into any one of them with equal probability. The k-semi-generalist is a neighbour of the generalist and of the l-specialist for $l = k$ and $k + 1$ (Fig. 9.2a,b) and can mutate into any of these with equal probability. Finally, the k-specialist species is a neighbour of the k-semi-generalist and can mutate into it. See Chapter 5 for experimental data of the mutation rates of *Pseudomonas fluorescens* in simple culture.

9.5. CLONAL SELECTION WITHOUT IMMUNE RESPONSE

Typical output from the model is shown in Figures 9.3, 9.4, and 9.5, with fixed parameters given in Table A1 in Appendix A, and an intermediate value for the preference-discrimination trade-off parameter $\delta = 3$. Simulations all begin with a uniform distribution of generalists over all sites. The small mutation probability (Table A1) allows other species to invade until a stable state is reached (as described in 9.3). Figure 9.3a shows the stable *inter-site* distribution $P_{g,i}$ of species g at site i as a proportion of the total population (using the colour coding in Fig. 9.2d). Figure 9.3b shows the *intra-site* distributions $Q_{g,i}$ of species g at site i as a proportion of the total population at site i. Figure 9.4a shows rank abundance of the twenty-one possible species; that is, the species ranked according to their total abundance. Figure 9.4b shows the rank ordering by degree of conservatism of the network nodes (measured by the self-transition probabilities c_{ii}, as described in section 1), with the colour coding associated to the specialist at each site. Finally, Figure 9.5 shows the stable inter-site distribution $P_{g,i}$ at various abundance scales (scales of order 10^{-s} for $s = 0$ to 5) to determine whether there is "subterranean" diversity of rare species.

The results show that the inter-site distribution (Fig. 9.3a) is dominated by a single specialist type (the site-10 specialist). This dominant type is predictable from the degree of conservatism at each site (Fig. 9.4): It is the specialist type associated with the site having the highest degree of conservatism. Almost all the distribution is concentrated at the site for which the dominant species is a specialist (site 10), with growth of species at other sites almost completely suppressed. Furthermore, there is little subterranean diversity at lower abundance scales (Fig. 9.5), largely represented by generalists and site-10 semi-generalists. This situation leads to a low level of diversity as measured by the Shannon entropy H (see Appendix B for details of how these

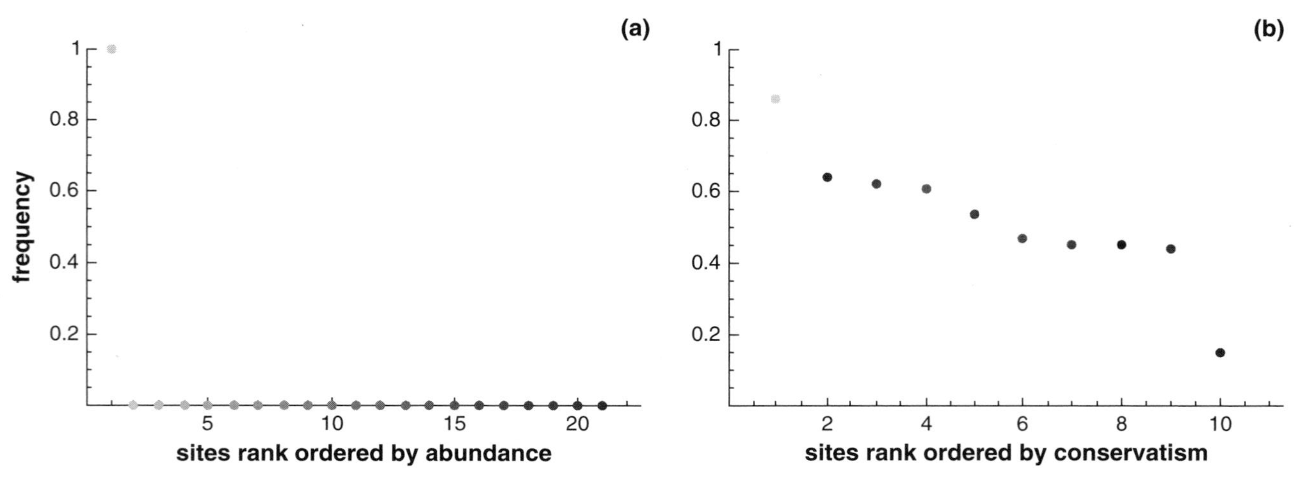

Figure 9.4. **(a)** Species by rank abundance for the stable state inter-site distribution from Figure 9.3a. **(b)** Specialist species ranked by degree of conservatism of the site at which they specialise. This measure is a good predictor of the dominant species.

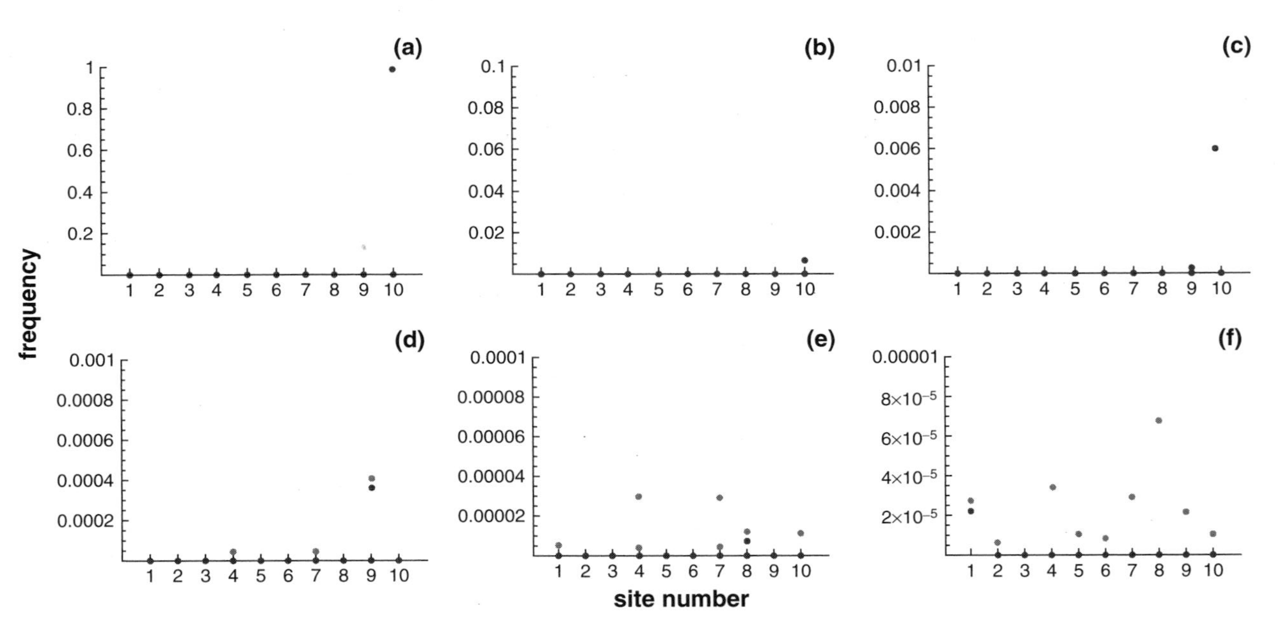

Figure 9.5. "Subterranean" diversity at various scales $P_{g,i} \leq 10^{-s}$ for $s = 0$ to 5. **(a)** $s = 0$. **(b)** $s = 1$. **(c)** $s = 2$. **(d)** $s = 3$. **(e)** $s = 4$. **(f)** $s = 5$. There is only a very little subterranean diversity at the lower scales, and these are largely represented by the small populations of the dominant species at sites other than 10.

Table 9.1. *Intra-site diversity indices H for each site, together with the mean and variance over all sites, associated with the simulation of Figure 9.3*

Site	1	2	3	4	5	6	7	8	9	10	mean	var
H	0.72	0.478	0.761	0.352	0.144	0.177	0.364	0.729	0.711	0.037	0.447	0.067

indices are computed). Similarly, the intra-site diversity is low, as illustrated in Figure 9.3b, and more quantitatively by the site-specific H-values of the intra-site distributions given in Table 9.1.

This situation is an extension to the network context of competitive exclusion in a competitive community. I now consider what might happen if a "discriminating predator" is introduced. The relevant analogy is with the immune system, which acts as a suppressor of the invading microorganisms.

9.6. MODELLING THE IMMUNE RESPONSE

The systemic immune response already affects migrating microorganisms through the fixed survival probability ν (section 9.3). In addition, I now assume there is a site-specific action that affects settled microorganisms. Thus, I model the immune system as acting to suppress the reproductive rates $\omega_{g,i}$ of the settled microorganisms. However, unlike a predator, I assume that the immune system does not have "fixed preferences" for particular microorganism species, but has the capacity to adapt to the particular properties of the invaders it detects. In this sense, the immune response is *adaptive*. Furthermore, this adaptation is *local*: It has a component that is specific to each site in the network, depending on what species are present.

I consider the immune response as having two components. The first is *systemic*, which responds to the total microorganism load at a site, but in a manner that is globally controlled. The second is *species specific* and allows the immune system to adjust its strength in response to the relative presence (or absence) of particular species at particular sites. That is, the ingredients of the local molecular and cellular "soups" recruited by the host at a particular site can be modulated by the local distribution of microorganisms at that site.

The formal manner in which this immune response is modelled is given by replacing constant microorganism growth rates $\omega_{g,i}$ by frequency-dependent growth rates:

$$\omega_{g,i} = \omega_{g,i}^{0} \exp\left\{-\beta \frac{P_{g,i}}{1 - P_i}\right\}. \tag{6}$$

Here, $\omega^0_{g,i}$ are the constant growth rates that would be realised in the absence of an immune response (section 9.4); β is a control parameter that measures the global *strength* of the systemic immune response. However, this strength grows rapidly in response to the microorganism load at site i through the factor $1 - P_i$, where $P_i = \sum_g P_{g,i}$ is the relative load at site i. Thus, $1 - P_i$ represents the load at sites other than i; if this is small (i.e., if P_i is large), then the immune system can direct more of its resources to site i, where most of the load is concentrated. Finally, the factor $P_{g,i}$ represents the proportion of the local response at site i that is specifically adapted to act against species g. This assumes that the "visibility" to the immune system of each species is proportional to its representation at the site. Thus, even if the total-load distribution P_i over the network sites remains fixed, the immune response can adapt to a change in the relative representation of species within this load.

The results of section 9.5 correspond to the no-immune response case $\beta = 0$. I now consider what happens with the immune response present; that is, when $\beta > 0$.

9.7. CLONAL SELECTION WITH IMMUNE RESPONSE

I illustrate the effect of the immune response by taking $\beta = 5$ and, as before, with an intermediate value of the trade-off parameter $\delta = 3$. This is a mid-strength response for the parameters given in Appendix A, Table A1. Figures 9.6, 9.7, and 9.8 correspond to Figures 9.4, 9.5, and 9.6 of section 9.5.

The effect of the immune response is dramatic. There are now several specialist species that dominate the system, each at relatively low frequency, that is, high evenness (Fig. 9.6a). Furthermore, the rank abundance of these dominant species is reasonably well correlated with the ranked degree of conservatism of the nodes at which they are specialist (Fig. 9.7). The subterranean diversity at lower abundance scales is even more dramatically increased compared with the no-immune response case (Fig. 9.8 compared with Fig. 9.5). However, the increase in mean intra-site diversity is very small, though the variance between sites is much greater, largely due to the fact that site 1 has a much higher intra-site diversity than the other sites (Fig. 9.6b and Table 9.2 – cf. Fig. 9.3b and Table 9.1). This appears to be because site 1 is the only site not dominated by its associated specialist, again apparently because site 1 has the lowest degree of conservatism of any of the sites (Fig. 9.7b). Because of this, site-1 specialists are more prone than other species to migrate to other sites where they fare badly. The effects of predators on population diversity is also discussed in Chapter 8.

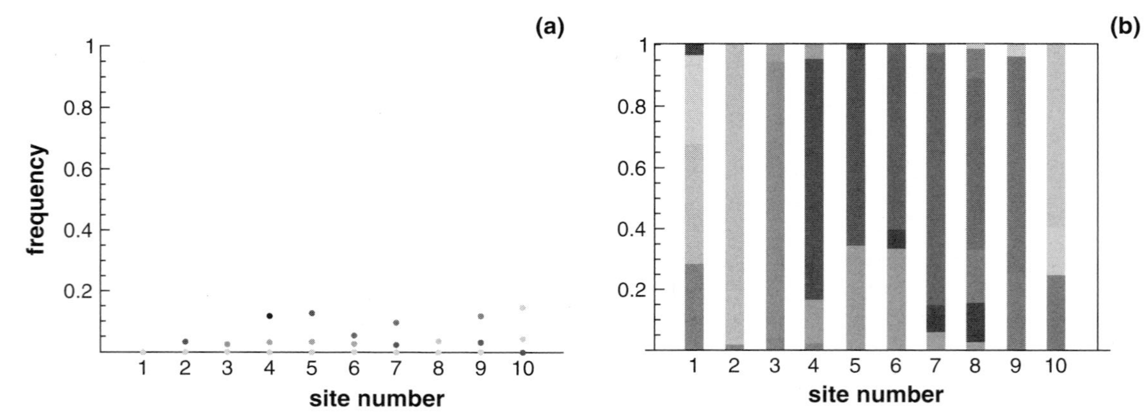

Figure 9.6. **(a)** The stable inter-site distribution $P_{g,i}$ after 250 time steps starting from a uniform distribution of generalists with preference-discrimination trade-off strength $\delta = 3$, and immune-response strength $\beta = 5$. The inter-site diversity is $H = 2.59$, which is high (maximum diversity is $H_{\max} = \ln(21) \approx 3$). **(b)** The intra-site stable distributions at each site. Most sites are dominated by their own specialist, with some representation of the corresponding semi-generalist, and smaller representation of other species. The exception is site 1, which is more diverse than the other sites.

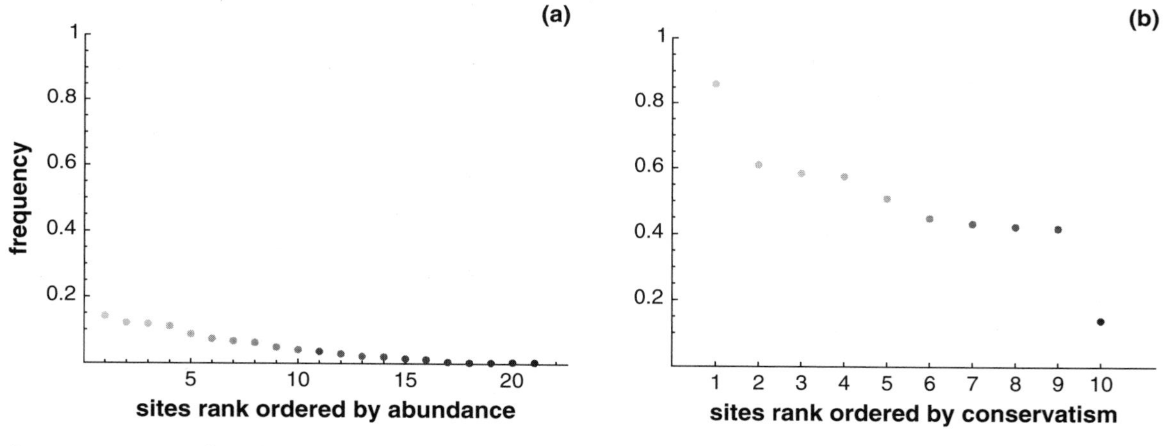

Figure 9.7. **(a)** Species by rank abundance for the stable state inter-site distribution from Figure 9.6a. **(b)** Specialist species ranked by degree of conservatism of the site at which they specialise. This measure predicts the dominant species with good accuracy.

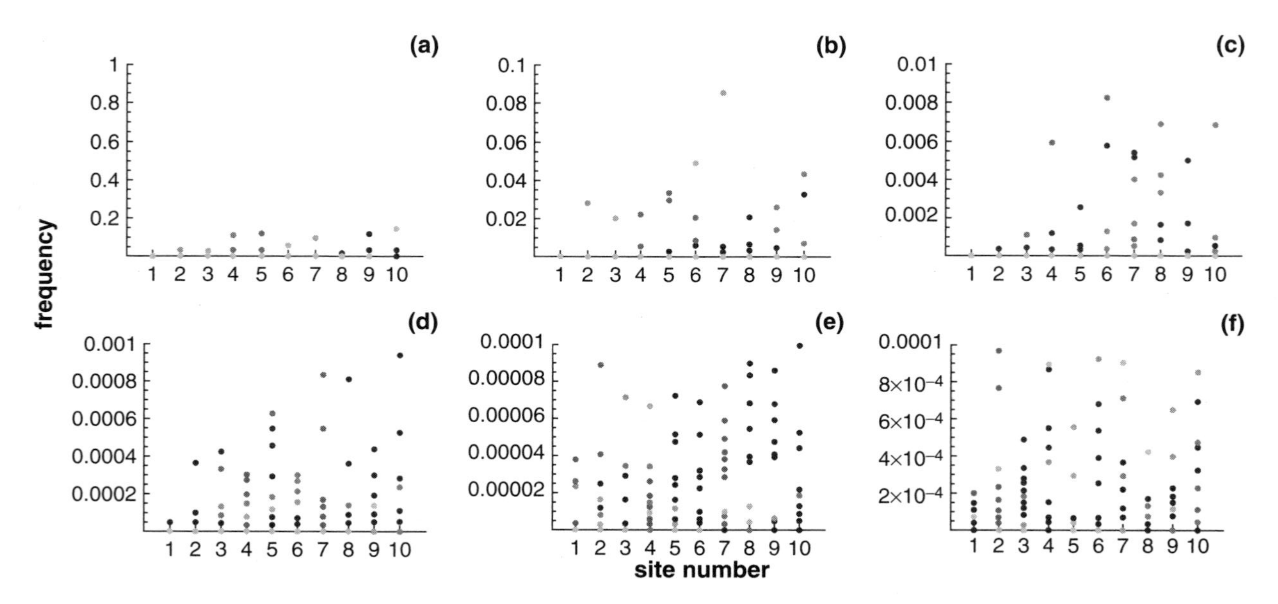

Figure 9.8. "Subterranean" diversity at various scales $P_{g,i} \leq 10^{-s}$ for $s = 0$ to 5. **(a)** $s = 0$. **(b)** $s = 1$. **(c)** $s = 2$. **(d)** $s = 3$. **(e)** $s = 4$. **(f)** $s = 5$. In contrast to the no-immune response case (Fig. 9.5), there is considerable diversity at smaller scales.

Table 9.2. *Intra-site diversity indices H for each site, together with the mean and variance over all sites, associated with the simulation of Figure 9.6*

Site	1	2	3	4	5	6	7	8	9	10	mean	var
H	1.682	0.132	0.449	0.712	1.02	1.228	0.768	1.464	0.962	1.056	0.947	0.19

9.8. DISTURBANCE

There are several kinds of disturbance that might be relevant to the diversity of microorganisms within the host. The first is a systemic disturbance to the immune response, which may arise from poor physiological condition or from some form of exogenous shock. This kind of disturbance affects all network sites equally, and can be modelled as a perturbation in the immune-response control parameter β. Second is a local disturbance of the immune response, which may affect its strength at a single site only. For example, a locally stable state of a cytokine network at a site may be perturbed by a transient invasion of alien microorganisms. Third, it may not be the immune response which is subject to disturbance, but rather the site-specific characteristics of the network nodes. For example, the gut or the mouth are under constant assault from the passage through them of material from the outside world. This may tend to homogenise the potential micro-niches within these environments. A microorganism which is a specialist at some specific site, say i, may then suddenly find that the site at which it has been thriving has taken on the characteristics of some other site, say j. The i-specialist will then find itself in a very unfavourable environment, whereas the formally disfavoured j-specialist will suddenly find that it has hit the jackpot.

The simplest disturbance is the first, systemic type. Assume the disturbance takes the form of reducing the strength parameter β from its current value to some perturbed value, say β_0 (which could be zero, thereby nullifying the immune response completely). I also allow for a recovery process from such a perturbation, which acts to restore β to its long-run average $\overline{\beta}$. Thus, at time t after the onset of the perturbation, and in the absence of any further perturbation, I assume that β is given by

$$\beta_t = \overline{\beta}(1 - e^{-rt}) + \beta_0 e^{-rt}, \tag{7}$$

where r is the recovery rate.

The effect of shocks of this kind is simply to allow the would-be dominant microorganism species to explode transiently (as in Fig. 9.3a) before it is brought under control again by the recovering immune response (as in Fig. 9.6a). This is clearly likely to be detrimental to the organism.

Furthermore, if these shocks occur sufficiently frequently, the diversity of the stable state (Fig. 9.6a) may be lost.

The second kind of disturbance is a localised version of the first. In this case, β in equation (2), which is the same for all sites, must be replaced by site-specific values β_i that can be chosen independently at each site. The "normal" situation is then identified with the long-run average, $\beta_i = \overline{\beta}$. A local perturbation at site i changes β_i, while leaving the β-values at other sites unaffected. Recovery is given by a local form of Equation (5).

Local shocks of this kind can occur with some (Poisson) frequency at randomly chosen sites. The effect is slight if the frequency is fairly high, though there is some reduction of diversity due to the average loss of effectiveness of the immune response over the long run. However, if the shock frequency is relatively low, then occasionally there may result the same dramatic explosion of a dominant specialist as occurs with systemic perturbations. This happens if the random site at which there is a compromise of the immune response happens to be the site associated with the previously suppressed dominant specialist. Nevertheless, on the whole, the system is much more robust against this kind of disturbance than against systemic disturbance.

Finally, I consider random disturbances of site characteristics. These are modelled as random permutations of site labels, so that an i-specialist growing at site i (at which it grows well) before a perturbation will find itself at site j (at which it grows poorly) after the perturbation. If such permutations happen frequently enough, then no species will have a systematic advantage at any site. In this case, the outcome is simply that the generalist species dominates the species distribution, leading to a loss of diversity. At low disturbance frequency, there are large random fluctuations in the representation of all the species, including generalists.

In summary, stochastic disturbances, either to the immune response or the site characteristics, lead to a loss of diversity, though this loss is greater for systemic immune disturbances than for local ones, and tend to lead to domination by generalists when site characteristics fluctuate.

9.9. NETWORK STRUCTURE, IMMUNE RESPONSE, AND DIVERSITY

The results illustrated in sections 9.5 and 9.7 are snapshots obtained from one particular small-worlds network with one particular set of connection probabilities and one particular value of the preference-discrimination trade-off parameter δ. In section 9.8, one particular value of the immune response, parameter β was used ($\beta = 5$). In this section, I consider the effect of systematically varying these factors.

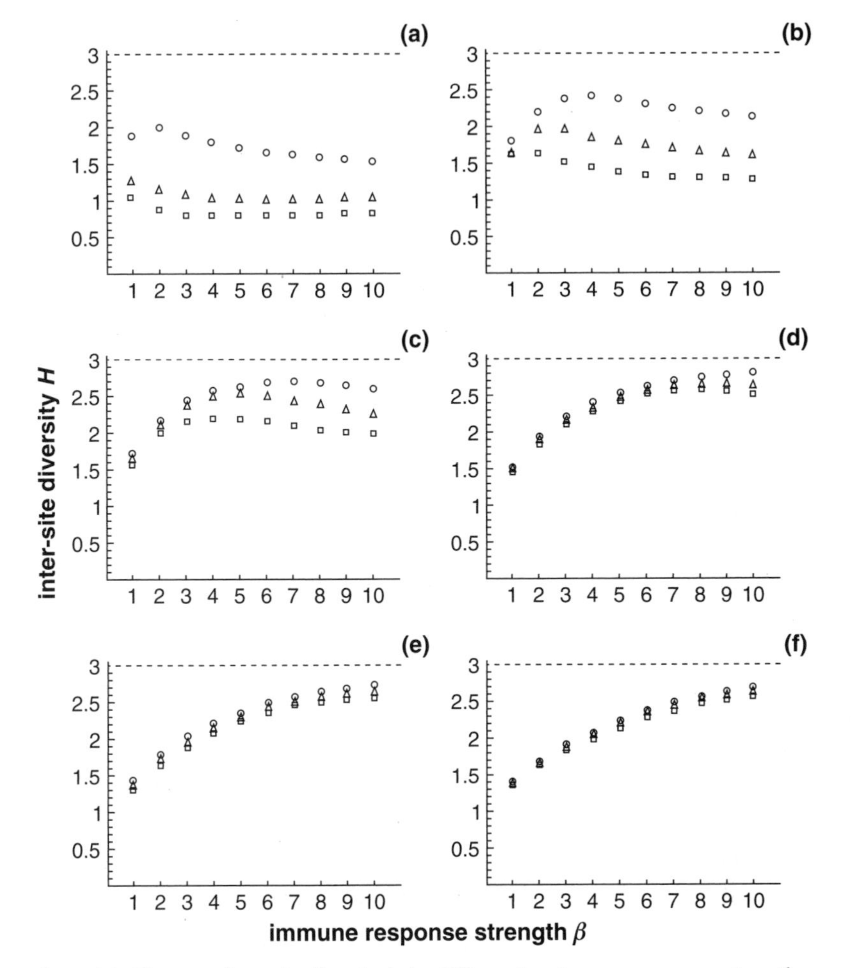

Figure 9.9. The mean inter-site diversity index H for various immune-response strengths $\beta = 0, 1, \ldots, 10$ and discrimination-preference trade-off strengths: **(a)** $\delta = 1$, **(b)** $\delta = 1.5$, **(c)** $\delta = 2$, **(d)** $\delta = 3$, **(e)** $\delta = 4$, **(f)** $\delta = 5$. The circles are local networks, the squares are random networks, and the triangles are small-worlds networks (with connection probability $p = 0.5$) – see Figure 9.1. The means are taken over ten randomly generated networks of each type. The dashed line indicates the maximum possible diversity $H_{\max} = \ln(21) \approx 3$.

The effects of increasing β in integer steps from 0 to 10 on the inter-site diversity index H for various values of δ are illustrated in Figure 9.9. For each δ, ten random local networks (Fig. 9.1a) were generated, and the stable-state inter-site diversity computed for each network and each value of β. The mean values of H over these networks was then obtained and plotted against β (Fig. 9.9, circles). A similar procedure was employed with ten randomly generated random networks (Fig. 9.1c) to obtain a plot of mean inter-site diversity against β (Fig. 9.9, squares). Finally, this procedure was repeated with ten randomly generated small-worlds networks (connection probability $p = 0.5$ – Fig. 9.1b), giving the triangles in Figure 9.9. Details of these computations are given in Appendix B. The results obtained may be summarised as follows:

1. For each network type and each value of δ, there is a positive value of β at which inter-site diversity is a maximum. This β value is largest for local networks, and smallest for random networks.
2. For each δ and $\beta > 0$, local networks have higher inter-site diversity than small-worlds networks, which have higher inter-site diversity than random networks. However, this difference decreases as δ increases.
3. For all network types, maximum inter-site diversity increases as δ increases.
4. For each β, the proportion of generalists in the population decreases as δ increases.
5. For each $\delta \geq 1$, the proportion of generalists in the population increases as β increases.

Conclusion 4 says that the generalist species does less well as the strength of the preference-discrimination trade-off increases. In these circumstances, the generalist pays a high price in terms of depressed growth rates for its lack of site discrimination. This is true whatever the immune response strength. However, conclusion 5 asserts that as the immune-response strength increases, generalists claw back some advantage, because the immune response acts more strongly to depress the more highly represented specialists and semi-specialists.

A similar procedure was adopted for intra-site diversity: The mean intra-site diversity index H_i over the ten networks of each type at each site i and for each value of β was calculated. Figure 9.10 plots the means of these over the ten sites against β (see Appendix B for details) and shows that the mean intra-site diversity increases somewhat to a low plateau as β increases. The intra-site diversity therefore responds only weakly to increasing immune-response strength, and diversity remains relatively low. This is the case for all network types. Again, the network structure effect, in which local networks have higher intra-site diversity than small-worlds networks, which have higher diversity than random networks, is pronounced at lower values of δ, but is actually reversed for high values of δ.

As observed in section 9.1, the Shannon entropy is relatively insensitive to rare species. However, it can be seen from Figure 9.8 that there

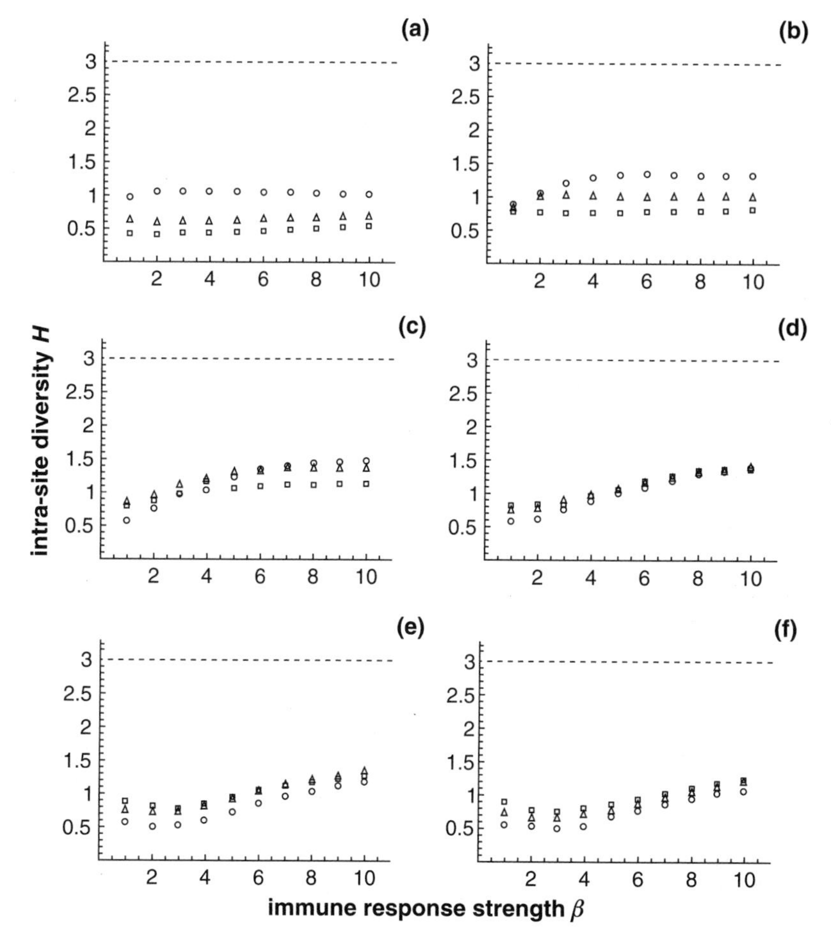

Figure 9.10. The mean intra-site diversity index H for various immune-response strengths $\beta = 0, 1, \ldots, 10$ and discrimination-preference trade-off strengths: **(a)** $\delta = 1$, **(b)** $\delta = 1.5$, **(c)** $\delta = 2$, **(d)** $\delta = 3$, **(e)** $\delta = 4$, **(f)** $\delta = 5$. The circles are local networks, the squares are random networks, and the triangles are small-worlds networks (with connection probability $p = 0.5$) – see Figure 9.1. The means are taken over ten randomly generated networks of each type at each site, and then over the ten sites. The dashed line indicates the maximum possible diversity $H_{\max} = \ln(21) \approx 3$.

is considerable diversity at various scales 10^{-s} for $s \geq 0$. To capture this quantitatively, I define in Appendix B a sequence of scaled inter-site diversity indices H^s, based on the part of the stable abundance distribution $P_{g,i}$ which satisfies $P_{g,i} \leq 10^{-s}$, and such that $H^0 = H$. Each of these indices has the same maximum value $H_{\max} = \ln(S)$. The means of these indices over ten

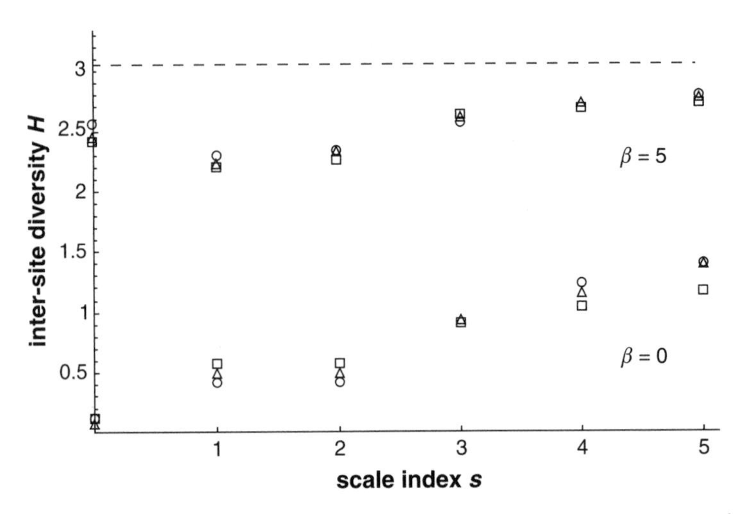

Figure 9.11. The mean scaled inter-site diversity index H^s (see Appendix B) associated with scale 10^{-s} for $s = 0$ to 5. The means of ten randomly generated networks of each type are shown in the no-immune-response case ($\beta = 0$), and for an intermediate immune-response strength ($\beta = 5$). In all cases, the preference-discrimination trade-off strength is $\delta = 3$. Circles are local networks, squares are random networks, and triangles are small-worlds networks (connection probability $p = 0.5$).

randomly generated networks with $\delta = 3$ and for two values of the immune-strength parameter, $\beta = 0$ and $\beta = 5$, are shown in Figure 9.11.

It can be seen from Figure 9.11 that for the medium immune-response case, $\beta = 5$, diversity is high at all scales, though there is a dip at scale 10^{-1} followed by a monotonic increase. For the case of no immune response, diversity is low at all scales, though it increases approximately monotonically at lower scales.

9.10. CONCLUSIONS

Competitive exclusion is a ubiquitous feature of communities in which no other forces intervene to moderate competition, and leads to low diversity communities dominated by one or a few superior competitors. In the context of the host-as-network considered here, the capacity of a site-specific specialist to grow at its specialist site with a low, but non-zero emigration rate (i.e., high "conservatism" as defined in section 9.1) allows the specialist to exclude almost all competitors in the network, even those which specialise at other sites, provided all other sites have lower conservatism (Figs. 9.3, 9.4, and 9.5).

For a community of microorganisms within the host-as-network, the immune response can act analogously to a predator in an ecological community, but one with adaptable prey preferences, depending on the proportional representation of particular species of microorganism within the population at particular sites. Inter-site diversity, as measured by the Shannon entropy index (Appendix B), increases rapidly as immune-response strength increases above zero. With low site preference–discrimination trade-off (i.e., low δ – section 9.2), the inter-site diversity increases to a maximum as immune-response strength increases, before declining to a (still comparatively high) plateau value (Figs. 9.6, 9.7, 9.8, and 9.9). High diversity is also sustained by the immune response at lower abundance scales, where species other than the dominant specialists, such as generalists and semi-generalists, can maintain a presence (Figs. 9.8 and 9.11).

The model simulations reported here were obtained from an initial uniformly distributed population of the generalist species, but with mutation (at a low rate) to neighbouring species. In the presence of a reasonably strong immune response and no systematic disturbance, the long-run stable outcomes were highly diverse communities dominated by several distinct species of specialist that do not directly mutate into each other. These outcomes, therefore, arise by speciation through niche specialisation.

Stochastic disturbance, whether in the form of a transitory compromise of the immune response, or random re-orderings of relative competitiveness at the network sites, tends to lead to loss of diversity. In the former case, there is a reversion to competitive exclusion – and hence to an explosion of the competitively dominant species – perhaps leading to host pathology – which can be countered by immune recovery after compromise. Site-preference disturbances sustained over time at a moderate Poisson frequency result in competitive dominance by the generalist species.

The immune response is modelled phenomenologically as acting to suppress microorganism reproductive rates, and is "adaptive" only in the sense that it is responsive to site-specific bacterial load and species composition (section 9.6). The model does not explicitly include molecular "cross-talk" amongst microorganism species that might mediate competition, or active evasion of the host immune response by microorganisms (Henderson and Seymour 2003). The model does include selection acting on microorganisms with fixed life-history trade-offs, mediated by a given immune response, but does not include co-evolution of microorganism life-history and immune-response strength. To do this would require a clearer idea than is currently available of what the costs and benefits are for the host of maintaining a diverse commensal microflora.

However, one aspect of this question is suggested by the model. Opportunistic pathogens perhaps may be represented by generalist species, which are relatively suppressed under high preference-discrimination trade-off conditions (section 9.6). Surprisingly, the representation of generalists *increases* with the immune-response strength. Thus, there may be a price to pay for the maintenance of a high-diversity commensal microflora by means of a strong immune response, in that the host is somewhat more vulnerable to invasion by contingently pathogenic generalists.

What about the "commensal paradox," that is, the lack of a chronic inflammatory response to the resident microbiota? High diversity promotes low but relatively even representation of most species, even the dominant specialists (Figs. 9.9 and 9.11). Perhaps this could be associated with the lack of inflammatory response, because the latter requires "above threshold" effects to perturb a normal (local) cytokine network to a pathological, pro-inflammatory stable state? This question could be investigated only by including the dynamics of cytokine networks more explicitly in the model (cf. Henderson et al. 1998; Seymour and Henderson 2001).

In conclusion, it seems likely that, however complex the molecular details of the interactions between a microorganism population and its host, the fact that animal hosts generally sustain a diverse community of microorganisms can be explained in terms of an intimate co-evolved collaboration between the host immune response and life-history features of microorganisms.

REFERENCES

Begon, M., Harper, J. L., and Townsend, C. R. (1990). *Ecology: Individuals, Populations and Communities*, 2nd ed. Oxford: Blackwell Scientific Publications.

Henderson, B., Seymour, R., and Wilson, M. (1998). The cytokine network in infectious diseases. *Journal of Immunology and Immunopharmacology* **18**, 7–14.

Henderson, B., Wilson, M., McNab, R., and Lax, A. J. (1999). *Cellular Microbiology: Bacteria-Host Interactions in Health and Disease*. Chichester: John Wiley & Sons.

Henderson, B. and Seymour, R. (2003). Microbial modulation of cytokine networks. In *Bacterial Evasion of Host Immune Responses*, eds. B. Henderson and P. F. C. Oyston, pp. 223–242. Cambridge: Cambridge University Press.

Johnson, C. R. and Seinen, I. (2002). Selection for restraint in competitive ability in spatial competition systems. *Proceeding of the Royal Society of London* B **269**, 655–663.

Kondoh, M. (2003). Foraging adaptation and the relationship between food-web complexity and stability. *Science* **299**, 1388–1391.

Krebs, C. J. (1985). *Ecology: The Experimental Analysis of Distribution and Abundance,* 3rd ed. New York: Harper & Row.

Mayr, E. (1963). *Animal Species and Evolution.* Cambridge, MA: Belknap Press.

Ofek, I. and Doyle, R. J. (1994). *Bacterial Adhesion to Cells and Tissues.* London: Chapman & Hall.

Pianka, E. R. (1988). *Evolutionary Ecology,* 4th ed. New York: Harper Collins.

Pielou, E. C. (1975). *Ecological Diversity.* New York: John Wiley and Sons.

Rosenzweig, M. L. (1999). Species diversity. In J. McGalde (ed): *Advanced Ecological Theory: Principles and Applications,* ed. J. McGalde, pp. 249–28 Oxford: Blackwell Science.

Seymour, R. M. and Henderson, B. (2001). Pro-inflammatory–anti-inflammatory cytokine dynamics mediated by cytokine-receptor dynamics in monocytes. *IMA Journal of Mathematics Applied in Medicine and Biology* **18**, 159–192.

Svanborg, C., Hedlund, M., Connell, H., Agace W., Duan, R. D., Nilsson, A., and Wullt, B. (1996). Bacterial adherence and mucosal cytokine responses: receptors and transmembrane. *Annals of the New York Academy of Science* **797**, 177–190.

Whittaker, C. J., Klier, C. M., and Kolenbrander, P. E. (1996). Mechanisms of adhesion by oral bacteria. *Annual Reviews of Microbiology* **50**, 513–552.

APPENDIX A: THE NETWORK MODEL

In this appendix, I describe the formal network model used in sections 9.3 through 9.9.

A network consists of n nodes (or sites) labelled $1, 2, \ldots, n$. Each site has two *nearest neighbours.* The nearest neighbours of site 1 are sites 2 and n, those of site n are 1 and $n-1$, and those of any other site i are $i \pm 1$. An edge is drawn between two nodes i and j if there is a positive probability c_{ij} of a microorganism migrating from i to j. In the semi-symmetric case considered here, if $c_{ij} > 0$, then also $c_{ji} > 0$, though they may have different values.

I assume that model microorganism life cycles can be split into two distinct phases: migration between sites and reproduction (with possible mutation) within sites.

The Conservative Migration Assumption

The connection probability matrix $C = (c_{ij})$ is constructed by taking $c_{ij} = C_{ij}/\sum_{k=1}^{n} C_{ik}$, where the C_{ij} are determined as follows:

$$\begin{aligned} C_{ii} &= \xi, \\ C_{ij} &= \eta\xi \qquad \text{if } i \text{ and } j \text{ are nearest neighbours}, \\ C_{ij} &= \sigma\eta\theta\xi \quad \text{otherwise}. \end{aligned} \tag{A.1}$$

Here, $0 < \eta, \sigma < 1$ are fixed parameters; ξ is a uniformly distributed random variable taking values between 0 and 1, chosen independently for each edge; and θ is a random variable taking integer values 0 or 1, chosen independently for each edge. The value $\theta = 1$ is taken with a given *connection probability* p. If $p = 0$, then there are only self- and nearest-neighbour connections, and in this case, the network is called *local*. If $p = 1$, then any path almost always has positive probability, and the network is called *random*. For intermediate p, the network is called *small world*, with self- and local-connections, but also a fraction of long-range connections having positive probability (see Fig. 9.1). Thus, (A.1) implies that in general, self-connections have the highest probability, nearest-neighbour connections the next highest, and long-range connections the lowest.

Migration

The migration phase is assumed to be split into a very large number of short search periods. There is a constant probability ν that a migrating organism will survive any search period, and a constant probability e that it will encounter a possible settlement site. There are S microorganism species, with species g having site preferences (settlement probabilities) $\lambda_{g,i}$ in the range $\lambda_{\min} \leq \lambda_{g,i} \leq \lambda_{\max}$. The settlement matrix for species g is $\Lambda_g = diag(\lambda_{g,1}, \lambda_{g,2}, \ldots, \lambda_{g,n})$. The non-settlement matrix is $I - \Lambda_g$, where I is the $n \times n$ identity matrix.

Let $N_{g,i}$ be the number of organisms of type g resident at site i just before migration, and let $E_{g,i}$ be the population of migrants from site i who have not yet reached another site. At the end of each search period, some fraction of the migrant populations, $S_{g,i}$, which arrive at site i settle there. If $\mathbf{E}_g^{(0)} = (E_{g,1}^{(0)}, \ldots, E_{g,n}^{(0)})$ is the initial migrant population vector for species g, then some simple algebra shows that, after k search periods, the vector of migrant and settled type-g organisms are:

$$\mathbf{E}_g^{(k)} = \nu^k \mathbf{E}_g^{(0)} \Omega_g^k, \quad \mathbf{S}_g^{(k)} = e\nu^k \mathbf{E}_g^{(0)} \Omega_g^{k-1} \Delta_g,$$

where $\Delta_g = C\Lambda_g$, $\Omega_g = (1-e)I + e\Gamma_g$, and $\Gamma_g = C(I - \Lambda_g)$. It follows that the population vector of type-g organisms that eventually settle *somewhere* is

$$\mathbf{S}_g^{(\infty)} = \sum_{k=1}^{\infty} e\nu^k \mathbf{E}_g^{(0)} \Omega_g^{k-1} \Delta_g = e\nu \mathbf{E}_g^{(0)} \left[\sum_{k=1}^{\infty} \nu^{k-1} \Omega_g^{k-1} \right] \Delta_g$$
$$= \gamma \mathbf{E}_g^{(0)} (I - \gamma \Gamma_g)^{-1} \Delta_g,$$

where $\gamma = e\nu/(1-(1-e)\nu) < 1$.

Now suppose the initial emigrant population was some fraction ρ of the resident population at time 0, $\mathbf{E}_g^{(0)} = \rho \mathbf{N}_g^{(0)}$. Then the resident population at the end of the migration-settlement round is:

$$\hat{\mathbf{S}}_g = \chi(1-\rho)\mathbf{N}_g^{(0)} + \rho\gamma \mathbf{N}_g^{(0)} (I - \gamma\Gamma_g)^{-1} \Delta_g,$$

where χ is the probability of survival of the remaining residents.

If $P_g = (P_{g,1}, \ldots, P_{g,n})$ is the distribution vector for species g just before the migration phase, and $\hat{\mathbf{P}}_g$ is the distribution vector just after settlement, then the above considerations can be expressed in terms of proportions:

$$\hat{\mathbf{P}}_g = \frac{1}{\hat{P}} \mathbf{P}_g \left\{ \chi(1-\rho)I + \rho\gamma(I - \gamma\Gamma_g)^{-1}\Delta_g \right\}, \qquad \text{(A.2)}$$

where $\hat{P}$ is a normalisation constant.

This gives proportions of settling organisms as a proportion of the *total* settling population at *all* sites. We also want the proportion $Q_{g,i}$ of organisms of each type who settle at a *given* site i. Then

$$Q_{g,i} = \hat{P}_{g,i} \Big/ \sum_h \hat{P}_{h,i}.$$

Mutation and Reproduction

Post settlement, the microorganisms undergo a reproductive phase, possibly preceded by mutation. The reproductive rate of species g at site i is $\omega_{g,i}$, and the probability of mutation from species h to species g is μ_g^h. In particular, the probability of *not* mutating is assumed to be the same for all species, $\mu_g^g = 1 - \mu$, where μ is a small mutation rate. Species g can mutate to any "neighbouring" species g' with equal probability $\mu_{g'}^g = \mu/m_g$, where m_g is the number of neighbouring species of g.

At the end of the growth phase, just before the next migration phase, the inter-site distribution is $P'_{g,i}$, given in terms of the previous generation distribution $P_{g,i}$ by

$$P'_{g,i} = \frac{w_{g,i}\hat{\mu}_{g,i}}{\overline{w}}, \quad \hat{\mu}_{g,i} = \sum_h \mu_g^h \hat{P}_{h,i}, \quad \overline{w} = \sum_{i=1}^{n} \sum_g \omega_{g,i}\hat{\mu}_{g,i}. \tag{A.3}$$

Equations (A.2) and (A.3) define the inter-site distribution dynamics.

Simulation Parameters

Table A1. *Fixed model parameters used in the simulations of sections 9.5 through 9.9*

Number of network nodes (sites)	Number of micro-organism species	Network conservatism parameters (eq. A.1)	Bounds for microorganism site preferences	Migration encounter-survival probability per search round	Resident migration-survival parameters	Mutation rate
n	S	(η, σ)	$(\lambda_{max}, \lambda_{min})$	γ	(ρ, χ)	μ
10	21	(0.5, 0.1)	(0.95, 0.05)	0.1	(0.8, 0.1)	0.01

APPENDIX B: DIVERSITY INDICES

To compute the *inter-site diversity*, take the stable species distribution over all sites, $P_g = \sum_i P_{g,i}$, and take the associated Shannon entropy: $H = -\sum_g P_g \ln(P_g)$. Because the number of possible species is fixed at $S = 21$, the maximum possible inter-site diversity is $H_{max} = \ln(21) \approx 3$.

In Figure 9.9, the mean inter-site diversities are computed as follows. First generate N random networks of a given type ($N = 10$ in Figs. 9.9 and 9.10). Now compute the diversities $H^u(\beta)$ for each value of the immune strength parameter $\beta = 0, 1, \ldots, 10$ and each network $u = 1, 2, \ldots, N$. Next, for each β, take the mean over the networks:

$$\overline{H}(\beta) = \tfrac{1}{N} \sum_{u=1}^{N} H^u(\beta).$$

Figure 9.9 plots $\overline{H}(\beta)$ against β for each of three network types.

To compute the *intra-site diversities*, begin with the intra-site distributions $Q_{g,i} = P_{g,i}/\sum_h P_{h,i}$, and compute the associated Shannon entropy:

$H_i = -\sum_g Q_{g,i} \ln(Q_{g,i})$. Again, the maximum diversity is $H_{\max}$. The *mean intra-site diversity* is $\overline{H}_{intra} = \frac{1}{n}\sum_i H_i$.

In Figure 9.10, the mean intra-site diversities are computed as follows. First compute $\overline{H}^u_{intra}(\beta)$ as above. Now take the mean over the class of networks to obtain $\hat{H}_{intra}(\beta) = \frac{1}{N}\sum_{u=1}^{N} \overline{H}^u_{intra}(\beta)$. Figure 9.10 plots $\hat{H}_{intra}(\beta)$ against β for each of three network types.

Scaled Diversity Indices

To take account of the fact that the Shannon entropy is relatively insensitive to rare species, define a sequence of scaled indices to quantify diversity at "subterranean" levels. Begin with the stable inter-site distribution $P_{g,i}$. For a given scale 10^{-s} with $s \geq 0$ an integer, define

$$\tilde{P}^s_{g,i} = P_{g,i} \quad \text{if } P_{g,i} \leq 10^{-s}; \quad \tilde{P}^s_{g,i} = 0 \quad \text{otherwise.} \tag{B.1}$$

Now normalise to define the *scale-s inter-site distribution*:

$$P^s_{g,i} = \frac{\tilde{P}^s_{g,i}}{\sum_{g,i} \tilde{P}^s_{g,i}}. \tag{B.2}$$

Clearly, $P^0_{g,i} = P_{g,i}$ is the total distribution. Now define the *scale-s inter-site diversity index* by

$$H^s = -\sum_g P^s_g \ln(P^s_g), \tag{B.3}$$

where $P^s_g = \sum_i P^s_{g,i}$. Clearly, $H^0 = H$ is the unrefined inter-site diversity index.

The intuition is that, in defining the scale-*s* distribution, we ignore the representation of species at coarser scales, and consider only the representation of the species at various sites on a scale at most as coarse as *s*. The means of these scaled indices over a family of ten randomly generated small-worlds networks ($p = 0.5$) are plotted in Figure 9.11.

III Cellular interactions at the bacteria–host interface

CHAPTER 10

Beneficial intracellular bacteria in the *Dryophthoridae*: Evolutionary and immunological features of a pathogenic-like relationship

Abdelaziz Heddi and Caroline Anselme

10.1. INTRODUCTION

Many associations between different species have been established over evolutionary time and have significantly contributed to biodiversity. Some associations are facultative, involving symbionts that are not represented in all populations, whereas other relationships involve partners that cannot survive independently and the entity is so-called obligatory (or integrated). Generally, associations are made between species that are phylogenetically very distant, such as bacteria–plant and bacteria–animal. Nevertheless, other types of associations can occur between eukaryotes, such as algae–fungi and algae–coral symbioses (see Chapter 6), or between prokaryotes such as soil bacteria (see Chapter 8). Even recently, an atypical association has been described in mealybugs where a γ-proteobacterium lives inside a β-proteobacterium (von Dohlen et al. 2001). "Aseptic biology" is not a common aspect in nature, and the latter example supports the occurrence of symbiotic associations at different levels of complexity, from bacteria to animals and plants.

In insects, perhaps the most diverse animal group, many species involve intracellular bacteria (endosymbionts) in their reproduction and development. Bacteria are hosted within specialized cells, named bacteriocytes, that sometimes group together to form an organ, the bacteriome (Nardon 1971). In these associations, bacteria interact with the host metabolic and cellular pathways by providing the insect with limited nutritional components (Douglas 1995; Heddi et al. 1993). Metabolic and physiological improvements result in the insect thriving on new and "empty" ecological niches that are nutritionally deficient or unbalanced, such as plant sap, wood, cereal grains, and mammalian blood. In return, endosymbionts become dependent on insect cells that provide them permanently with basal nutritive molecules such as sugars. In this syntrophy context, selection pressure deletes bacterial genes encoding redundant metabolic pathways present in the host (Shigenobu et al. 2000) or genes become "unnecessary" to the new association, such as virulence genes or genes encoding the bacterial cell wall. Serial gene deletions lead subsequently to bacterial genome size reduction (Charles and Ishikawa 1999; van Ham et al. 2003), which results in "bacterial domestication" by the host. Moreover, endosymbionts are transmitted vertically through maternal germ cells throughout insect generations, which leads to the co-evolution between partners that may end in co-speciation (Baumann et al. 1995). See Chapter 7 for a detailed discussion of the *Wolbachia*-insect paradigm.

However, although physiological and evolutionary aspects of insect symbiosis have been well dissected over the past two decades, very little is known about molecular mechanisms and the partners' dialogue that permit the establishment and the evolution of such beneficial associations. One striking question relates to insect immunity, particularly in the first evolutionary steps, during which time the partners may select a "molecular consensus" for their being together. In this chapter, we focus on a relatively recently established endosymbiosis that involves the weevil *Sitophilus* (Dryophthoridae) and a γ-proteobacterium. In addition, we review current data on insect innate immunity and discuss its potential impact on endosymbiosis.

10.2. THE BIOLOGY OF INSECTS AND THEIR INTRACELLULAR BACTERIA

The Dryophthoridae weevils are phytophagous insects that thrive on a large host-plant spectrum. Around 500 species (previously named Rhynchophoridae) have been described so far (Alonzo-Zarazaga and Lyal 1999). Some have a worldwide distribution (species such as *Sitophilus oryzae* and *Sitophilus zeamais*), whereas others are limited geographically, such as *Trigonotarsus rugosus* Boisduval (Australia). Feeding diets and ecological niches are also very variable, ranging from the trunk base of banana trees and palm trees (*Cosmopolytes sordidus* and *Rhynchophorus palmarum*) to cereal seeds (*S. oryzae, S. zeamais, Sitophilus granarius*).

However, in spite of the huge economic importance of these pests, only the genus *Sitophilus*, which destroys up to forty percent of cereal products, has been extensively studied with regard to intracellular symbiosis. Pierantoni first described this model (Pierantoni 1927), then Mansour, Tiegs, and Murray provided details of the endosymbiont behavior throughout the insect's embryonic and postembryonic development (Mansour 1934, 1935; Tiegs and Murray 1938). Recently, we have used fluorescence in situ

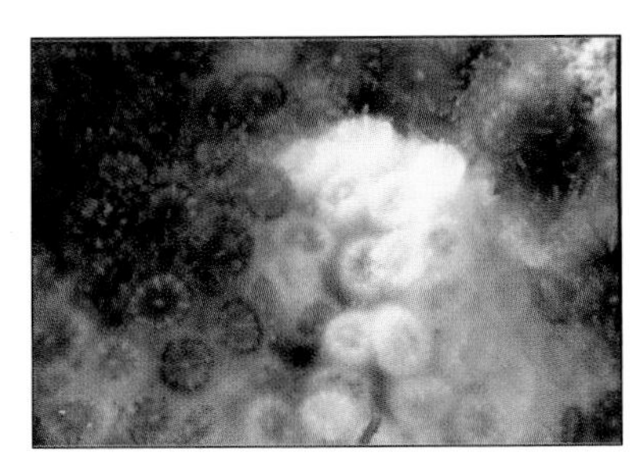

Figure 6.1. A colony of *Oculina patagonica* showing bleached and healthy tissues.

Figure 6.2. Photograph of the fireworm *H. caruncolata* feeding on a coral colony.

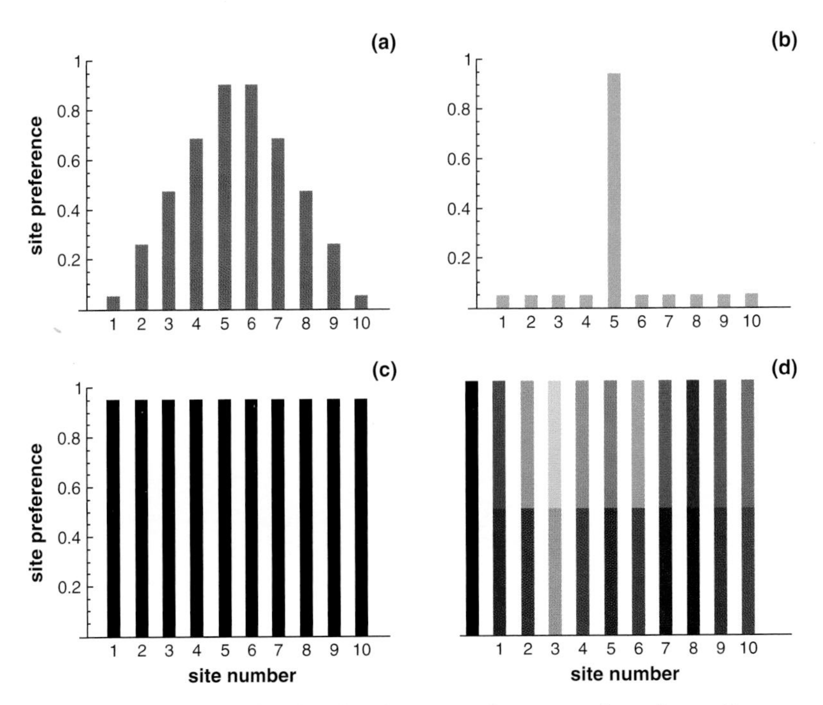

Figure 9.2. Preference profiles for selected species with corresponding colour coding. **(a)** A site-5 semi-generalist. **(b)** A site-5 specialist. **(c)** An optimal generalist. **(d)** Colour coding for species: black is a generalist; light colours are the specialist at the indicated site; darker shades are the semi-generalists at the indicated sites. The maximum and minimum site-preference values are $\lambda_{max} = 0.95$, $\lambda_{min} = 0.05$.

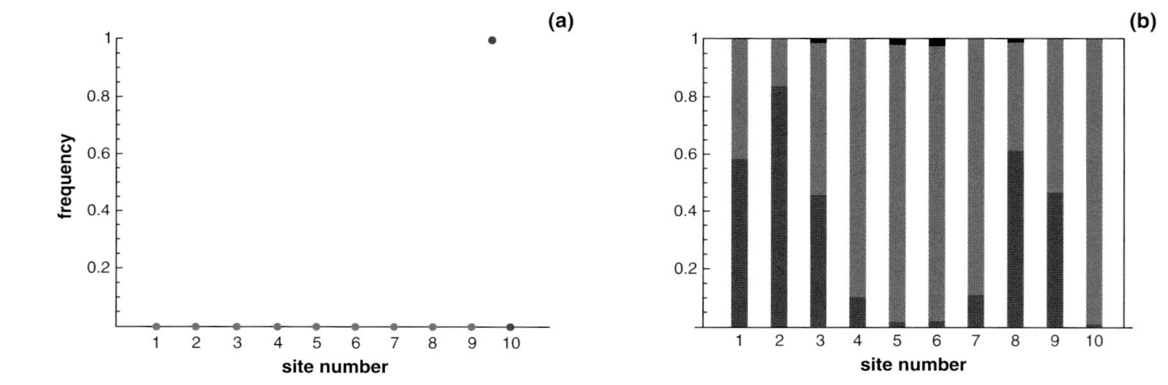

Figure 9.3. **(a)** The stable inter-site distribution $P_{g,i}$ after 250 time steps starting from a uniform distribution of generalists with preference-discrimination trade-off strength $\delta = 3$ and no immune response ($\beta = 0$). The inter-site diversity is $H = 0.039$, which is very low (maximum diversity is $H_{\max} = \ln(21) \approx 3$). **(b)** The intra-site stable distributions at each site. These are dominated by the specialist at site 10, with some representation of the site-8 semi-generalist, and a little of the true generalist. See Appendix A for model and parameters.

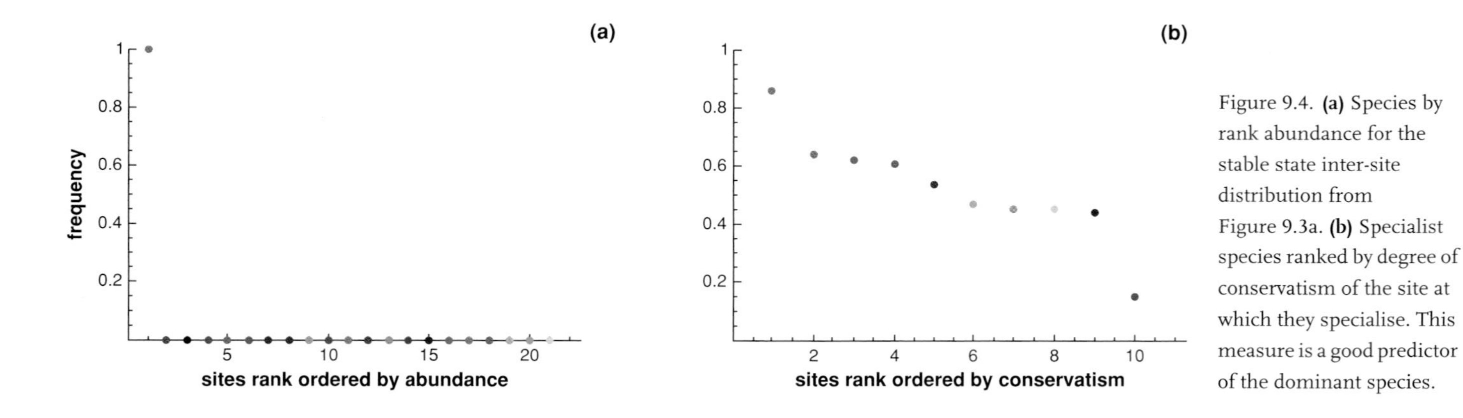

Figure 9.4. **(a)** Species by rank abundance for the stable state inter-site distribution from Figure 9.3a. **(b)** Specialist species ranked by degree of conservatism of the site at which they specialise. This measure is a good predictor of the dominant species.

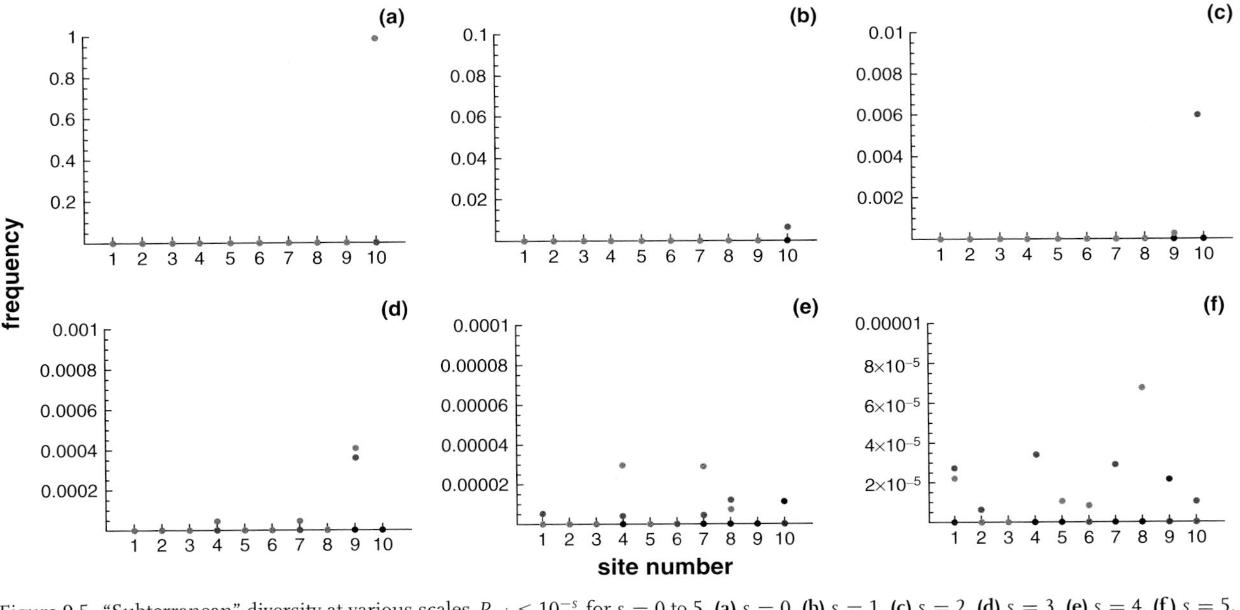

Figure 9.5. "Subterranean" diversity at various scales $P_{g,i} \leq 10^{-s}$ for $s = 0$ to 5. **(a)** $s = 0$. **(b)** $s = 1$. **(c)** $s = 2$. **(d)** $s = 3$. **(e)** $s = 4$. **(f)** $s = 5$. There is only a very little subterranean diversity at the lower scales, and these are largely represented by the small populations of the dominant species at sites other than 10.

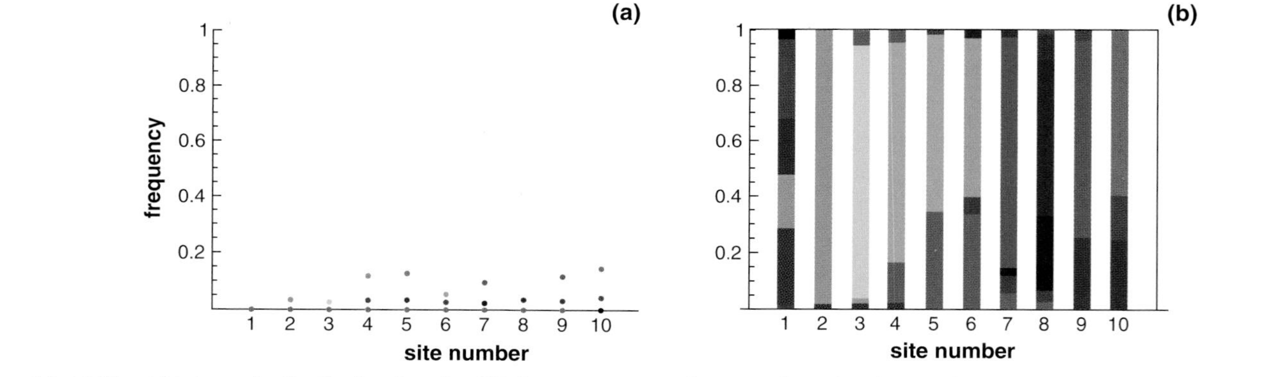

Figure 9.6. **(a)** The stable inter-site distribution $P_{g,i}$ after 250 time steps starting from a uniform distribution of generalists with preference-discrimination trade-off strength $\delta = 3$, and immune-response strength $\beta = 5$. The inter-site diversity is $H = 2.59$, which is high (maximum diversity is $H_{\max} = \ln(21) \approx 3$). **(b)** The intra-site stable distributions at each site. Most sites are dominated by their own specialist, with some representation of the corresponding semi-generalist, and smaller representation of other species. The exception is site 1, which is more diverse than the other sites.

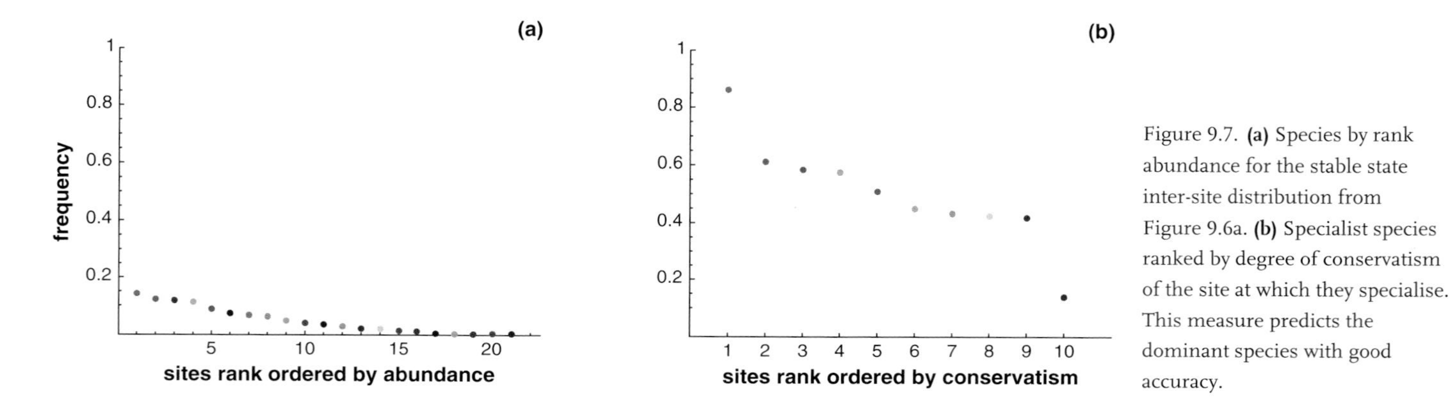

Figure 9.7. **(a)** Species by rank abundance for the stable state inter-site distribution from Figure 9.6a. **(b)** Specialist species ranked by degree of conservatism of the site at which they specialise. This measure predicts the dominant species with good accuracy.

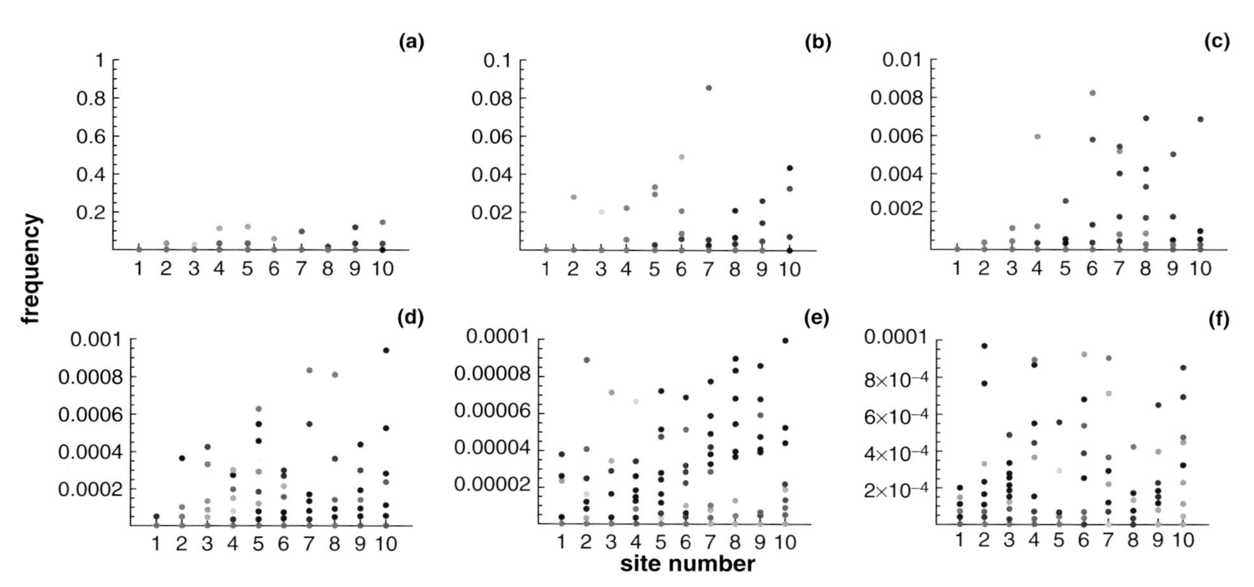

Figure 9.8. "Subterranean" diversity at various scales $P_{g,i} \leq 10^{-s}$ for $s = 0$ to 5. **(a)** $s = 0$. **(b)** $s = 1$. **(c)** $s = 2$. **(d)** $s = 3$. **(e)** $s = 4$. **(f)** $s = 5$. In contrast to the no-immune response case (Fig. 9.5), there is considerable diversity at smaller scales.

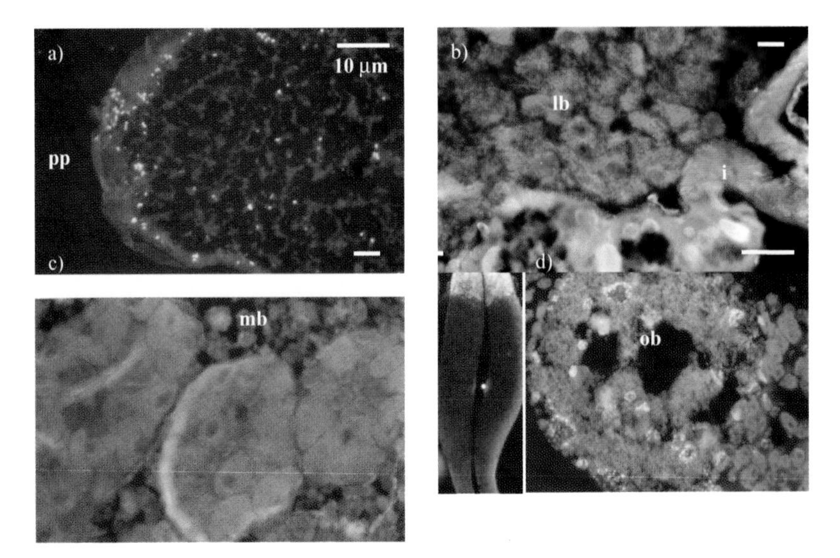

Figure 10.1. Fluorescent in situ hybridization (FISH) of *Sitophilus oryzae* intracellular bacteria. Specific oligonucleotide probes were designed for *S. oryzae primary endosymbiont* (SOPE) 16S rDNA. The SOPE probe was 5′-end labeled with rhodamine. Hybridization was performed as described by Heddi et al. (1999). Slides were mounted in Vectarshield medium containing DAPI. a, Oocyte; b, Larval bacteriome (lb); c, mesenteric bacteriome (mb), d, ovary bacteriome (ob). pp, posterior pole. Scale bars = 10 μm.

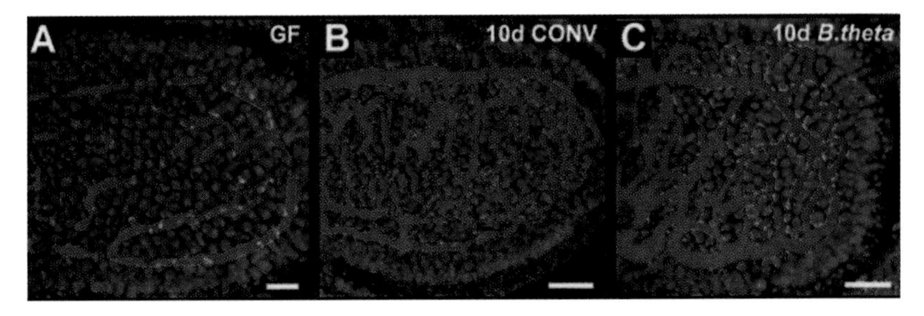

Figure 12.4. Bacterial induction of intestinal angiogenesis. Confocal scans of the capillary network in the upper third of small intestinal villi of (**A**) a germ-free mouse, (**B**) a mouse associated with a conventional (conv) microflora for 10 days, and (**C**) a mouse colonized for 10 days with *B. thetaiotaomicron* (*B. theta*). Blood vessels were detected by retro-orbital injection of FITC-conjugated dextran, followed by visualization of villi by confocal microscopy (bars = 25 μm) (Stappenbeck et al. 2002). Note that the colonized mice show a higher-density capillary network. (Figure originally published in the *Proceedings of the National Academy of Sciences USA* [Stappenbeck et al. 2002]).

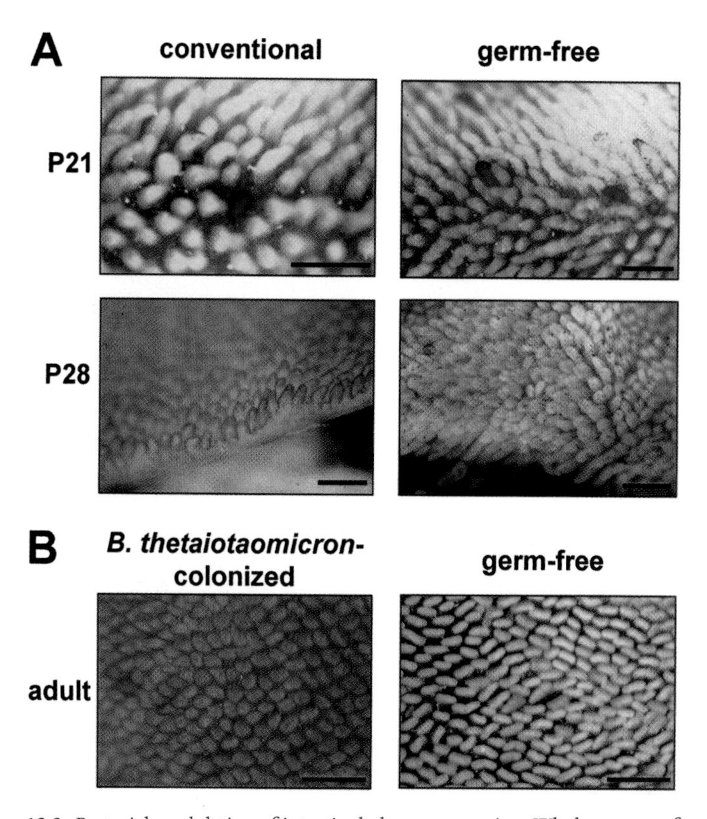

Figure 12.2. Bacterial modulation of intestinal glycan expression. Wholemounts of small intestines from postnatal day 21 (P21), P28, and adult mice were stained with peroxidase-conjugated *Ulex europeaeus* agglutinin. This lectin reacts with terminal α1,2-linked fucose produced by epithelial cells overlying finger-like villi, producing a brown staining pattern. Note that mice undergo weaning during P14–21. Scale bars = 0.5 mm. (**A**) Intestinal bacteria are required to initiate the developmentally regulated program of fucosylated glycan production, which is complete by P28. (**B**) Fucosylated glycans are induced in adult germ-free mice upon colonization with a single member of the indigenous gut flora, *B. thetaiotaomicron* (Bry et al. 1996; Hooper et al. 1999). (Figure originally published in *Glycobiology* [Oxford University Press] [Hooper and Gordon 2001]).

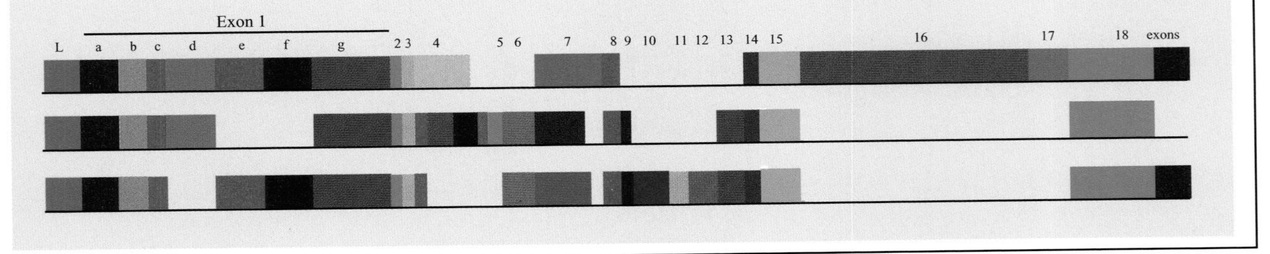

Figure 14.6. Alignment of three full-length *185/333* cDNA sequences. The amino acids deduced from the cDNA sequences were used to generate the alignment with Bioedit. Colored blocks represent exons, and missing blocks represent regions of deleted sequence due to alternative splicing. A leader and eighteen exons were identified from 81 full-length sequences. Exon 1 is shown as a set of subexons, "a" through "g," based on alternative splicing from within the exon as identified by alignments and by cryptic splice sites (see Fig. 14.7). Exons 4 and 7 are shown in multiple colors to indicate significant sequence variability. Exon 18 is shown in two colors of blue to denote a nucleotide variation that introduces an early stop codon.

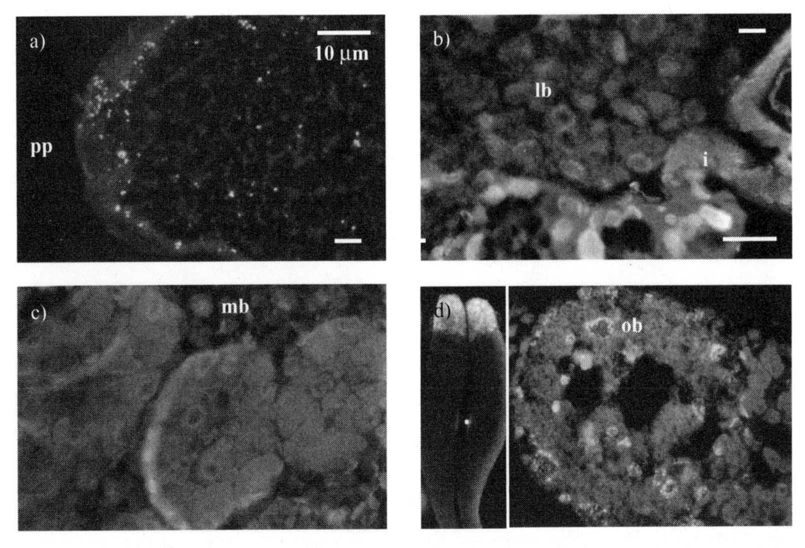

Figure 10.1. Fluorescent in situ hybridization (FISH) of *Sitophilus oryzae* intracellular bacteria. Specific oligonucleotide probes were designed for *S. oryzae* primary endosymbiont (SOPE) 16S rDNA. The SOPE probe was 5′-end labeled with rhodamine. Hybridization was performed as described by Heddi et al. (1999). Slides were mounted in Vectarshield medium containing DAPI. a, Oocyte; b, Larval bacteriome (lb); c, mesenteric bacteriome (mb), d, ovary bacteriome (ob). pp, posterior pole.
Scale bars = 10 µm.

hybridization (FISH) in order to define the behavior of this endosymbiont (Heddi et al. 1999). The bacteria are concentrated at the posterior pole of oocytes and young embryos (Fig. 10.1a), in close association with the germ cells. As embryo cleavage begins, the endosymbionts start moving to the central part of the embryo. Bacteriome differentiation occurs on the third day, and the complete organ is structured by the first instar larva (Fig. 10.1b). Bacteriomes are intimately attached to the intestine but without any direct communication with the gut. From this developmental phase, meticulous FISH screening throughout the whole larval body failed to reveal any bacterial cell outside the bacteriocytes, which suggests that the endosymbiont is recognized as "pathogenic" outside the bacteriocytes and predicts as well the existence of specific immune regulation inside these cells. The bacteriome organ remains attached to the junction of the fore- and midgut during the fourth larval instars and increases in size. In the young nymph, through an unknown phenomenon but probably linked to the metamorphosis, the larval bacteriome dissociates and bacteriocytes (and perhaps free bacteria) migrate back to the midgut where they form in young adults new mesenteric ceca bacteriomes (Fig. 10.1c) that disappear during third week, suggesting that the endosymbiont role is essential during insect growth and development. Nevertheless, symbionts are permanent in the female germ cells (Fig. 10.1d), from which they are transmitted to the offspring.

10.2.1. Physiological Impact of the Endosymbionts

Nardon (and collaborators) have extensively studied the biological impact of the endosymbiont on weevils. He first succeeded in obtaining an aposymbiotic strain from a wild-type one by treating the insect at 35°C for one month (Nardon 1973), which opened the field of comparative physiology of this endosymbiont. The endosymbiont elimination was shown to result in diminished fertility, increased development time, and flying ineptitude of the aposymbiotic adult (Nardon and Grenier 1988; Grenier et al. 1994). Soon after, biochemical studies reported that these physiological impairments were related to nutrition. Aposymbiotic and symbiotic strain comparisons have shown that symbiotic bacteria provide their host with pantothenic acid, biotin, and riboflavin (Wicker 1983), data that have been recently confirmed by using microarray technology with *Escherichia coli* (Rio et al. 2003). As these vitamins are involved in mitochondrial metabolism (respiratory chain and Krebs cycle), experiments were investigated to determine the impact of symbiosis on energy production. As a result, the *Sitophilus* endosymbiont was shown to increase mitochondrial respiratory control and ADP/O ratios (Heddi et al. 1991), as well as several mitochondrial activities (Heddi et al. 1993; Heddi et al. 1999). We have concluded therefore that the endosymbiont represents a third genome implicated in ATP production in weevil cells (Heddi et al. 2001). By increasing mitochondrial capacities, the bacteria interact with the physiology (development and fertility) and the behavior (flight ability) of the insect. The flight ability of aposymbiotic adults is indeed reduced by eighty percent (as compared with that of wild-type insects), which may have a significant consequence on insect diversification and invasive power (Heddi 2003).

10.2.2. Evolution of Endosymbiosis within the *Dryophthoridae* Family

The first molecular phylogeny aimed at placing the *Sitophilus oryzae* primary endosymbiont (SOPE) among free-living bacteria and several insect intracellular bacteria. The phylogenetic analysis based on bacterial 16S rDNA sequencing has indicated that SOPE belongs to the γ-proteobacteria group, closely related to the free-living bacteria *Erwinia herbicola* and *E. coli*, and to *Buchnera* and *Sodalis*, the primary endosymbiont of aphids and the secondary endosymbiont of tsetse, respectively (Heddi et al. 1998). Moreover, genomic G+C content analysis has revealed that SOPE exhibits an "extracellular molecular profile" in that fifty-four percent of its genomic bases are G+C, whereas most of intracellular bacteria are A+T biased.

Therefore, at least two aspects indicate that SOPE may have established recently as a symbiotic association with *Sitophilus*: (i) *molecular*, the relatively high genomic G+C content and (ii) *physiological*, the possibility of obtaining viable aposymbiotic insects; most symbiotic insects, including aphids, mealybugs, tsetse, and ants, cannot survive without their primary endosymbionts.

To gain more precise information about the age of *Sitophilus* symbiosis and to investigate the occurrence of bacterial intracellular symbiosis through the Dryophthoridae, the family of *Sitophilus*, we have collected worldwide and analyzed cytologically and phylogenetically nineteen Dryophthoridae species (Nardon et al. 2002; Lefèvre et al. 2004). All except *Sitophilus linearis* were shown to bear symbiotic bacteria within the bacteriome structure, which does not exhibit any diversity with regard to the shape and the location, always being located in the anterior part of larvae (Nardon et al. 2002). Phylogenetically, three endosymbiotic clades, named R, S, and D, were defined by a heterogeneous model of DNA evolution (Fig. 10.2). The R clade exhibits a much higher relative rate of substitution, has a relatively high A+T bias (40.5 percent G+C on the 16S rDNA), and contains six out of nine genera studied, including the putative ancestral species *Yaccaborus frontalis*. It was therefore assigned as the oldest clade, on which the ancestral bacterial infection occurred around 100 million years ago. The genus name *Candidatus* Nardonella was ascribed to this clade of endosymbionts (Lefèvre et al. 2004). Conversely, the S and the D clades are restricted to only one and two genera, respectively. Furthermore, these clades do not show any A+T bias or significantly high rates of evolution. These determinations, along with the dating estimation (O'Meara and Farrell, personal communication), suggest that the S and the D clade endosymbionts established symbiosis with the Dryophthoridae recently, probably by endosymbiont replacement (Lefèvre et al. 2004).

Interestingly, both the S and the D clades are closely related to the free-living parasitic bacteria *Haemophilus* and *Pasteurella* (Fig. 10.2). This indicates that endosymbiosis may have evolved from an ancestor with a parasitic lifestyle within these clades. In keeping with these findings, Dale et al. (2002)

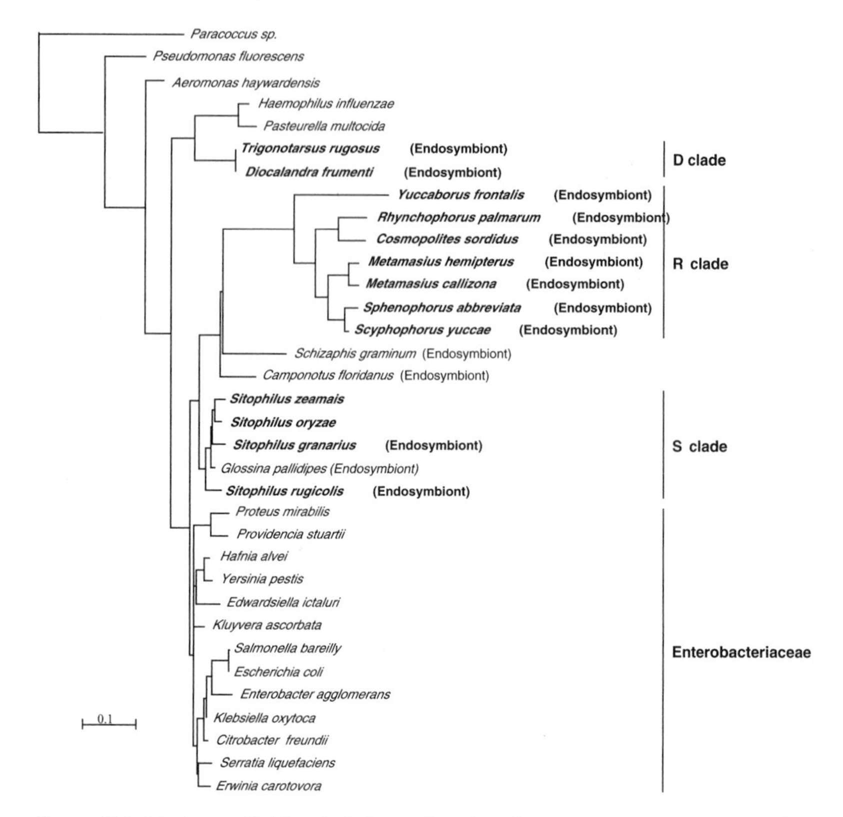

Figure 10.2. Maximum likelihood phylogeny based on the 16S rDNA gene sequence of the Dryophthoridae endosymbionts, other insect endosymbionts, and free-living bacteria. The phylogenetic tree was constructed as described in Lefèvre et al. (2004). *Paracoccus* sp. was taken as an outgroup. Three endosymbiont clades are represented (S, D, and R). Modified from Lefèvre et al. (2004).

have recently identified in SOPE, the endosymbiont of the S clade, a type-three secretion system (TTSS) similar to those of the pathogenic bacteria *Salmonella* and *Shigella*. In Gram-negative bacteria, such secretion systems are required for protein export and toxin delivery into the host cells (see Chapter 11 for more detailed discussion of TTSSs). We must emphasize here that all insect endosymbiont genomes sequenced so far (i.e., *Buchnera, Wigglesworthia, Blochmannia, Baumannia*), except *Sodalis* (Dale et al. 2001), the closely related bacteria to SOPE, do not encode such a secretion system. Therefore, the question that arises is how does such a beneficial relationship become established between potentially defensive hosts and potentially virulent bacteria in the initial stage of symbiogenesis?

10.3. INNATE IMMUNITY IN INSECTS

The renewal of interest in innate immunology is mainly a result of the recent data obtained from *Drosophila* genome sequencing as well as its role as the first line of defense against microbial invaders of mammals (see Chapter 15 for details of the innate host bacterial recognition systems and Chapter 17 for discussion of the NF-κB transcriptional control system).

Insect innate immunity is expressed in two principal ways against microorganisms: cellular and humoral. The former consists mainly of phagocytosis by plasmatocytes (macrophage-like cells) and encapsulation by the lamellocytes. The latter involves proteolytic cascades that result in local production of melanin and the synthesis and secretion in the hemolymph of antimicrobial peptides (Hoffmann and Reichhart 2002).

10.3.1. Bacterial Recognition by the Insect Humoral System

The *Drosophila* humoral system does not constitutively synthesize most of its antibacterial peptides, and the antimicrobial response is now recognized to be selective, depending on the nature of the invading pathogen (Gram-negative bacteria, Gram-positive bacteria or fungi). This "specific" recognition is the result of the existence of pattern-recognition receptors (PRRs) that are associated with pathogen-associated molecular patterns (PAMPs – see Chapters 2, 14, and 15 for more details). Two receptor types have been identified: the Gram-negative bacteria-binding proteins (GNBPs) that are specific to invertebrates and bind to lipopolysaccharide (LPS) as well as to β-1,3-glucans (Kim et al. 2000), and peptidoglycan recognition proteins (PGRPs) that recognize both bacterial peptidoglycan and LPS (Werner et al. 2003; Kurata 2004).

During the last decade, much information on the PGRPs has become available. This gene family encodes proteins well conserved throughout the animal kingdom. *PGRP* genes are expressed in immunocompetent organs of insects (fat body, intestine, hemocytes) but also in the thymus, lung, and liver of mammals (Kang et al. 1998). In the lepidopteran *Trichoplusia ni*, the *PGRP* gene exhibits twenty-eight percent sequence identity and fifty percent similarity with the bacteriophage T3 lysozyme (N-acetylmuramoyl-L-alanine amidase). Nevertheless, and in contrast to several insect *PGRPs* (Kim et al. 2003) and to the phage lysozyme, *T. ni* PGRP protein does not seem to possess the amidase function that cleaves the peptidoglycan (Kang et al. 1998). Therefore, invertebrate *PGRPs* may arise from an ancestral phage sequence, which integrated into the insect genome and then duplicated into several gene copies. Throughout evolution, some PGRPs have conserved, but some have lost, the amidase activity (Mellroth et al. 2003).

In the *Drosophila* genome, 13 *PGRP* genes were identified and all share a PGRP domain of approximately 160 amino acids, conserved from insects to humans (Werner et al. 2000; Werner et al., 2003). Outside of this domain, there is little sequence similarity. Based on their general structure, *PGRPs* have been so far separated into two groups (Werner et al. 2000): i- *PGRPs* with long transcripts (1 to 2.5 kb) and without any typical signal peptide (*PGRP-LA, -LB, -LC, -LD, -LE,* and *-LF*), and *PGRPs* with short transcripts (0.6 to 1.25 kb) that are very similar to the genes previously found in moths and mammals (*PGRP-SA, -SB1, -SB2, -SC1A, -SC1B, SC2,* and *SD*). Some of the long *PGRPs* (*PGRP-LA, -LC,* and *-LD*) exhibit a predicted transmembrane region, indicating that they are likely to encode membrane proteins, whereas short *PGRPs* are preceded by a typical signal peptide and the mature products are predicted to be exported proteins.

Thus far, three genes, *PGRP-SA, PGRP-LC,* and *PGRP-LE,* have been studied genetically. The *PGRP-SA* that is associated with the *semmelweis* (*seml*) mutation is involved in the activation of the Toll pathway, an intracellular cascade that triggers antibacterial peptide synthesis principally by Gram-positive bacteria (Michel et al. 2001). *PGRP-LC* and *PGRP-LE* act instead upstream of the IMD (Immune Deficiency) pathway, another cellular cascade that is induced primarily by Gram-negative bacteria and that regulates antibacterial peptide production (Gottar et al. 2002; Takehana et al. 2002). Moreover, *PGRP-LC* is involved in Gram-negative bacterial phagocytosis (Ramet et al. 2002), and the *PGRP-LE* activates the prophenoloxidase cascade in insect larvae (Takehana et al. 2002).

Taken together, these data suggest that an insect-specific response to a variety of microorganisms may result partly from the specificity of these proteins, which evokes a functional analogy to the immunoglobins that are involved in the adaptive immunity of mammals. A detailed discussion of marine invertebrate immune responses is to be found in Chapter 14.

10.3.2. Insect Signaling Pathways

Two signaling pathways have thus far been described in *Drosophila*, Toll and IMD (Fig. 10.3) (Hoffmann and Reichhart 2002). When the invading microorganisms are Gram-positive bacteria or fungi, the transcription of genes encoding the anti–Gram-positive and antifungal peptides is triggered by the Toll pathway, which was initially identified through mutations that affect dorsoventral patterning in the early embryo of *Drosophila* (Lemaitre

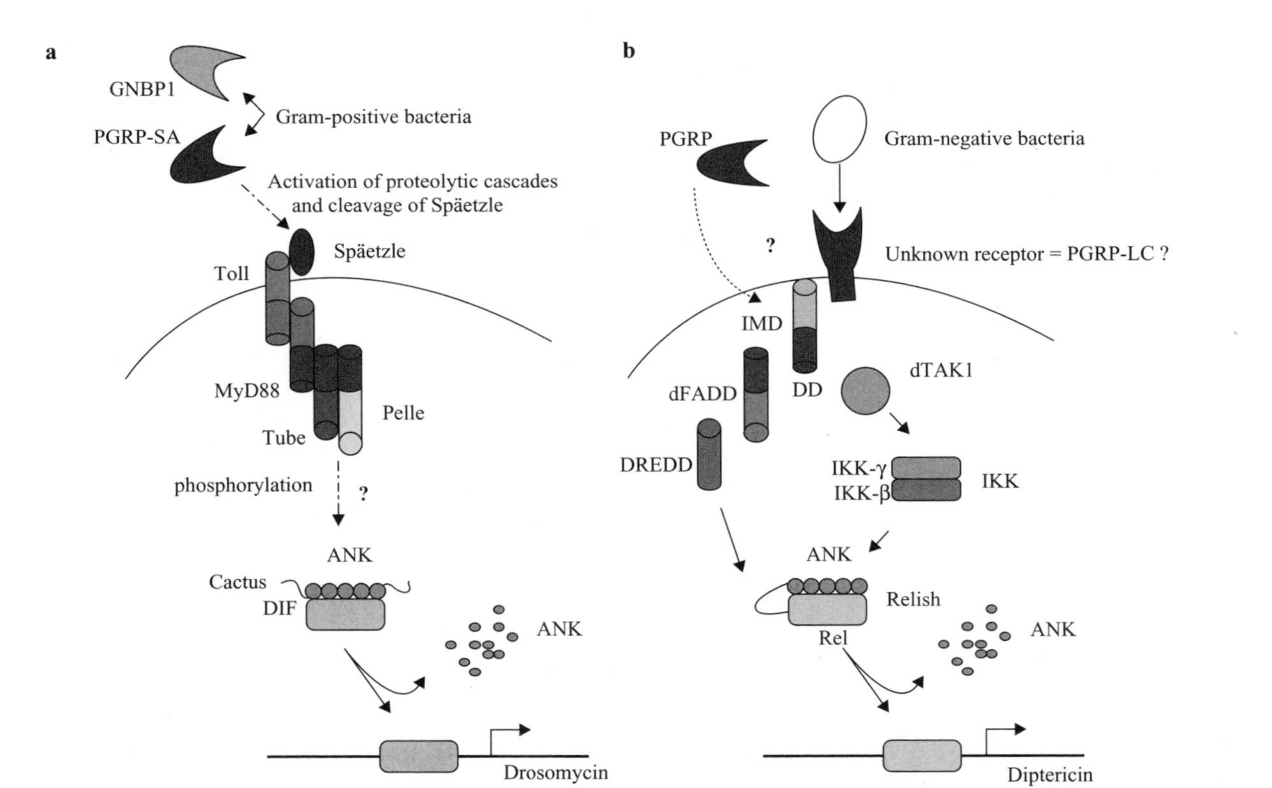

Figure 10.3. *Drosophila* Toll (a) and IMD (b) involved in antimicrobial peptide synthesis. See the text for legend and comments.

et al. 1996). Toll is a transmembrane receptor containing an extracellular leucine-rich repeat (LRR) domain and an intracytoplasmic domain similar to the corresponding region of the interleukin-1 receptor (IL-1R). In *Drosophila*, nine Toll-like receptors have been identified (*Toll*, *18-Wheeler*, and *Toll 3 to 9*), but only *Toll*, *Toll 5*, and *18-Wheeler* were shown to be involved in immunity (Vasselon and Detmers 2002). Challenging *Drosophila* with Gram-positive bacteria or fungi induces an extracellular proteolytic cascade that ends with the cleavage of a protein called Spätzle, which in turn triggers the activation of the Toll receptor. Through a set of phosphorylating reactions involving a complex adaptor (Sun et al. 2002) formed with dMyD88 (Horng and Medzhitov 2001), Tube and Pelle (a protein kinase), an inhibitor protein called Cactus (similar to mammalian I-kB) is degraded, releasing therefore its associated nuclear factor protein called DIF (Dorsal-related Immunity Factor)(Fig. 10.3a). DIF, which is an analogue to the mammalian cytoplasmic transcription factor, NF-κB, and to the *Drosophila* Dorsal factors, translocates into the nucleus and induces antimicrobial peptide genes. See Chapter 17 for more details on NF-κB and inflammatory signaling.

The second insect signaling pathway was identified while studying *Drosophila imd* mutants (Lemaitre et al. 1995). This mutant fails to produce antibacterial peptides after insect infection with Gram-negative bacteria. IMD protein has a death domain similar to that of mammalian RIP (Receptor Interacting Protein) that interacts with the TNF-α (tumor necrosis factor alpha) receptor (Georgel et al. 2001). This may confer on IMD an intermediate role between intracellular cascade proteins and a transmembrane receptor, although no receptor and no extracellular pathway have been characterized so far. Furthermore, considering the intracellular and the transmembrane location of the PGRP-LE and the PGRP-LC proteins (involved in Gram-negative bacterial recognition), respectively, two nonexclusive hypotheses can be put forward:

The IMD pathway does not possess any transmembrane receptor other than the PGRP protein.

The IMD pathway responds to intracellular signals after bacterial invasion into the cell. In this case, *PGRP* genes acting in this way may play a crucial role in insect intracellular symbioses.

As for the intracellular pathway, evidence from the mammalian TNF-α receptor pathway suggests that the IMD protein would be upstream of two separate pathways:

The first would involve a protein from the mitogen-activated protein (MAP) 3 kinase family, the dTAK1 (*Drosophila* transforming growth factor β activated kinase 1), that appears as a good candidate for activating the IKK (inhibitor kB kinase) signalosome proteins, which in turn phosphorylate the Relish (Rel family) transcriptional factor (Fig. 10.3b). In contrast to Dorsal and DIF, Relish contains Ankyrin (ANK) domains that are similar to the inhibitor Cactus in addition to the Rel domain (Stoven et al. 2000). As a consequence, Relish may undergo an autoinhibition, and phosphorylation may represent one step of transcriptional factor release.

The second pathway would control the cleavage of Relish. The "*Drosophila* Fas-associated death-domain-containing protein" (dFADD), which is homologous to the mammalian adaptor protein that interacts with the complex "tumor necrosis factor receptor 1" (TNF-R1) to recruit procaspase-8, would link IMD to the caspase "death-related ced-3/Nedd2-like" (DREDD) in order to build the "adaptor" complex allowing the activation of caspases (Leulier et al. 2000; Leulier et al. 2002). This pathway may end with a proteasome-independent proteolytic cleavage of Relish, probably by the DREDD protein (Stoven et al. 2000; Stoven et al. 2003). The Relish cleavage dissociates the Rel and the Ankyrins and allows the processing of the nuclear transcriptional factor (Fig. 10.3b). See Chapter 17 for more details.

10.3.3. Antibacterial Peptides, the Target Molecules

Regardless of the signaling pathway being induced, the cascade reactions end in the synthesis by the insect fat body of the humoral effectors that are antibacterial peptides. Seven types of antibacterial peptides are known in *Drosophila*, and many others have been found in invertebrates (Mitta et al. 2000). They are small peptides with twenty to fifty amino acids and are separated according to their activity spectrum: antibacterial Gram positive (Defensin), antibacterial Gram negative (Diptericin, Attacin, Drosocin, Cecropin), and antifungal (Metchnikowin, Drosomycin) (Tzou et al. 2002; Hoffmann and Reichhart 2002).

10.4. CONCLUSIONS

Intracellular symbiosis in the Dryophthoridae family, and particularly in *Sitophilus* spp. weevils, in which the association is recent, provides an appropriate model for studying the role of innate immunity in bacterial control by animals as well as for dissecting the early cell–cell molecular dialogue that evolves toward a beneficial relationship.

First, as emphasized above, endosymbionts are transmitted to the offspring through oocytes and are involved as an epigenetic factor in the developmental program of the insect. In *Sitophilus* spp., SOPE interacts with the embryonic cells and "induces" by the third day after the egg laying the differentiation of the bacteriocytes that form the bacteriome. From this developmental time, which may correspond to the induction of insect immune genes, bacteria are eliminated from the whole larvae, except from the germ cells and the bacteriocytes. Before nymphosis, the larval bacteriome dissociates and many bacteriocytes degenerate while some migrate to the mesenteric caeca. In three-week-old adults, the caeca bacteriomes dissociate and again bacteria are eliminated. Moreover, genetic crosses and back-crosses have demonstrated that the host controls genetically the bacterial density in *Sitophilus* weevils (Nardon et al. 1998). Taken together, these findings suggest the occurrence of an embryonic cell–cell interaction, leading to the differentiation of the bacteriocyte, and the existence of a postembryonic regulation of the endosymbiont density and location, probably through a selected immunological program (Heddi 2003).

Second, among insect endosymbionts, only SOPE and *Sodalis* genomes (continue to) encode the pathogenic island of the TTSS (Dale et al. 2001; Dale et al. 2002), which is probably related to their relatively recent integration into the host (around 20 million years ago) (Lefèvre et al. 2004). Moreover, the recent data on the Dryophthoridae endosymbiont phylogeny show that SOPE is closely related to the parasitic free-living bacteria, such as *Haemophilus* and *Pasteurella* (Lefèvre et al. 2004) (Fig. 10.2). Hence, SOPE seems to possess infectious abilities and virulence faculties that allow it to penetrate cells and perhaps cause cell damage as well (Heddi 2003). However, one might conjecture that SOPE virulence genes should be down-regulated (or depressed), at least at some critical phases, in order to permit the establishment and the maintenance of such a "mutualistic association" with the weevils. This condition seems to occur within the bacteriome as no evidence of host cell damage has been reported so far except for the work of Dale et al. (2002) describing the induction of virulence genes (*invA* and *spaPQ*) during weevil nymphosis.

Although SOPE virulence gene repression (or noninduction) looks crucial for intracellular symbiosis, the current data from *Sitophilus* do not resolve whether the inhibition is exerted by the bacterium itself or by the host system. Nevertheless, an analysis based on the bacterial quorum sensing (QS), which is the ability of a microorganism to perceive and respond to microbial population density, and its manipulation by the host system supports the latter hypothesis (see Chapter 8 for more details). For instance, symbiotic bacteria in the weevil's bacteriocyte reach such a high density that if bacterial QS is functional, and if it triggers virulence, it would induce pathogenesis. In this case, one of the possible host manipulations may consist of the catabolism of the bacterial signal molecules (like acyl-homoserine lactone (AHLs) involved in the QS in order to maintain their concentration below the critical threshold. Such a QS manipulation by the host system was recently reported in the legume model *Medicago truncatula*, in which the plant responds to bacterial AHLs by the accumulation of more than 150 proteins (Mathesius et al. 2003).

The other side of host–bacterial interactions involves the host response, to the bacteria. To form a successful symbiotic relationship, the host may turn down its immune response, particularly in the symbiotic structures (i.e., the bacteriome). Immunodepression could be driven by the host itself or by a cocktail of molecules secreted by the bacteria and directed toward the immune reactions that occur within the bacteriocytes. Nevertheless, one should emphasize that total inhibition of the immunity may result in an uncontrolled bacterial proliferation, which is basically not the case in the weevils as the bacterial density is genetically controlled (Nardon et al. 1998). Regulation may occur at different levels of the Toll and IMD pathway reactions. For example, it has been shown in *Drosophila* that some bacteria are able to inhibit the antibacterial peptide synthesis by the LPS (lipopolysaccharide), upstream of the Relish factor activation. The molecular mechanism is currently unknown but requires a minimum of twenty bacteria per cell (Lindmark et al. 2001). Another mechanism was reported by Orth (2002) in *Yersinia enterocolitica*. This bacterium secretes through the TTSS a virulence factor called YopJ (a cysteine protease that cleaves a reversible posttranslational modification in the form of ubiquitin or a ubiquitin-like protein), which inhibits the host immune response and induces apoptosis by blocking multiple signaling pathways, including the MAP kinase and NF-κB pathways in the infected cell. Chapter 17 discusses in more detail the ability of bacteria to interfere with inflammatory signaling via NF-κB.

The immune processes involved in the regulation of insect intracellular symbioses are currently unknown. In the Dryophthoridae weevils, we have obtained phylogenetic data that support a pathogenic-like origin of symbiotic bacteria (Lefèvre et al. 2004). Moreover, the molecular data from the "recent" Dryophthoridae endosymbiotic clade (i.e., *Sitophilus*) have shown that the bacteria (i.e., SOPE) encode the TTSS (Dale et al. 2002) and that the host system induces a humoral response against the bacteria (Heddi et al., unpublished result). Therefore, one of the crucial and striking questions would deal with the establishment of cooperative relationships between pathogenic-like bacteria and potentially defensive host insects. We are currently investigating this field by the use of comparative and functional genomics in order to understand the origin of pathogenicity through mutualism and vice versa.

ACKNOWLEDGMENTS

The authors would like to thank the Rockefeller Foundation for supporting and encouraging this field of research, Cédric Lefèvre for the phylogenetic construction, and Margaret McFall-Ngai for the critical reading of the manuscript.

This work was supported by the "Institut National des Sciences Appliquées" (INSA de Lyon) and the Institut National de la Recherche Agronomique (INRA).

REFERENCES

Alonso-Zarazaga, M. A. and Lyal, C. H. C. (1999). A world catalogue of family and genera of Curculionoidea (Insecta, Coleoptera) (Excepting Scolytidae and Platipodidae). Barcelona: Entomopraxis. p. 315.

Baumann, P., Baumann, L., Lai, C.-Y., Rouhbakhsh, D., Moran, N. A., and Clark, M. A. (1995). Genetics, physiology, and evolutionary relationships of the genus *Buchnera*: Intracellular symbionts of aphids. *Annual Review of Microbiology* **49**, 55–94.

Charles, H. and Ishikawa, H. (1999). Physical and genetic map of the genome of *Buchnera*, the primary endosymbiont of the pea aphid *Acyrthosiphon pisum*. *Journal of Molecular Evolution* **48**, 142–150.

Dale, C., Plague, G. R., Wang, B., Ochman, H., and Moran, N. A. (2002). Type III secretion systems and the evolution of mutualistic endosymbionsis. *Proceedings of the National Academy of Sciences USA* **99**, 12397–12402.

Dale, C., Young, S. A., Haydon, D. T., and Welburn, S. C. (2001). The insect endosymbiont *Sodalis glossinidius* utilizes a type III secretion for cell invasion. *Proceedings of the National Academy of Sciences USA* **98**, 1883–1888.

Douglas, A. E. (1995). Reproductive failure and the free amino acid pools in pea aphids (*Acyrthosiphon pisum*) lacking symbiotic bacteria. *Journal of Insect Physiology* **42**, 247–255.

Georgel, P., Naitza, S., Kappler, C., Ferrandon, D., Zachary, D., Swimmer, C., Kopczynski, C., Duyk, G., Reichhart, J. M., and Hoffmann, J. A. (2001). *Drosophila* immune deficiency (IMD) is a death domain protein that activates antibacterial defense and can promote apoptosis. *Developmental Cell* **1**, 503–514.

Gottar, M., Gobert, V., Michel, T., Belvin, M., Duyk, G., Hoffmann, J. A., Ferrandon, D., and Royet, J. (2002). The *Drosophila* immune response against Gram-negative bacteria is mediated by a peptidoglycan recognition protein. *Nature* **416**, 640–644.

Grenier, A. M., Nardon, C., and Nardon, P. (1994). The role of symbiotes in flight activity of *Sitophilus* weevils. *Entomology Experimental and Applied* **70**, 201–208.

Heddi, A. (2003). Endosymbiosis in the weevil of the genus *Sitophilus*: Genetic, physiological, and molecular interactions among associated genomes. In *Insect Symbiosis,* eds. K. Bourtzis and T. Miller, pp. 67–82. Boca Raton/London/New York/Washington DC: CRC Press.

Heddi, A., Charles, H., and Khatchadourian, C. (2001). Intracellular bacterial symbiosis in the genus *Sitophilus:* the 'biological individual' concept revisited. *Research Microbiology* **152**, 431–437.

Heddi, A., Charles, H., Khatchadourian, C., Bonnot, G., and Nardon, P. (1998). Molecular characterization of the principal symbiotic bacteria of the weevil *Sitophilus oryzae*: A peculiar G – C content of an endocytobiotic DNA. *Journal of Molecular Evolution* **47**, 52–61.

Heddi, A., Grenier, A. M., Khatchadourian, C., Charles, H., and Nardon, P. (1999). Four intracellular genomes direct weevil biology: Nuclear, mitochondrial, principal endosymbionts, and *Wolbachia*. *Proceedings of the National Academy of Sciences USA* **96**, 6814–6819.

Heddi, A., Lefebvre, F., and Nardon, P. (1991). The influence of symbiosis on the respiratory control ratio (RCR) and the ADP/O ratio in the adult weevil *Sitophilus oryzae* (Coleoptera, Curculionidae). *Endocytobiosis & Cell Research* **8**, 61–73.

Heddi, A., Lefebvre, F., and Nardon, P. (1993). Effect of endocytobiotic bacteria on mitochondrial enzymatic activities in the weevil *Sitophilus oryzae* (Coleoptera, Curculionidae). *Insect Biochemistry and Molecular Biology* **23**, 403–411.

Hoffmann, J. A. and Reichhart, J. M. (2002). *Drosophila* innate immunity: an evolutionary perspective. *Nature Immunology* **3**, 121–126.

Horng, T. and Medzhitov, R. (2001). *Drosophila* MyD88 is an adapter in the Toll signalling pathway. *Proceedings of the National Academy of Sciences USA* **98**, 12654–12658.

Kang, D., Liu, G., Lundstrom, A., Gelius, E., and Steiner, H. (1998). A peptidoglycan recognition protein in innate immunity conserved from insects to humans. *Proceedings of the National Academy of Sciences USA* **95**, 10078–10082.

Kim, M.-S., Byun, M., and Oh, B.-H. (2003). Crystal structure of peptidoglycan recognition protein LB from *Drosophila melanogaster. Nature Immunology* **4**, 787–793.

Kim, Y. S., Ryu, J. H., Han, S. J., Choi, K. H., Nam, K. B., Jang, I. H., Lemaitre, B., Brey, P. T., and Lee, W. J. (2000). Gram-negative bacteria-binding protein, a pattern recognition receptor for lipopolysaccharide and beta-1,3-glucan that mediates the signaling for the induction of innate immune genes in *Drosophila melanogaster* cells. *Journal of Biological Chemistry* **275**, 32721–32727.

Kurata, S. (2004). Recognition of infectious non-self and activation of immune responses by peptidoglycan recognition protein (PGRP)-family members in *Drosophila. Developmental and Comparative Immunology* **28**, 89–95.

Lefèvre, C., Charles, H., Vallier, A., Delobel, B., Farrell, B., and Heddi, A. (2004). Endosymbiont phylogenesis in the Dryophthoridae weevils: Evidence for bacterial replacement. *Molecular Biology and Evolution*, **21**, 965–973.

Lemaitre, B., Kromer-Metzger, E., Michaut, L., Nicolas, E., Meister, M., Georgel, P., Reichhart, J. M., and Hoffmann, J. A. (1995). A recessive mutation, immune deficiency (*imd*), defines two distinct control pathways in the *Drosophila* host defense. *Proceedings of the National Academy of Sciences USA* **92**, 9465–9469.

Lemaitre, B., Nicolas, E., Michaut, L., Reichhart, J. M., and Hoffmann, J. A. (1996). The dorsoventral regulatory gene cassette spatzle/Toll/cactus controls the potent antifungal response in *Drosophila* adults. *Cell* **86**, 973–983.

Leulier, F., Rodriguez, A., Khush, R. S., Abrams, J. M., and Lemaitre, B. (2000). The *Drosophila* caspase Dredd is required to resist gram-negative bacterial infection. *EMBO Reports* **1**, 353–358.

Leulier, F., Vidal, S., Saigo, K., Ueda, R., and Lemaitre, B. (2002). Inducible expression of double-stranded RNA reveals a role for dFADD in the regulation of the antibacterial response in *Drosophila* adults. *Current Biology* **12**, 996–1000.

Lindmark, H., Johansson, K. C., Stoven, S., Hultmark, D., Engstrom, Y., and Soderhall, K. (2001). Enteric bacteria counteract lipopolysaccharide induction of antimicrobial peptide genes. *Journal of Immunology* **167**, 6920–6923.

Mansour, K. (1930). Preliminary studies on the bacterial cell mass (accessory cell mass) of *Calandra oryzae*: the rice weevil. *Quarterly Journal Microscopical Science* **73**, 421–436.

Mansour, K. (1934). On the so-called symbiotic relationship between Coleopterous insects and intracellular micro-organisms. *Quarterly Journal of Microscopical Science* **77**, 255–272.

Mansour, K. (1935). On the microorganisms free and the infected *Calandra granaria. Bulletin of the Entomological Society of Egypt* **19**, 290–306.

Mathesius, U., Mulders, S., Gao, M. S., Teplitski, M., Caetano Anolles, G., Rolfe, B. G., and Bauer, W. D. (2003). Extensive and specific responses of a eukaryote to bacterial quorum-sensing signals. *Proceedings of the National Academy of Sciences USA* **100**, 1444–1449.

Mellroth, P., Karlsson, J., and Steiner, H. (2003). A scavenger function for a *Drosophila* peptidoglycan recognition protein. *Journal of Biological Chemistry* **278**, 7059–7064.

Michel, T., Reichhart, J.-M., Hoffmann, J. A., and Royet, J. (2001). *Drosophila* Toll is activated by Gram-positive bacteria through a circulating peptidoglycan recognition protein. *Nature* **414**, 756–759.

Mitta, G., Hubert, F., Dyrynda, E. A., Boudry, P., and Roch, P. (2000). Mytilin B and MGD2, two antimicrobial peptides of marine mussels: gene structure and expression analysis. *Developmental and Comparative Immunology* **24**, 381–393.

Nardon, C., Lefèvre, C., Delobel, B., Charles, H., and Heddi, A. (2002). Occurence of endosymbiosis in Dryophthoridae weevils: Cytological insights into bacterial symbiotic structures. *Symbiosis* **33**, 227–241.

Nardon, P. (1971). Contribution à l'étude des symbiotes ovariens de *Sitophilus saskii*: localisation, histochimie et ultrastructure chez la femelle adulte. *Comptes Rendus de l'Académie des Sciences de Paris* **272D**, 2975–2978.

Nardon, P. (1973). Obtention d'une souche asymbiotique chez le charançon *Sitophilus sasakii* Tak: différentes méthodes d'obtention et comparaison avec la souche symbiotique d'origine. *Comptes Rendus de l'Académie des Sciences de Paris* **277D**, 981–984.

Nardon, P. and Grenier, A. M. (1988). Genetical and biochemical interactions between the host and its endosymbiotes in the weevil *Sitophilus* (Coleoptera Curculionidae) and other related species. In *Cell to Cell Signals in Plant, Animal and Microbial Symbiosis*, ed. S. Scannerini et al., pp. 255–270. Berlin: Springer-Verlag.

Nardon, P., Grenier, A. M., and Heddi, A. (1998). Endocytobiote control by the host in the weevil *Sitophilus oryzae*, Coleoptera, curculionidae. *Symbiosis* **25**, 237–250.

Orth, K. (2002). Function of the Yersinia effector YopJ. *Current Opinion Microbiology* **5**, 38–43.

Pierantoni, U. (1927). L'organo simbiotico nello sviluppo di *Calandra oryzae. Rend Reale Acad Sci Fis mat Napoli* **35**, 244–250.

Ramet, M., Manfruelli, P., Pearson, A., Mathey-Prevot, B., and Ezekowitz, R. A. (2002). Functional genomic analysis of phagocytosis and identification of a *Drosophila* receptor for *E. coli*. *Nature* **416**, 644–648.

Rio, R. V. M., Lefèvre, C., Heddi, A., and Aksoy, S. (2003). Comparative genomics of insect-symbiotic bacteria: Influence of host environmental on microbial genome composition. *Appl Environ Microbiol* **69**, 6825–6832.

Shigenobu, S., Watanabe, H., Hattori, M., Sakaki, Y., and Ishikawa, H. (2000). Genome sequence of the endocellular bacterial symbiont of aphids *Buchnera sp.* APS. *Nature* **407**, 81–86.

Stoven, S., Ando, I., Kadalayil, L., Engstrom, Y., and Hultmark, D. (2000). Activation of the *Drosophila* NF-kappaB factor Relish by rapid endoproteolytic cleavage. *EMBO Reports* **1**, 347–352.

Stoven, S., Silverman, N., Junell, A., Hedengren-Olcott, M., Erturk, D., Engstrom, Y., Maniatis, T., and Hultmark, D. (2003). Caspase-mediated processing of the *Drosophila* NF-kappaB factor Relish. *Proceedings of the National Academy of Sciences USA* **100**, 5991–5996.

Sun, H., Bristow, B. N., Qu, G., and Wasserman, S. A. (2002). A heterotrimeric death domain complex in Toll signaling. *Proceedings of the National Academy of Sciences USA* **99**, 12871–12876.

Takehana, A., Katsuyama, T., Yano, T., Oshima, Y., Takada, H., Aigaki, T., and Kurata, S. (2002). Overexpression of a pattern-recognition receptor, peptidoglycan-recognition protein-LE, activates imd/relish-mediated antibacterial defense and the prophenoloxidase cascade in *Drosophila* larvae. *Proceedings of the National Academy of Sciences USA* **99**, 13705–13710.

Tiegs, O. N. and Murray, F. U. (1938). Embryonic development of *Calandra oryzae. Quarterly Journal of Microscopical Science* **80**, 159–284.

Tzou, P., De Gregorio, E., and Lemaitre, B. (2002). How *Drosophila* combats microbial infection: a model to study innate immunity and host-pathogen interactions. *Current Opinion in Microbiology* **5**, 102–110.

van Ham, R. C., Kamerbeek, J., Palacios, C., Rausell, C., Abascal, F., Bastolla, U., Fernandez, J. M., Jimenez, L., Postigo, M., Silva, F. J., et al. (2003). Reductive genome evolution in *Buchnera aphidicola. Proceedings of the National Academy of Sciences USA* **100**, 581–586.

Vasselon, T. and Detmers, P. A. (2002). Toll receptors: a central element in innate immune responses. *Infection and Immunity* **70**, 1033–1041.

von Dohlen, C. D., Kohler, S., Alsop, S. T., and McManus, W. R. (2001). Mealybug beta-proteobacterial endosymbionts contain gamma-proteobacterial symbionts. *Nature* **412**, 433–436.

Werner, T., Borge-Renberg, K., Mellroth, P., Steiner, H., and Hultmark, D. (2003). Functional diversity of the *Drosophila* PGRP-LC gene cluster in the response to lipopolysaccharide and peptidoglycan. *Journal of Biological Chemistry* **278**, 26319–26322.

Werner, T., Liu, G., Kang, D., Ekengren, S., Steiner, H., and Hultmark, D. (2000). A family of peptidoglycan recognition proteins in the fruit fly *Drosophila melanogaster*. *Proceedings of the National Academy of Sciences USA* **97**, 13772–13777.

Wicker, C. (1983). Differential vitamin and choline requirements of symbiotic and aposymbiotic *Sitophilus oryzae* (Coleoptera: Curculionidae). *Comparative Biochemistry and Physiology* **76A**, 177–182.

CHAPTER 11

Type III secretion in *Bordetella*–host interactions

Seema Mattoo and Jeffrey F. Miller

11.1. INTRODUCTION

The respiratory system of a healthy immunocompetent host has, it is believed, evolved to eliminate pathogenic microbes and restrict colonizing flora to the upper airways. This assertion is supported by the fact that serious respiratory infections are rare, and the assumption is that an equilibrium is maintained between the host and its normal "respiratory" microbiota. The Gram-negative respiratory pathogen *Bordetella bronchiseptica* has evolved sophisticated mechanisms to overcome host defenses in order to successfully colonize the ciliated respiratory epithelial surfaces of a variety of mammalian hosts. *B. bronchiseptica* belongs to a cluster of four closely related subspecies that share a nearly identical virulence control system encoded by the *bvgAS* locus. They also express a common set of surface and secreted molecules involved in colonization and virulence. They differ, however, in a variety of characteristics, including host specificity, severity of disease, ability to establish persistent infection, and perhaps, even pathways for transmission. Differential regulation of gene expression appears to be a primary determinant of subspecies phenotypes. As a result of their extremely high degree of genetic relatedness, comparative studies of the similarities and differences in the infectious cycles of *B. bronchiseptica* and the other *Bordetella* subspecies serve as a guide to understanding fundamental features of bacterial–host interactions and the evolution of pathogenesis. Differences in the expression of type III secretion may be particularly important in this regard.

11.2. PHYLOGENETIC RELATIONSHIPS BETWEEN *B. BRONCHISEPTICA* AND OTHER *BORDETELLA* SUBSPECIES

Bordetella bronchiseptica infects a wide range of mammals, causing respiratory diseases that include atrophic rhinitis in pigs, bronchopneumoniae

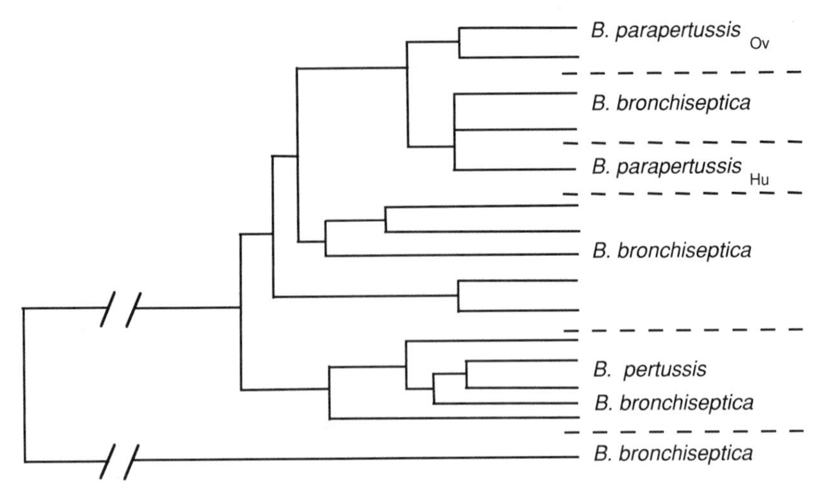

Figure 11.1. Phylogenetic relationships among *Bordetella* subspecies. The dendrogram suggests that *B. pertussis* and *B. parapertussis* descended from a *B. bronchiseptica*–like ancestor. It also shows that *B. parapertussis*$_{\mathrm{hu}}$ and *B. parapertussis*$_{\mathrm{ov}}$ evolved as distinct clonal lineages.

in rabbits, and kennel cough in dogs (Deeb et al. 1990; Keil et al. 1999; Magyar et al. 1988). Human infections are rare, occurring only in severely immunocompromised individuals (Dworkin et al. 1999; Tamion et al. 1996). As with many pathogenic bacteria, *disease is the exception rather than the rule.* Epidemiological and laboratory studies show that infections established by *B. bronchiseptica* are chronic and often asymptomatic. The ability of this organism to establish long-term asymptomatic infections seems to be adaptive, representing a balance between immunostimulatory events associated with infection and immunomodulatory events mediated by the bacterium (Harvill et al. 1999; Mattoo et al. 2001; Yuk et al. 2000).

The dendrogram in Figure 11.1 depicts the phylogeny of various *Bordetella* strains based on a combination of multilocus enzyme electrophoresis (MLEE), insertion sequence (IS) polymorphisms, sequence data, and microarray-based comparative genome hybridization (van der Zee et al. 1997; Gerlach et al. 2001; von Wintzingerode et al. 2002; Parkhill et al. 2003; Cummings et al. 2004). It demonstrates that *B. pertussis* and *B. parapertussis* are most closely related to *B. bronchiseptica. Bordetella pertussis* is a strictly human pathogen with no known animal or environmental reservoir. It is the etiological agent of whooping cough (or pertussis), an acute respiratory disease in children, and a major cause of respiratory infections in adolescents and adults. Classical pertussis is characterized by severe paroxysmal coughing accompanied by leukocytosis and lymphocytosis and with dangerous, even lethal, complications such as encephalopathy (Cherry and Heininger 1999; Yeh 2003). The disease is highly contagious, and transmission is thought to occur primarily by respiratory droplets (Mattoo et al. 2001). Despite extensive vaccination programs, whooping cough is still endemic in many countries and caused an estimated 285,000 deaths in 2001 (Parkhill et al. 2003). *B. parapertussis* infects both humans and sheep. *B. parapertussis*$_{hu}$ is restricted to humans and causes a milder pertussis-like respiratory illness not accompanied by lymphocytosis (Heininger et al. 1994). *B. parapertussis*$_{ov}$ causes a chronic infection of the ovine respiratory tract. There appears to be little or no transmission between the sheep and human reservoirs. Whereas *B. parapertussis*$_{ov}$ isolates are genetically diverse, *B. parapertussis*$_{hu}$ isolates are genetically uniform, and all evidence suggests that human and ovine strains of *B. parapertussis* represent distinct clonal lineages that diverged independently (Parkhill et al. 2003). The remarkably limited genetic diversity among *B. pertussis, B. parapertussis,* and *B. bronchiseptica* strains has led to their reclassification as "subspecies" of a single species with different host adaptations. *B. pertussis* seems to have made the earliest switch toward human adaptation, followed more recently by *B. parapertussis*$_{hu}$. In both cases, *B. bronchiseptica* or a *B. bronchiseptica*–like organism is the likely evolutionary progenitor. *B. pertussis* and *B. parapertussis*$_{hu}$ are, therefore, considered to be human-adapted lineages of *B. bronchiseptica*. These subspecies together comprise the "*B. bronchiseptica* cluster" (Gerlach et al. 2001).

The genomes of three *Bordetella* subspecies (*B. pertussis* strain Tohama 1, *B. parapertussis*$_{hu}$ strain 12822, and *B. bronchiseptica* strain RB50) were recently sequenced by the Sanger Center (Parkhill et al. 2003; Preston et al. 2004). Comparative analysis of the genomes supports the notion that *B. pertussis* and *B. parapertussis* recently and independently evolved from *B. bronchiseptica*–like ancestors by a process that included significant reductions in genome size. The genome of RB50 is 5.34 Mb, whereas that of Tohama 1 and 12822 is 4.09 Mb and 4.77 Mb, respectively. In comparison with Tohama 1 and 12822, a large portion of the extra DNA in RB50 is attributed to prophage and prophage remnants (Parkhill et al. 2003). Other genes lost by *B. pertussis* and *B. parapertussis*$_{hu}$ include loci involved in small-molecule metabolism, membrane transport, and biosynthesis of surface structures. In addition to this substantial gene deletion, *B. pertussis* and *B. parapertussis*$_{hu}$ contain pseudogenes, many of which have been inactivated by insertion of IS elements, in-frame stop codons, or frame shift mutations. Interestingly, very few genes known or suspected to be involved in pathogenicity are missing in the genomes of human-adapted bordetellae.

11.3. THE *BORDETELLA* VIRULENCE REGULON

Bordetella bronchiseptica, *B. pertussis*, and *B. parapertussis* (human and ovine) share a nearly identical virulence control system encoded by the *bvgAS* locus (Weiss and Falkow 1984). BvgA and BvgS are members of a two-component signal transduction system. BvgA is a 23 kd DNA-binding response regulator, and BvgS is a 135 kd transmembrane sensor kinase containing a periplasmic domain, a linker region (L), a transmitter (T), a receiver (R), and a histidine phosphotransfer domain (HPD) (Uhl and Miller 1994 – Fig. 11.2). BvgA and BvgS from *B. pertussis* and *B. bronchiseptica* share one-hundred percent and ninety-six percent amino acid sequence identity, respectively, and the loci are functionally interchangeable (Martinez de Tejada et al. 1996).

BvgAS is environmentally responsive; it can be activated by growth at 37°C in the relative absence of $MgSO_4$ or nicotinic acid. Bordetellae grown under such "nonmodulating" conditions are referred to as Bvg^+ phase bacteria. Signal inputs detected by the periplasmic domain of BvgS are relayed through the membrane as a four-step His-Asp-His-Asp phosphotransfer signaling mechanism that eventually leads to the phosphorylation (and thus activation) of BvgA (Uhl and Miller 1996a,b – Fig. 11.2). BvgA~P promotes transcription of Bvg^+ phase-specific genes called *vags* [*vir* activated genes (*bvgAS* was originally termed *vir*)] by binding to *cis*-acting activation (A) sequences in their promoter regions (Boucher et al. 1997, 2003). An additional class of genes, termed *vrg* (*vir* repressed genes), is repressed by the products of the *bvgAS* locus. The repression of these genes is mediated via a 32 kd BvgAS-activated cytoplasmic repressor protein called BvgR (Merkel et al. 1998, 2003). In addition, *bvgA* can repress transcription of certain genes by binding to *cis*-acting repression sites (R) in their promoter regions (Cotter and Jones 2003). The BvgAS phosphorelay can be inactivated by growing bordetellae under "modulating" conditions, such as at 25°C, or at 37°C in the presence of ≥10 mM nicotinic acid or ≥40 mM $MgSO_4$. Under these Bvg^- phase conditions, BvgAS is unable to activate transcription of *vags* and repress *vrgs*. Although a variety of conditions modulate BvgAS activity *in vitro*, the true signals recognized *in vivo* are unknown.

Experiments with phase-locked and ectopic expression mutants have demonstrated that the Bvg^+ phase is necessary and sufficient for respiratory tract colonization by *B. pertussis* and *B. bronchiseptica*, and that inappropriate expression of Bvg^- phase factors is, in fact, detrimental to the infectious process (Cotter and Miller 1994; Akerley et al. 1995). The Bvg^- phase of *B. bronchiseptica* is necessary and sufficient for survival under nutrient-limiting conditions, suggesting the existence of an environmental reservoir

(Cotter and Miller 1997). It was recently demonstrated that, instead of controlling a biphasic transition between the Bvg^+ and Bvg^- states, BvgAS controls expression of a spectrum of phenotypic phases in response to quantitative differences in environmental cues (Cotter and Miller 1997). Wild-type *Bordetella* grown in the presence of submodulating conditions, such as concentrations of 0.4 to 2 mM nicotinic acid for *B. bronchiseptica*, express a phenotypic phase that is characterized by the absence of Bvg-repressed phenotypes, the presence of a subset of Bvg-activated virulence factors, and the expression of several polypeptides that are expressed maximally or exclusively in this phase. This phase, designated as Bvg-intermediate (Bvg^i), appears to be conserved between *B. pertussis* and *B. bronchiseptica* and is predicted to play a role in aerosol transmission of these strains (Fuchslocher et al. 2003).

From a phylogenetic perspective, Bvg-regulated genes fall into two categories (Fig. 11.2). Some loci are commonly expressed by *B. pertussis*, *B. parapertussis* (human and ovine), and *B. bronchiseptica*. Their products are highly similar and, in some cases, interchangeable between different subspecies. In contrast, other loci, which are present in the genomes of all four subspecies, appear to be differentially expressed. These genes provide important clues for understanding the fundamental differences in the modes of *Bordetella*–host interactions.

A type III secretion system (TTSS) has been identified in *Bordetella* subspecies (Yuk et al. 1998). Type III genes are intact and highly conserved in members of the *B. bronchiseptica* cluster; however, only *B. bronchiseptica* and *B. parapertussis*$_{ov}$ readily display type III secretion (TTS)-associated phenotypes *in vitro* (Mattoo et al. 2004). The TTSS appears to play a pivotal role in promoting persistent infection of the respiratory epithelium by modulating host immunity (Yuk et al. 1998, 2000). Our recent analysis of the control of TTS in *Bordetella* subspecies provides insight into the dynamics of virulence gene regulation and its implications for host adaptation (Mattoo et al. 2004). For additional discussions of $TTSS_s$ in other organisms, see Chapters 3 and 10.

11.4. THE *BORDETELLA* TYPE III SECRETION SYSTEM

Type III secretion systems allow Gram-negative bacteria to inject protein effector molecules into the cytoplasm or plasma membrane of eukaryotic cells (Cornelis and Van Gijsegem 2000; Cornelis 2002b; Ramamurthi and Schneewind 2002). Secretion is sometimes triggered by direct contact between the bacterium and host cell; however, host cell contact is not a universal requirement. The myriad roles these systems play in the pathogenic strategies of bacteria that express them are as varied as the pathogens themselves

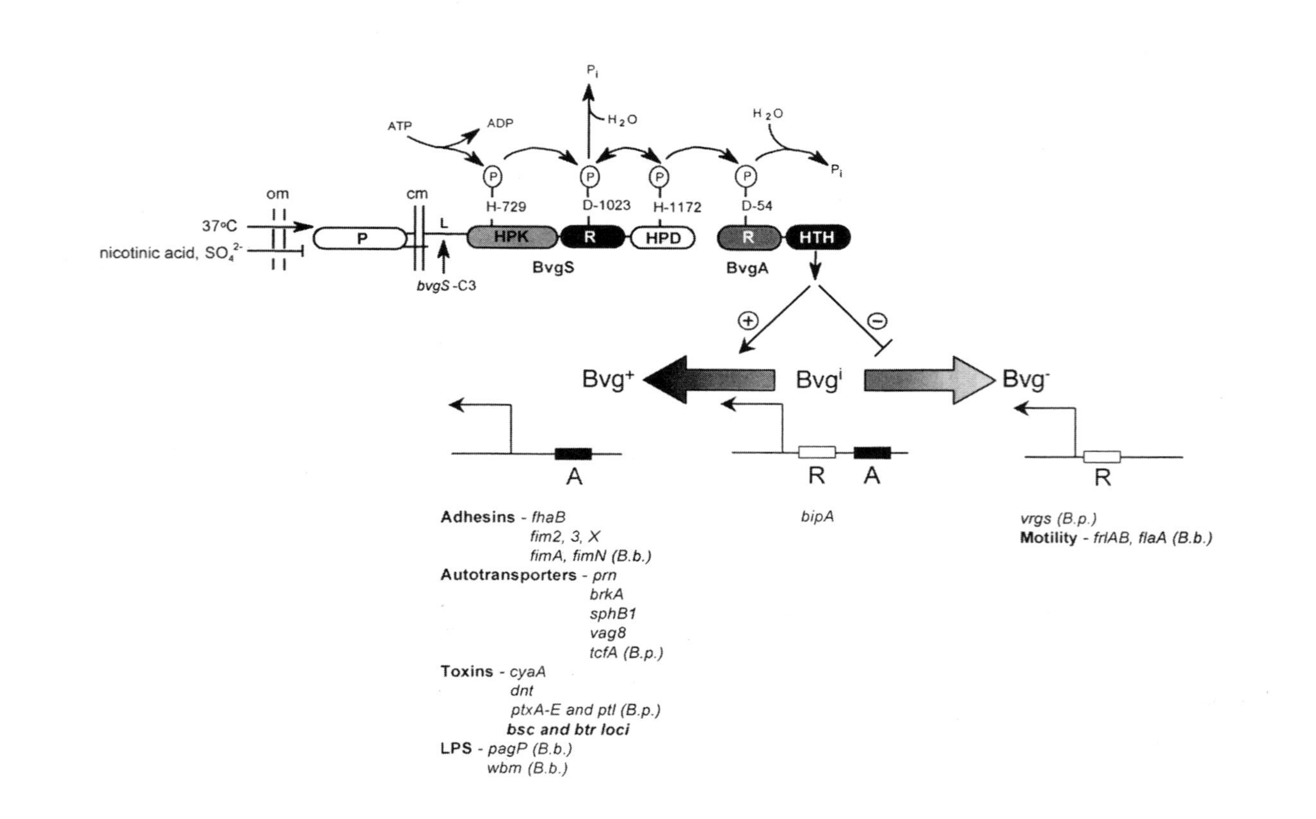

ATP
ADP
P_i
H_2O
H_2O
P_i
P
P
P
P
H-729
D-1023
H-1172
D-54
om
cm
37°C
nicotinic acid, SO_4^{2-}
P
L
HPK
R
HPD
R
HTH
BvgS
BvgA
bvgS-C3
+
−
Bvg+
Bvgi
Bvg-
A
R
A
R
Adhesins - fhaB
fim2, 3, X
fimA, fimN (B.b.)
Autotransporters - prn
brkA
sphB1
vag8
tcfA (B.p.)
Toxins - cyaA
dnt
ptxA-E and ptl (B.p.)
bsc and btr loci
LPS - pagP (B.b.)
wbm (B.b.)
bipA
vrgs (B.p.)
Motility - frlAB, flaA (B.b.)

(Galan and Collmer 1999; Cornelis 2002a). It is safe to say that for any given bacterium, a TTSS, if expressed, is likely to occupy a pivotal place in the organism's infectious cycle.

Type III secretion systems typically require more than twenty genes, which can be grouped into discrete functional categories (Hueck 1998; Kimbrough and Miller 2002). Export apparatus loci encode proteins that form a secretion complex that spans the bacterial cytoplasmic and outer membranes. In several cases, "needle complexes" have been observed to be associated with the secretion apparatus (Kimbrough and Miller 2002). Translocon genes encode proteins that form a pore in the eukaryotic cell membrane, and effector loci encode proteins that enter eukaryotic cells. Type III secreted effectors have been shown to alter an impressive array of host-cell functions, including activation and repression of signal transduction pathways, rearrangement of cytoskeletal structures and redirection of vesicle trafficking (Galan and Collmer 1999; Cornelis 2002a). Apparatus-, translocon-, and TTS-specific regulatory loci are usually linked, they are often present in pathogenicity islands, and they display easily recognized sequence similarities between different bacterial species (Cornelis 2002c). In contrast, effector genes are often unlinked to the apparatus locus and different pathogens encode different arrays of effectors. The close linkage between apparatus genes, and the availability of a "universe" of effectors that can potentially be mixed and matched

Figure 11.2 The *Bordetella* virulence regulon. The BvgAS phosphorelay controls expression of Bvg$^+$, Bvg^i, and Bvg$^-$ phase-specific genes. Upon detecting inducing signals with its periplasmic domain (P), BvgS autophosphorylates at His-729 of its transmitter domain (T). His-729 then donates the phosphoryl group to Asp-1023 of the receiver (R). Asp-1023 can donate the phosphoryl group to His-1172 of the HPD or to water to form inorganic phosphate. The HPD can then transfer the phosphate back to BvgS or, alternatively, to BvgA at Asp-54 and thus activate it. Phosphorylated BvgA (BvgA~P) can bind to high-affinity activation sites (A; black rectangles) in the promoter regions of Bvg$^+$ phase-specific genes, which include those encoding adhesins, toxins (including the *bsc* and *btr* genes for the TTSS), autotransporters, and LPS biosynthesis machinery. BvgA~P can also bind to repression sites (R; white rectangles) in the promoters of Bvg$^-$ phase genes, such as those required for motility in *B. bronchiseptica*. The promoters of Bvg^i phase genes, such as *bipA*, are predicted to contain both high-affinity A sites and low-affinity R sites. At intermediate concentrations of BvgA~P, such as in the Bvg^i phase, the A site(s) gets bound by BvgA~P and *bipA* is transcribed. However, at high BvgA~P concentrations, such as in the Bvg$^+$ phase, the R site(s) also gets bound by BvgA~P, thereby shutting off *bipA* transcription. Brackets denote genes that are expressed exclusively by *B. bronchiseptica* (*B.b.*) or *B. pertussis* (*B.p.*)

in various combinations, has profound implications for the evolution of bacterial pathogens.

Type III secretion represents one of the most complex mechanisms of protein translocation yet discovered in biology (Thomas and Finlay 2003). It also provides a striking example of the ability of pathogenic bacteria to manipulate their eukaryotic hosts. Understanding the regulation of TTS greatly facilitates functional studies, both in vitro and in vivo. For *Bordetella* subspecies, it may provide a key to understanding host adaptation.

11.4.1. Discovery and Characterization of the *bsc* Locus

The *Bordetella* TTSS was discovered using a differential display technique to identify BvgAS-activated transcripts in the *B. bronchiseptica* strain, RB50 (Yuk et al. 1998). The analysis yielded a Bvg$^+$ phase-specific cDNA product that mapped to an open reading frame (ORF) with sixty-four percent identity to the *yscN* gene product from *Yersinia* spp. YscN is an integral component of the *Yersinia* TTS machinery and is postulated to provide energy for secretion by hydrolyzing ATP (Cornelis 2002c). Isolation of chromosomal fragments adjacent to the *B. bronchiseptica yscN* homolog (*bscN*) led to the discovery of the *Bordetella* TTS locus, termed the *bsc* locus.

The *bsc* locus includes twenty-two genes that encode components of a TTS apparatus, secreted proteins, and putative chaperones (Fig. 11.3). Recently, a cluster of four ORFs (*btrS*, *btrU*, *btrW*, and *btrV*) adjacent to the *bsc* locus was identified and shown to encode regulatory proteins for the TTSS (Mattoo et al. 2004). Reverse-transcriptase polymerase chain reaction (RT-PCR), *lacZ* fusion, and microarray analyses indicate that all the genes shown in Figure 11.3 are transcriptionally activated by BvgAS. In *Bordetella*, type III secreted polypeptides can be readily detected in the culture supernatants of Bvg$^+$ phase bacteria without manipulating salt concentrations and in the absence of contact with host cells. Further, a deletion of *bscN*, which encodes the ATPase required for TTS, eliminates secretion of type III proteins. A comparison of supernatant protein profiles of wild-type and Δ*bscN* mutants revealed approximately twelve polypeptides that are secreted by this pathway, four of which were identified as BopB, BopD, BopN, and Bsp22 (Yuk et al. 1998). Western blot analysis using sera from animals infected with wild-type *B. bronchiseptica* further revealed that these proteins are antigenic and expressed during infection. BopB and BopD are orthologues of the *Yersinia* YopB and YopD translocation proteins, and are predicted to form part of the type III translocation pore (Cornelis 2002c). BopN is orthologous to YopN, which functions to prevent Yop secretion in *Yersinia* in the absence

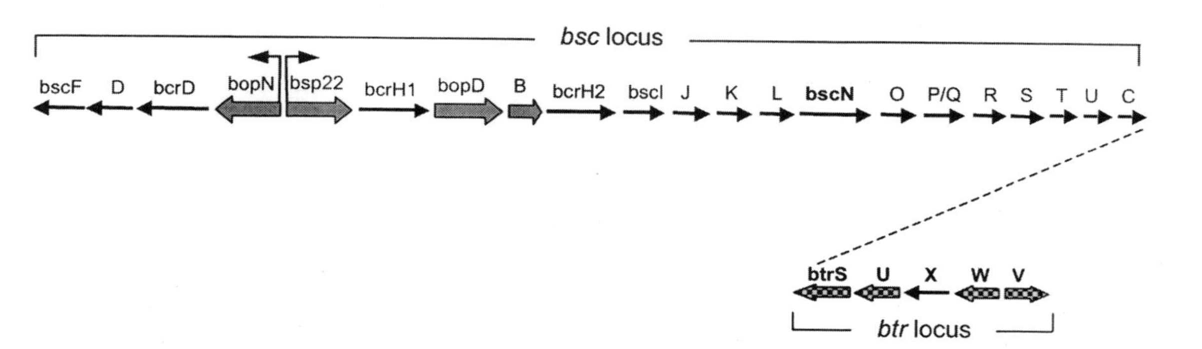

Figure 11.3. Genomic organization and open reading frames of type III secretion genes in *B. bronchiseptica*. The stretch of genes from *bscF* to *bscC* constitutes the *bsc* locus. *bopN*, *bsp22*, *bopD*, and *bopB* (shown as gray block arrows) encode secreted proteins. *bscN* encodes the putative ATPase. The genes of the *btr* type III regulatory locus, *btrS*, *btrU*, *btrW*, and *btrV*, are represented as checkered arrows.

of host cell contact (Cheng et al. 2001). Bsp22, the most abundantly secreted polypeptide, is unique. Although an in-frame deletion in *bsp22* has no effect on secretion of other polypeptides, all TTS-dependent phenotypes detected so far are dependent on the presence of an intact *bsp22* locus, suggesting that it too may be a component of the translocation apparatus (Yuk et al. 2000). Despite extensive efforts to identify them, *Bordetella* type III effector proteins have not yet been described in the literature.

11.4.2. In vivo Phenotypes Associated with Type III Secretion

In vivo, the *B. bronchiseptica* TTSS contributes to persistent colonization of the trachea in both rat and mouse models of respiratory infection (Yuk et al. 1998, 2000). Comparisons between a wild-type strain and a Δ*bscN* mutant in a rat model of respiratory infection revealed that the defect in colonization resulting from the elimination of type III secretion depended on two parameters: time and anatomical location. At 14 days post inoculation, no phenotype was resolved. By 35 days post inoculation, however, the Δ*bscN* mutants showed a striking defect in colonization of the trachea, but not the nasal cavity. Similar results were observed in C57BL/6 and BALB/c mice infected with Δ*bscN* or Δ*bsp22* mutant strains. The inflammatory cells that infiltrate the lungs during infection undergo apoptosis in mice infected with a wild-type strain but not with a mutant strain deficient in TTS (Yuk et al. 2000). These data indicate a role for TTS in promoting persistent infection in the lower respiratory tract, which is normally maintained in a state of near sterility. The defect in tracheal colonization is consistent with the notion that at least one function of TTS effectors is to down-regulate immune functions that selectively protect the lower respiratory tract against bacterial infection. Indeed, mice infected with the TTS-deficient strain elicit higher titers of anti-*Bordetella* antibodies (specifically serum IgA) than animals infected with a wild-type strain (Yuk et al. 2000). Consistent with this, animals infected with the TTS-deficient strain are completely protected against superinfection with wild-type *B. bronchiseptica* (Mattoo et al., manuscript in preparation). Taken together, these data suggest the *B. bronchiseptica* TTSS may be involved in modulating the host immune response and could contribute to the typically chronic nature of *B. bronchiseptica* infections. It further raises the possibility that TTS-deficient *Bordetella* strains could serve as live vaccine delivery vehicles.

Comparative analysis of wild-type and TTS mutants using immunocompromised SCID and SCID-beige mice revealed that deletions in *bscN* or *bsp22* consistently resulted in a hypervirulent phenotype (Yuk et al. 2000). This intriguing phenotype probably reflects the lack of the ability of TTSS mutants to modulate innate immune responses, and a corresponding increase in the inflammatory response to bacterial infection. Thus, in immunocompromised animals, the exacerbated inflammatory response is pathological whereas in immunocompetent animals, the result is a qualitative and/or quantitative increase in acquired immunity.

11.4.3. In vitro Phenotypes Associated with Type III Secretion

Although the type III effector proteins secreted by the *B. bronchiseptica* TTSS have not yet been identified, *B. bronchiseptica* causes a variety of in vitro phenotypes that are dependent on an intact and functional TTSS (Yuk et al. 1998, 2000; Stockbauer et al. 2003). The Bsc TTSS induces cytotoxicity in several cultured cell lines, dephosphorylation of specific host cell proteins, and activation of the MAP kinases ERK1 and ERK2. Additionally, the TTSS causes aberrant localization of the transcription factor NF-κB into aggregates within the host cell cytoplasm; the NF-κB of cells infected with wild-type *B. bronchiseptica* does not translocate to the nucleus, even upon stimulation with TNFα (Yuk et al. 2000). See Chapter 17 for detailed discussion of the NF-κB transcriptional control system. *B. bronchiseptica* also causes very rapid apoptosis in macrophage and epithelial cell lines (Yuk et al. 2000). The cell death pathway induced by RB50 is distinct from the apoptotic pathway induced by the *Yersinia* Ysc TTSS or the caspase-1–dependent death pathways induced by the *Shigella* or *Salmonella* Spi-1 TTSSs. Death does not require caspase-1. Procaspase-3, procaspase-7, and PARP (poly(ADP)-ribose Polymerase) remain uncleaved, and death is not blocked by the pancaspase inhibitor zVAD (Stockbauer et al. 2003). Dying cells morphologically resemble necrotic rather than apoptotic cells, and death is efficiently blocked by the addition of exogenous glycine, which functions as a nonspecific cytoprotectant. These and other data suggest that the *Bordetella* TTSS induces a necrotic form of cell death (Stockbauer et al. 2003). Although inhibition of NF-κB combined with activation of Toll-like receptor (TLR)4 (see Chapter 15 for details of TLR receptors) has been suggested to account for *Yersinia* TTS-mediated cell death, a significantly different mechanism appears to be operating in the case of *Bordetella*. Although diverse cell types are affected in vitro, the cellular targets for Bsc-dependent killing in vivo are presently unknown.

11.5. THE *btr* (*BORDETELLA* TYPE III REGULATION) LOCUS

As mentioned earlier, the type III regulatory locus consists of *btrS*, *btrU*, *btrW*, and *btrV*, which are located adjacent to the *bsc* TTS locus (Mattoo et al.

2004; Fig. 11.3). BtrS bears sequence similarity to the ECF (extra-cytoplasmic function) family of alternative sigma factors, which include HrpL, an activator of TTS in the phytopathogen *Pseudomonas syringae*, and σE, which activates heat shock responses in *E. coli* (Xiao et al. 1994; Xiao and Hutcheson 1994; De Las Penas et al. 1997; Missiakis et al. 1997). ECF sigma factors typically regulate functions related to extracytoplasmic compartments, and their activity is often controlled by cognate membrane bound anti-sigma factors (Missiakis and Raina 1998; Campbell et al. 2003).

Sequence similarities displayed by BtrU, BtrW, and BtrV are quite surprising (Fig. 11.4). These proteins contain an array of domains that define "partner switching" complexes characterized in Gram-positive bacteria. In *Bacillus subtilis*, different sets of partner switches regulate σB activity in the stress response pathway and σF activity during sporulation (Hughes and Mathee 1998; Helmann 1999). These systems typically consist of a phosphatase, a switch protein/serine protein kinase that can function as an anti-sigma factor, and an antagonist protein that can function as an anti-anti-sigma. They represent a new paradigm for signal transduction, which was, until recently, assumed to be confined to Gram-positive organisms. As indicated in Figure 11.4, BtrU is orthologous to the RsbU and SpoIIE serine phosphatases and contains a PP2C-like serine phosphatase domain (Kang et al. 1996; Schroeter et al. 1999; Voelker et al. 1995). BtrU also contains two predicted transmembrane sequences and a HAMP (histidine kinases, adenylyl cyclases, methylbinding proteins and phosphatases) domain (Aravind and Ponting 1999). BtrW bears sequence similarity to RsbW and SpoIIAB, which act as anti-sigma factors for σB and σF, respectively (Benson and Haldenwang 1993; Garsin et al. 1998). Like its orthologues, BtrW contains an HPK (histidine protein kinase)-like serine kinase domain. BtrV bears sequence similarity to RsbV and SpoIIAA, which act as antagonists of the RsbW and SpoIIAB anti-sigma factors (Garsin et al. 1998; Dufour and Haldenwang 1994). RsbV, SpoIIAA, and BtrV all contain anti-sigma factor antagonist (STAS) domains (Kovacs et al. 1998).

A summary of the RsbUVW pathway of *B. subtilis* is shown in Figure 11.5 (Yang et al. 1996). RsbW functions as an anti-sigma factor that controls the activity of σB. This interaction is countered by the RsbV anti-anti-sigma factor that binds to the RsbW-σB complex, freeing σB to interact with RNA polymerase and promote transcription. RsbW also functions as a serine kinase, and phosphorylation of RsbV dissociates the RsbVW complex. RsbU can dephosphorylate RsbV, restoring its ability to bind RsbW. The phosphorylation state of RsbV is therefore controlled by opposing kinase (RsbW) and phosphatase (RsbU) activities. RsbU represents a key point for signal

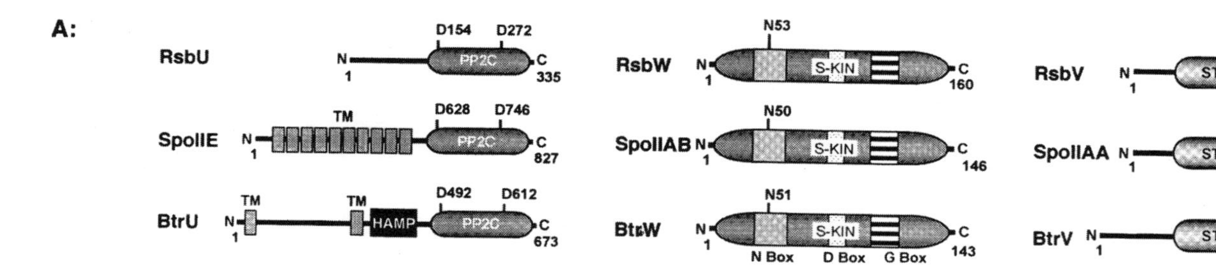

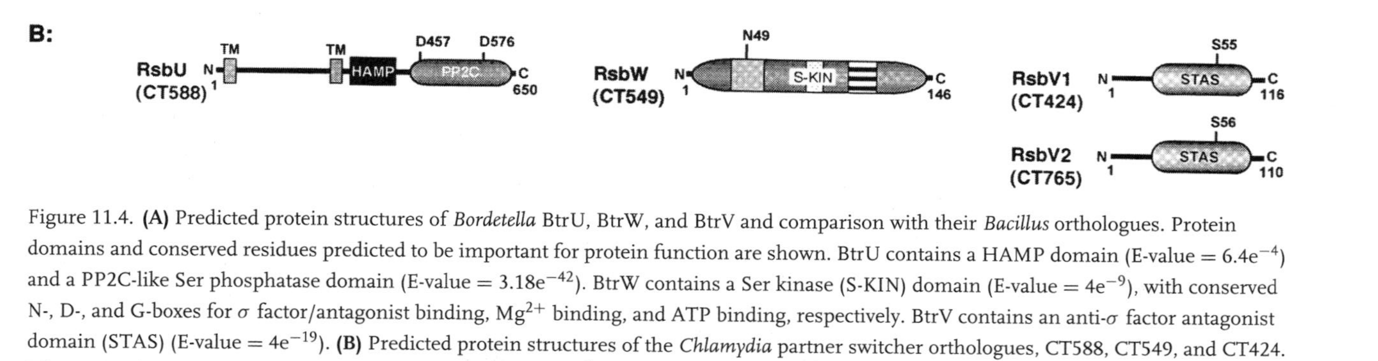

Figure 11.4. **(A)** Predicted protein structures of *Bordetella* BtrU, BtrW, and BtrV and comparison with their *Bacillus* orthologues. Protein domains and conserved residues predicted to be important for protein function are shown. BtrU contains a HAMP domain (E-value = $6.4e^{-4}$) and a PP2C-like Ser phosphatase domain (E-value = $3.18e^{-42}$). BtrW contains a Ser kinase (S-KIN) domain (E-value = $4e^{-9}$), with conserved N-, D-, and G-boxes for σ factor/antagonist binding, Mg^{2+} binding, and ATP binding, respectively. BtrV contains an anti-σ factor antagonist domain (STAS) (E-value = $4e^{-19}$). **(B)** Predicted protein structures of the *Chlamydia* partner switcher orthologues, CT588, CT549, and CT424. Like BtrU, CT588 contains a HAMP domain (E-value = $4.35e^{-5}$) and a PP2C-like Ser phosphatase domain (E-value = $3.09e^{-72}$). CT549 contains a Ser kinase domain (E-value = $8e^{-60}$). *Chlamydia* have two RsbV orthologues, CT424 and CT765, each with a STAS domain (E-values = $1.7e^{-28}$ and 1.2^{-26}, respectively).

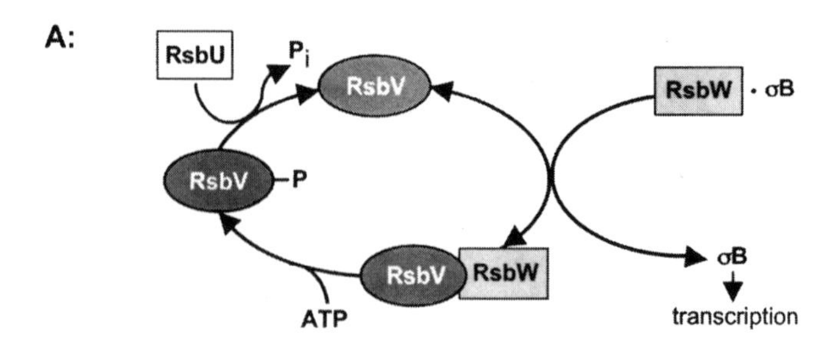

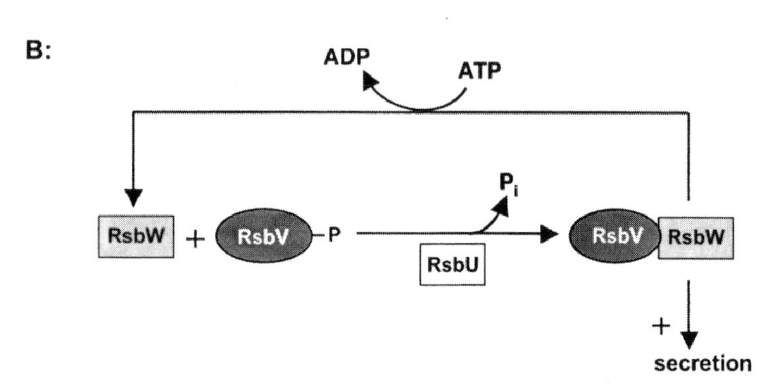

Figure 11.5. **(A)** The *Bacillus subtilis* RsbUVW partner switching cycle that regulates the stress responsive σB (see text). **(B)** A model for the partner switching cycle in *Bordetella*.

integration; extrinsic signals control the output of the system by regulating its phosphatase activity (Helmann 1999; Hughes and Mathee 1998; Yang et al. 1996). The partner switching complex that controls σF activity during sporulation operates through a similar mechanism.

The *btr* locus is unique, as it is predicted to encode both an ECF family member, which typically controls extracytoplasmic events in Gram-negative bacteria, and a complete partner switching complex, which regulates sigma factors in Gram-positive bacteria. Quantitative RT-PCR analysis and fusions to *lacZ* confirmed that, like *bsc* genes, the *btr* genes are also positively regulated by BvgAS (Mattoo et al. 2004). Deletion analysis of the *btr* genes revealed an interesting set of phenotypes associated with these loci (Mattoo et al. 2004). BtrS was found to be necessary and sufficient for transcription of type III genes, as a deletion in *btrS* eliminated expression of all *bsc* genes as well as *btrU* and *btrW*. Further, ectopic expression of *btrS* by expressing it under the control of a constitutive promoter in a Bvg$^-$ phase-locked strain restored expression of these genes. Thus, BvgAS exerts control over TTS by regulating *btrS*. The Δ*btrU* and Δ*btrW* mutants displayed a particularly interesting phenotype that appears to involve an uncoupling of protein expression from secretion; these strains expressed type III proteins such as Bsp22 and BopD in nearly normal amounts, but were unable to secrete them. The phenotype associated with a Δ*btrV* mutation was even more striking, as it eliminated Bsp22 and BopD expression altogether. Because *bsp22, bopD*, and other type III loci are transcribed at normal levels in Δ*btrV* strains, lack of detectable protein could result from a defect in translation and/or a marked increase in the rate of degradation. The above phenotypes correlated with the inability of the Δ*btrS*, Δ*btrU*, Δ*btrW*, or Δ*btrV* mutants to induce cytotoxicity in HeLa cells (Mattoo et al. 2004).

The above data support the hypothesis that although BtrS is required for expression of type III loci, BtrU and BtrW are required for secretion and BtrV is essential for translation and/or stability. This is the first example of a functionally active partner switching complex in a Gram-negative bacterium.

11.6. COMPARATIVE ANALYSIS OF VIRULENCE GENE REGULATION AND TYPE III SECRETION

The *bsc* and *btr* loci are highly conserved (ninety-seven to one hundred percent amino acid sequence identity) between *B. bronchiseptica* and *B. pertussis* (Parkhill et al. 2003; Mattoo et al. 2004). The *bsc* and *btr* genes of *B. pertussis* are predicted to be transcribed and translated to yield full-length polypeptides. In vitro, however, *B. pertussis* and *B. parapertussis*$_{hu}$ strains were not cytotoxic to mammalian cells, whereas *B. bronchiseptica* and *B. parapertussis*$_{ov}$ were highly toxic (Mattoo et al. 2004). Immunoblot analysis confirmed the lack of Bsp22 and BopD expression in *B. pertussis* and *B. parapertussis*$_{hu}$ strains, leading to the hypothesis that the lack of TTS in vitro by human-adapted strains might be caused by the lack of expression of *bsc* and/or *btr* genes. Surprisingly, RT-PCR analysis of *B. pertussis* and *B. parapertussis* showed that all *bsc* loci were actively transcribed and Bvg regulated (Mattoo et al. 2004). The same was true for the *btrS, btrU, btrW*, and *btrV* TTS regulatory loci. The *B. pertussis* BtrS protein is predicted to be identical to BtrS expressed by *B. bronchiseptica* and ectopic expression of *btrS* in a Bvg^{-} phase *B. pertussis* strain fully restored expression as was previously observed in *B. bronchiseptica* (Mattoo et al. 2004). These results strongly suggest that the BtrS regulon is intact and functional in *B. pertussis*, and the same may be true for human-adapted strains of *B. parapertussis*.

The multiple events occurring during *Bordetella*–host interactions are controlled to a significant extent by BvgAS. Different patterns of gene expression required to produce the various phenotypic phases are predicted

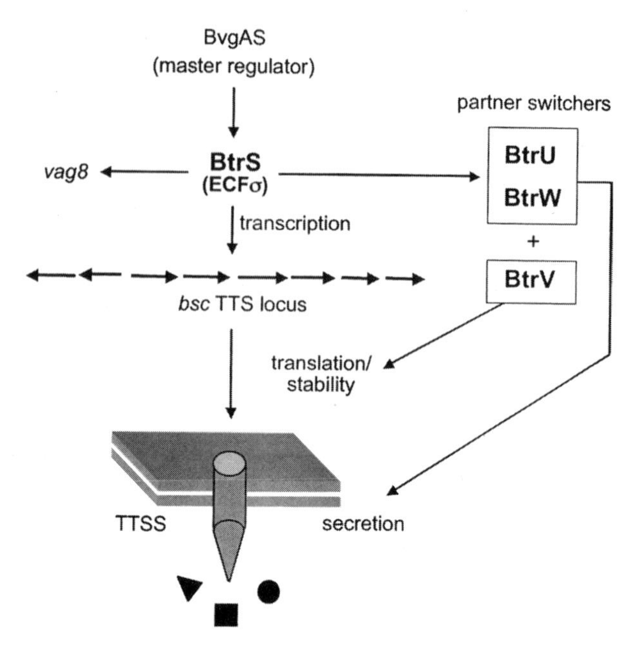

Figure 11.6. A model for type III secretion regulation in *Bordetella* (see text).

to occur in response to variations in phosphorylated BvgA levels, which are in turn regulated by the ability of BvgS to sense its environment and mediate subsequent phosphorylation/dephosphorylation events (Uhl and Miller 1996a). Results obtained from a detailed analysis of the TTSS demonstrate that BvgAS-controlled regulatory circuits are subject to further fine-tuning by downstream regulatory factors that can respond to an additional collection of environmental cues (Mattoo et al. 2004). The *Bordetella* type III regulatory cascade shown in Figure 11.6 demonstrates how multiple regulatory systems can be integrated to control a virulence function. The master virulence regulatory locus, BvgAS, sits at the top of a hierarchy and activates *btrS* expression, either directly or indirectly. BtrS then activates expression of genes encoding components of the TTS apparatus; secreted factors such as Bsp22, BopD, and BopB; and members of the partner switching complex. Expression of BtrS alone is sufficient for transcription of genes encoded by the *bsc* locus. However, secretion of type III proteins specifically requires the partner switching orthologues BtrU and BtrW. Translation and/or stability of secreted factors require a third orthologue, BtrV. Additional loci subject to BtrS control include, but are probably not limited to, the autotransporter gene *vag8*. Thus, TTS can be monitored at the levels of transcription, protein expression, and secretion.

It is suspected that, like other ECF sigma factors, BtrS function may be controlled by a cognate anti-sigma factor that remains to be identified. By analogy with *B. subtilis,* the partner switching complex could function to regulate an additional sigma factor required for expression of loci governing secretion. The sequence of BtrU includes two predicted transmembrane motifs followed by a HAMP domain (Fig. 11.4). HAMP domains are typically involved in transmitting signals between input and output modules in signaling proteins (Aravind and Ponting 1999). As such, BtrU could function as a membrane-bound sensor and transmit signals to regulate TTS directly to the partner switching complex. In an alternative model, BtrU and BtrW may themselves be components of the TTS apparatus or may regulate one or more apparatus components. It is unknown if the TTS apparatus is even assembled in strains lacking BtrU or BtrW. The role of BtrV in the type III regulatory hierarchy is intriguing and is reminiscent of data from other systems demonstrating that TTS and translation of secreted proteins can be coupled (Karlinsey et al. 2000; Trcek et al. 2002). It seems likely that BtrU, BtrW, and BtrV comprise a partner switching complex that operates in basic accordance with the predicted biochemical activities of the various conserved domains. The phenotypes of Δ*btrU,* Δ*btrW,* and Δ*btrV* mutants, however, clearly indicate significant differences in comparison to the *B. subtilis* Rsb system, as summarized in Figure 11.5A. For example, on the basis of the Rsb paradigm, a deletion in *btrW* would be expected to have the *opposite* phenotype as a deletion in *btrU*; however, this is not the case in *Bordetella,* in which both the *btrU* and *btrW* deletion mutants are defective for TTS (Mattoo et al. 2004). One alternative mechanism is that nonphosphorylated BtrV binds BtrW, and the resulting BtrVW complex is required for TTS (Fig. 11.5B).

The use of what had previously been considered to be a Gram-positive system of gene regulation to control TTS is intriguing, and *Bordetella* species may not be the only example. Homologues of the partner switching complex have been identified in *Chlamydia pneumoniae* and *Chlamydia trachomatis,* which also encode TTSSs (Kalman et al. 1999; Fig. 11.4). Interestingly, the *Chlamydia* BtrU homologue is more closely related to *Bordetella* BtrU than to their *Bacillus* counterparts (Mattoo et al. 2004). Its predicted protein sequence suggests that it also contains two transmembrane domains separated by a hydrophilic stretch of 274 aa followed by a HAMP domain. Understanding the *Bordetella* type III regulatory cascade may therefore have implications for understanding how this process is regulated in *Chlamydia* species for which tractable genetic systems have not yet been developed.

Although relatively little is known about the mechanisms of regulation for many TTS-competent bacteria, the hierarchical nature of pathways controlling TTS in general has become evident in recent years. In many cases, environmentally responsive global virulence control systems analogous to BvgAS are positioned at the top of the hierarchy (Pegues et al. 1995; Garcia et al. 1996), and signals are often channeled to AraC family members (Mellies et al. 1999; Schechter et al. 1999; Darwin and Miller 2001). Some bacteria, such as *Pseudomonas, Erwinia,* and *Ralstonia* species, utilize alternative sigma factors belonging to the ECF or RpoN families at various points in their TTS regulatory cascades (Hendrickson et al. 2000; Uhl and Miller 1996a; Wei and Beer, 1995). Regulation is often linked to secretion, as illustrated by a particularly interesting negative feedback loop that operates in *Yersinia* species. Secretion of a negative regulator, LcrQ, is triggered by target-cell contact, resulting in the derepression of TTS gene expression (Pettersson et al. 1996; Cambronne et al. 2000). A hallmark of TTS is the involvement of a novel class of specialized, energy-independent cytosolic chaperones with a variety of proposed functions (Wattiau et al. 1996; Parsot et al. 2003). These include stabilization of specific substrate proteins, protein targeting to the TTS apparatus, maintenance of substrate proteins in a secretion-competent state, translocation and transcriptional control of TTS genes (Wattiau et al. 1996; Francis et al. 2002; Parsot et al. 2003). Finally, additional levels of control can be exerted by type III apparatus components themselves. In *Yersinia enterocolitica,* YscP monitors Yop secretion by controlling elongation of the type III needle (Journet et al. 2003). This protein presumably anchors its N- and C-termini at the proximal and distal ends of the needle. Upon reaching the maximum needle length, as determined by the size of the YscP molecular ruler, YscP is secreted, thereby allowing secretion of Yop proteins. A similar mechanism is predicted to be involved in the control of TTS in *Salmonella* and *Shigella* species by InvJ and Spa32, respectively (Kubori et al. 2000; Magdalena et al. 2002; Thomas and Finlay 2004). Although the *Bordetella* TTSS shares some features of TTS regulation with the Gram-negative pathogens described above, the putative partner switching complex and its control of protein expression and secretion in a manner independent of transcriptional control of the type III apparatus sets it in a class apart.

While the *bsc* and *btr* loci are highly conserved and fully transcribed in the various *Bordetella* subspecies, human-adapted strains do not readily express phenotypes associated with TSS during growth under standard laboratory conditions (Mattoo et al. 2004). Interestingly, the block in TTS observed with *B. pertussis* in vitro is apparently post-transcriptional, similar to the phenotype observed in a *B. bronchiseptica* Δ*btrV* mutant. In comparing *B. pertussis* to *B. bronchiseptica*, BtrW and BtrV are identical whereas their BtrU proteins differ at six residues (Mattoo et al. 2004). Of the nucleotide differences observed in the genes encoding Btr or TTSS proteins, substitutions are dramatically skewed toward those that are silent or conservative. This implies evolutionary pressure for maintaining TTS in *B. pertussis*. On the basis of this and other data, we suspect that *B. pertussis* does express its TTSS, but in a manner that is regulated differently from that observed in *B. bronchiseptica*. Signal recognition, mediated by unidentified factors, is a possible source of divergence between subspecies. Alternatively, human-adapted strains may have lost genes required for stable expression of type III proteins. If so, then why would these strains transcribe the various type III genes if they no longer require the TTSS for human adaptation? From an evolutionary perspective, it would be more economical to mutate *btrS* and eliminate transcription of all type III genes. Multiple lines of evidence suggest that the BtrS regulon is conserved in *B. pertussis* and the comparative analysis of this regulon is likely to reveal new mechanisms of pathogenesis. If the hypothesis that TTS occurs in human-adapted strains in response to unidentified environmental cues is correct, it will dramatically change the way we view and potentially vaccinate against *B. pertussis*-host interactions.

11.7. CONCLUSIONS

The *Bordetella* TTSS is one example of how bacteria can modulate host immune responses to establish chronic infections. It is especially interesting to consider that, in *B. bronchiseptica*, it functions to promote persistence as opposed to disease. Understanding the fundamental mechanisms that pathogens employ to modulate the host response to infection may lead to new methods for the treatment and prevention of bacterial disease. Although historically TTS has been considered a feature specific to pathogenic bacteria, recent discoveries of such systems in endosymbionts and invertebrate pathogens have changed our perspective. TTSSs have been identified in nitrogen-fixing plant endosymbionts such as *Rhizobium* spp., as well as in endosymbionts of insects such as *Sodalis glossinidius* and of nematodes pathogenic to insects such as *Photorhabdus luminescens* (Viprey et al. 1998; Dale et al. 2002; ffrench-Constant et al. 2003). These observations suggest that a common yet underappreciated function of TTS is to enable bacteria to establish cell–cell communications across species in order to persist and replicate within a host *without* causing disease.

REFERENCES

Akerley, B. J., Cotter, P. A., and Miller, J. F. (1995). Ectopic expression of the flagellar regulon alters development of the *Bordetella*-host interaction. *Cell* **80**, 611–620.

Aravind, L. and Ponting, C. P. (1999). The cytoplasmic helical linker domain of receptor histidine kinase and methyl-accepting proteins is common to many prokaryotic signalling proteins. *FEMS Microbiology Letters* **176**, 111–116.

Benson, A. K. and Haldenwang, W. G. (1993). *Bacillus subtilis* sigma B is regulated by a binding protein (RsbW) that blocks its association with core RNA polymerase. *Proceedings of the National Academy of Sciences USA* **90**, 2330–2334.

Boucher, P. E., Maris, A. E., Yang, M. S., and Stibitz, S. (2003). The response regulator BvgA and RNA polymerase alpha subunit C-terminal domain bind simultaneously to different faces of the same segment of promoter DNA. *Molecular Cell* **11**, 163–173.

Boucher, P. E., Murakami, K., Ishihama, A., and Stibitz, S. (1997). Nature of DNA binding and RNA polymerase interaction of the *Bordetella pertussis* BvgA transcriptional activator at the *fha* promoter. *Journal of Bacteriology* **179**, 1755–1763.

Cambronne, E. D., Cheng, L. W., and Schneewind, O. (2000). LcrQ/YscM1, regulators of the *Yersinia* yop virulon, are injected into host cells by a chaperone-dependent mechanism. *Molecular Microbiology* **37**, 263–273.

Campbell, E. A., Tupy, J. L., Gruber, T. M., Wang, S., Sharp, M. M., et al. (2003). Crystal structure of *Escherichia coli* sigma E with the cytoplasmic domain of its anti-sigma RseA. *Molecular Cell* **11**, 1067–1078.

Cheng, L. W., Kay, O., and Schneewind, O. (2001). Regulated secretion of YopN by the type III machinery of *Yersinia enterocolitica*. *Journal of Bacteriology* **183**, 5293–5301.

Cherry, J. D. and Heininger, U. (1999). Pertussis. In *Textbook of Pediatric Infectious Diseases*, ed. R. D. Feigin and J. D. Cherry, pp. 1423–1440. Philadelphia: Saunders.

Cornelis, G. R. (2002a). *Yersinia* type III secretion: send in the effectors. *Journal of Cell Biology* **158**, 401–408.

Cornelis, G. R. (2002b). The *Yersinia* Ysc-Yop 'type III' weaponry. *Nature Reviews Molecular Cell Biology* **3**, 742–752.

Cornelis, G. R. (2002c). The *Yersinia* Ysc-Yop virulence apparatus. *International Journal of Medical Microbiology* **291**, 455–462.

Cornelis, G. R. and Van Gijsegem, F. (2000). Assembly and function of type III secretory systems. *Annual Review of Microbiology* **54**, 735–774.

Cotter, P. A. and Jones, A. M. (2003). Phosphorelay control of virulence gene expression in *Bordetella*. *Trends Microbiology* **11**, 367–373.

Cotter, P. A. and Miller, J. F. (1994). BvgAS-mediated signal transduction: analysis of phase-locked regulatory mutants of *Bordetella bronchiseptica* in a rabbit model. *Infection and Immunity* **62**, 3381–3390.

Cotter, P. A. and Miller, J. F. (1997). A mutation in the *Bordetella bronchiseptica* bvgS gene results in reduced virulence and increased resistance to starvation, and identifies a new class of Bvg-regulated antigens. *Molecular Microbiology* **24**, 671–685.

Cummings, C. A., Brinig, M. M., Lepp, P. W., van de Pas, S., and Relman, D. A. (2004). *Bordetella* species are distinguished by patterns of substantial gene loss and host adaptation. *Journal of Bacteriology* **186**, 1484–1492.

Dale, C., Plague, G. R., Wang, B., Ochman, H., and Moran, N. A. (2002). Type III secretion systems and the evolution of mutualistic endosymbiosis. *Proceedings of the National Academy of Sciences USA* **99**, 12397–12402.

Darwin, K. H. and Miller, V. L. (2001). Type III secretion chaperone-dependent regulation: activation of virulence genes by SicA and *InvF in Salmonella typhimurium. EMBO Journal* **20**, 1850–1862.

De Las Penas, A., Connolly, L., and Gross, C. A. (1997). The sigmaE-mediated response to extracytoplasmic stress in *Escherichia coli* is transduced by RseA and RseB, two negative regulators of sigmaE. *Molecular Microbiology* **24**, 373–385.

Deeb, B. J., DiGiacomo, R. F., Bernard, B. L., and Silbernagel, S. M. (1990). *Pasteurella multocida* and *Bordetella bronchiseptica* infections in rabbits. *Journal of Clinical Microbiology* **28**, 70–75.

Dufour, A. and Haldenwang, W. G. (1994). Interactions between a *Bacillus subtilis* anti-sigma factor (RsbW) and its antagonist (RsbV). *Journal of Bacteriology* **176**, 1813–1820.

Dworkin, M. S., Sullivan, P. S., Buskin, S. E., Harrington, R. D., Olliffe, J., et al. (1999). *Bordetella bronchiseptica* infection in human immunodeficiency virus-infected patients. *Clinical Infectious Diseases* **28**, 1095–1099.

ffrench-Constant, R., Waterfield, N., Daborn, P., Joyce, S., Bennett, H., et al. (2003). *Photorhabdus:* towards a functional genomic analysis of a symbiont and pathogen. *FEMS Microbiology Reviews* **26**: 433–456.

Francis, M. S., Wolf-Watz, H., and Forsberg, A. (2002). Regulation of type III secretion systems. *Current Opinion in Microbiology* **5**, 166–172.

Fuchslocher, B., Millar, L. L., and Cotter, P. A. (2003). Comparison of bipA alleles within and across *Bordetella* species. *Infection and Immunity* **71**, 3043–3052.

Galan, J. E. and Collmer, A. (1999). Type III secretion machines: bacterial devices for protein delivery into host cells. *Science* **284**, 1322–1338.

Garcia Vescovi, E., Soncini, F. C., and Groisman, E. A. (1996). Mg2+ as an extracellular signal: environmental regulation of *Salmonella* virulence. *Cell* **84**, 165–174.

Garsin, D. A., Paskowitz, D. M., Duncan, L., and Losick, R. (1998). Evidence for common sites of contact between the antisigma factor SpoIIAB and its partners SpoIIAA and the developmental transcription factor sigmaF in *Bacillus subtilis*. *Journal of Molecular Biology* **284**, 557–568.

Gerlach, G., von Wintzingerode, F., Middendorf, B., and Gross, R. (2001). Evolutionary trends in the genus *Bordetella*. *Microbes and Infection* **3**, 61–72.

Harvill, E. T., Cotter, P. A., Yuk, M. H., and Miller, J. F. (1999). Probing the function of *Bordetella bronchiseptica* adenylate cyclase toxin by manipulating host immunity. *Infection and Immunity* **67**, 1493–1500.

Heininger, U., Stehr, K., Schmittgrohe, S., Lorenz, C., Rost, R., et al. (1994). Clinical characteristics of illness caused by *Bordetella parapertussis* compared with illness caused by *Bordetella pertussis*. *Pediatric Infectious Disease Journal* **13**, 306–309.

Helmann, J. D. (1999). Anti-sigma factors. *Current Opinion in Microbiology* **2**, 135–141.

Hendrickson, E. L., Guevera, P., Ausubel, F. M. (2000). The alternative sigma factor RpoN is required for hrp activity in *Pseudomonas syringae* pv. maculicola and acts at the level of hrpL transcription. *Journal of Bacteriology* **182**, 3508–3516.

Hueck, C. J. (1998). Type III protein secretion systems in bacterial pathogens of animals and plants. *Microbiology and Molecular Biology Reviews* **62**, 379–433.

Hughes, K. T. and Mathee, K. (1998). The anti-sigma factors. *Annual Review of Microbiology* **52**, 231–286.

Journet, L., Agrain, C., Broz, P., and Cornelis, G. R. (2003). The needle length of bacterial injectisomes is determined by a molecular ruler. *Science* **302**, 1757–1760.

Kalman, S., Mitchell, W., Marathe, R., Lammel, C., Fan, J., et al. (1999). Comparative genomes of *Chlamydia pneumoniae* and *C. trachomatis*. *Nature Genetics* **21**, 385–389.

Kang, C. M., Brody, M. S., Akbar, S., Yang, X., and Price, C. W. (1996). Homologous pairs of regulatory proteins control activity of *Bacillus subtilis* transcription factor sigma(b) in response to environmental stress. *Journal of Bacteriology* **178**, 3846–3853.

Karlinsey, J. E., Lonner, J., Brown, K. L., and Hughes, K. T. (2000). Translation/secretion coupling by type III secretion systems. *Cell* **102**, 487–497.

Keil, D. J. and Fenwick, B. (1999). Strain- and growth condition-dependent variability in outer membrane protein expression by *Bordetella bronchiseptica* isolates from dogs. *American Journal of Veterinary Research* **60**, 1016–1021.

Kimbrough, T. G. and Miller, S. I. (2002). Assembly of the type III secretion needle complex of *Salmonella typhimurium. Microbes and Infection* **4**, 75–82.

Kovacs, H., Comfort, D., Lord, M., Campbell, I. D., and Yudkin, M. D. (1998). Solution structure of SpoIIAA, a phosphorylatable component of the system that regulates transcription factor sigmaF of *Bacillus subtilis. Proceedings of the National Academy of Sciences USA* **95**, 5067–5071.

Kubori, T., Sukhan, A., Aizawa, S. I., and Galan, J. E. (2000). Molecular characterization and assembly of the needle complex of the *Salmonella typhimurium* type III protein secretion system. *Proceedings of the National Academy of Sciences USA* **97**, 10225–10230.

Magdalena, J., Hachani, A., Chamekh, M., Jouihri, N., Gounon, P., et al. (2002). Spa32 regulates a switch in substrate specificity of the type III secreton of *Shigella flexneri* from needle components to Ipa proteins. *Journal of Bacteriology* **184**, 3433–3441.

Magyar, T., Chanter, N., Lax, A. J., Rutter, J. M., and Hall, G. A. (1988). The pathogenesis of turbinate atrophy in pigs caused by *Bordetella bronchiseptica. Veterinary Microbiology* **18**, 135–146.

Martinez de Tejada, G., Miller, J. F., and Cotter, P. A. (1996). Comparative analysis of the virulence control systems of Bordetella *pertussis* and *Bordetella bronchiseptica. Molecular Microbiology* **22**, 895–908.

Mattoo, S., Foreman-Wykert, A. K., Cotter, P. A., and Miller, J. F. (2001). Mechanisms of *Bordetella* pathogenesis. *Frontiers Bioscience* **6**, E168–186.

Mattoo, S., Yuk, M. H., Huang, L. L., and Miller, J. F. (2004). Regulation of type III secretion in *Bordetella. Molecular Microbiology* **52**, 1201–1214.

Mellies, J. L., Elliott, S. J., Sperandio, V., Donnenberg, M. S., and Kaper, J. B. (1999). The Per regulon of enteropathogenic *Escherichia coli*: identification of a regulatory cascade and a novel transcriptional activator, the locus of enterocyte effacement (LEE)-encoded regulator (Ler). *Molecular Microbiology* **33**, 296–306.

Merkel, T. J., Boucher, P. E., Stibitz, S., and Grippe, V. K. (2003). Analysis of bvgR expression in *Bordetella pertussis. Journal of Bacteriology* **185**: 6902–6912.

Merkel, T. J., Stibitz, S., Keith, J. M., Leef, M., and Shahin, R. (1998). Contribution of regulation by the bvg locus to respiratory infection of mice by *Bordetella pertussis. Infection and Immunity* **66**, 4367–4373.

Missiakas, D., Mayer, M. P., Lemaire, M., Georgopoulos, C., and Raina, S. (1997). Modulation of the *Escherichia coli* sigmaE (RpoE) heat-shock transcription-factor activity by the RseA, RseB and RseC proteins. *Molecular Microbiology* **24**, 355–371.

Missiakas, D. and Raina, S. (1998). The extracytoplasmic function sigma factors: role and regulation. *Molecular Microbiology* **28**, 1059–1066.

Parkhill, J., Sebaihia, M., Preston, A., Murphy, L. D., Thomson, N., et al. (2003). Comparative analysis of the genome sequences of *Bordetella pertussis, Bordetella parapertussis* and *Bordetella bronchiseptica. Nature Genetics* **35**, 32–40.

Parsot, C., Hamiaux, C., and Page, A. L. (2003). The various and varying roles of specific chaperones in type III secretion systems. *Current Opinion in Microbiology* **6**, 7–14.

Pegues, D. A., Hantman, M. J., Behlau, I., and Miller, S. I. (1995). PhoP/PhoQ transcriptional repression of *Salmonella typhimurium* invasion genes: evidence for a role in protein secretion. *Molecular Microbiology* **17**, 169–181.

Pettersson, J., Nordfelth, R., Dubinina, E., Bergman, T., Gustafsson, M., et al. (1996). Modulation of virulence factor expression by pathogen target cell contact. *Science* **273**, 1231–1233.

Preston, A., Parkhill, J., and Maskell, D. J. (2004). The *Bordetellae*: Lessons from genomics. *Nature Reviews Microbiology* **2**, 379–390.

Ramamurthi, K. S. and Schneewind, O. (2002). Type III protein secretion in *Yersinia* species. *Annual Review of Cell and Developmental Biology* **18**, 107–133.

Schechter, L. M., Damrauer, S. M., and Lee, C. A. (1999). Two AraC/XylS family members can independently counteract the effect of repressing sequences upstream of the hilA promoter. *Molecular Microbiology* **32**, 629–642.

Schroeter, R., Schlisio, S., Lucet, I., Yudkin, M., and Borriss, R. (1999). The *Bacillus subtilis* regulator protein SpoIIE shares functional and structural similarities with eukaryotic protein phosphatases 2C. *FEMS Microbiology Letteres* **174**, 117–123.

Stockbauer, K. E., Foreman-Wykert, A. K., and Miller, J. F. (2003). *Bordetella* type III secretion induces caspase 1-independent necrosis. *Cellular Microbiology* **5**, 123–132.

Tamion, F., Girault, C., Chevron, V., Pestel, M., and Bonmarchand, G. (1996). *Bordetella bronchiseptica* pneumonia with shock in an immunocompetent patient. *Scandinavian Journal of Infectious Diseases* **28**, 197–198.

Thomas, N. A. and Finlay, B. B. (2003). Establishing order for type III secretion substrates–a hierarchical process. *Trends in Microbiology* **11**, 398–403.

Thomas, N. A. and Finlay, B. B. (2004). Pathogens: bacterial needles ruled to length and specificity. *Current Biology* **14**, R192–194.

Trcek, J., Wilharm, G., Jacobi, C. A., and Heesemann, J. (2002). *Yersinia enterocolitica* YopQ: strain-dependent cytosolic accumulation and post-translational secretion. *Microbiology* **148**, 1457–1465.

Uhl, M. A. and Miller, J. F. (1994). Autophosphorylation and phosphotransfer in the *Bordetella pertussis* BvgAS signal transduction cascade. *Proceedings of the National Academy of Sciences USA* **91**, 1163–1167.

Uhl, M. A. and Miller, J. F. (1996a). Central role of the BvgS receiver as a phosphorylated intermediate in a complex two-component phosphorelay. *Journal of Biological Chemistry* **271**, 33176–33180.

Uhl, M. A. and Miller, J. F. (1996b). Integration of multiple domains in a two-component sensor protein: the *Bordetella pertussis* BvgAS phosphorelay. *EMBO Journal* **15**, 1028–1036.

van der Zee, A., Mooi, F., Van Embden, J., and Musser, J. (1997). Molecular evolution and host adaptation of *Bordetella* spp.: phylogenetic analysis using multilocus enzyme electrophoresis and typing with three insertion sequences. *Journal of Bacteriology* **179**, 6609–6617.

Viprey, V., Del Greco, A., Golinowski, W., Broughton, W. J., and Perret, X. (1998). Symbiotic implications of type III protein secretion machinery in *Rhizobium*. *Molecular Microbiology* **28**, 1381–1389.

Voelker, U., Dufour, A., and Haldenwang, W. G. (1995). The *Bacillus subtilis* rsbU gene product is necessary for RsbX-dependent regulation of sigma B. *Journal of Bacteriology* **177**, 114–122.

von Wintzingerode, F., Gerlach, G., Schneider, B., and Gross, R. (2002). Phylogenetic relationships and virulence evolution in the genus *Bordetella*. *Current Topics in Microbiology and Immunology* **264**, 177–199.

Wattiau, P., Woestyn, S., and Cornelis, G. R. (1996). Customized secretion chaperones in pathogenic bacteria. *Molecular Microbiology* **20**, 255–262.

Wei, Z. M. and Beer, S. V. (1995). hrpL activates *Erwinia amylovora* hrp gene transcription and is a member of the ECF subfamily of sigma factors. *Journal of Bacteriology* **177**, 6201–6210.

Weiss, A. A. and Falkow, S. (1984). Genetic analysis of phase change in *Bordetella pertussis*. *Infection and Immunity* **43**, 263–239.

Xiao, Y., Heu, S., Yi, J., Lu, Y., and Hutcheson, S. W. (1994). Identification of a putative alternate sigma factor and characterization of a multicomponent regulatory cascade controlling the expression of *Pseudomonas syringae* pv. syringae Pss61 hrp and hrmA genes. *Journal of Bacteriology* **176**, 1025–1036.

Xiao, Y. and Hutcheson, S. W. (1994). A single promoter sequence recognized by a newly identified alternate sigma factor directs expression of pathogenicity and host range determinants in *Pseudomonas syringae*. *Journal of Bacteriology* **176**, 3089–3091.

Yang, X., Kang, C. M., Brody, M. S., and Price, C. W. (1996). Opposing pairs of serine protein kinases and phosphatases transmit signals of environmental stress to activate a bacterial transcription factor. *Genes and Development* **10**, 2265–2275.

Yeh, S. H. (2003). Pertussis: persistent pathogen, imperfect vaccines. *Expert Reviews on Vaccines* **2**, 113–127.

Yuk, M. H., Harvill, E. T., Cotter, P. A., and Miller, J. F. (2000). Modulation of host immune responses, induction of apoptosis and inhibition of NF-kappaB activation by the bordetella type III secretion system. *Molecular Microbiology* **35**, 991–1004.

Yuk, M. H., Harvill, E. T., and Miller, J. F. (1998). The BvgAS virulence control system regulates type III secretion in *Bordetella bronchiseptica. Molecular Microbiology* **28**, 945–959.

CHAPTER 12

Resident bacteria as inductive signals in mammalian gut development

Lora V. Hooper

12.1. INTRODUCTION

As is discussed in previous chapters, we humans harbor a vast and complex society of bacteria that is present with us from the moment of our birth, and whose relationship with us has evolved over millions of years of coexistence. Many of these microbes reside in the gut, where they are in continuous contact with our mucosal tissues. The size of this population and its intimate contact with our own cells points to a profound intertwining of human and microbial biology. However, the extent to which indigenous microbes influence our physiology and development is only just beginning to come to light.

The study of animal development has traditionally focused on understanding how interactions among animal cells trigger developmental pathways. A frequent implicit assumption has been that all steps in the development of a complex multicellular organism are genetically preordained. However, the field of ecological developmental biology (abbreviated "Eco-Devo") has recently begun to focus on the idea that some developmental triggers can come from the environment (Dusheck 2002). A major goal of Eco-Devo is thus to understand how much of our development is hardwired in our genes and what proportion derives from our interactions with microbes, which are virtually ubiquitous in the environment.

Given the fact that the gut mucosal tissues are continuously awash with bacteria starting at birth (Savage 1977), the mammalian intestine is an arena where bacterial contributions to animal development are likely to be especially pronounced. This chapter will focus on growing evidence that these bacteria drive key postnatal developmental events in the gut, including shifts in the glycoconjugate repertoire, microbicidal protein expression, and angiogenesis. Despite the diverse nature of these processes, common themes emerge from each of these bacterially driven developmental transitions.

12.2. MUTUALISTIC HOST–MICROBIAL INTERACTIONS IN THE MAMMALIAN GUT

The evolutionary forces underlying our relationships with our prokaryotic partners are still poorly understood. However, a growing body of evidence indicates that, for the most part, we enjoy a mutually beneficial relationship with our resident gut bacteria. One important host benefit derives from the fact that gut bacterial communities collectively constitute a metabolically active entity. Intestinal bacteria play a critical role in mammalian nutrition by degrading dietary substances that are otherwise indigestible by the host (Savage 1986). By recruiting communities of resident microbes with diverse metabolic capabilities, allowing breakdown of numerous dietary compounds, mammals may have been relieved of the need to evolve such functions. This arrangement affords the host a degree of metabolic adaptability that may help it cope with acute changes in diet and nutrient availability. In return, gut bacteria benefit from having a protected, nutrient-rich habitat in which to multiply (Hooper et al. 2002).

The mutually beneficial tone of this host–bacterial interaction underscores an important semantic point. The term *commensalism* is frequently invoked to describe the relationship between mammals and their resident gut bacteria. This term derives from Latin roots meaning "at table together," and generally connotes a benign but passive relationship in which neither partner harms the other but for which there are no overt benefits. However, given the contributions of the gut microflora to mammalian nutrition, *mutualism* is arguably a much better descriptor of the nature of this gut–microbe interaction (Darveau et al. 2003).

12.3. THE SQUID–*VIBRIO* PARTNERSHIP: INSIGHTS FROM A SIMPLE ASSOCIATION

The pioneering analysis of the association between a species of marine squid (*Euprymna scolopes*) and its bacterial symbiont (*Vibrio fischeri*) was among the first to reveal that bacteria can induce morphological phenotypes in their animal partners. These studies showed that bacteria can play a critical inductive role in the normal development of animal organs by affecting fundamental developmental processes such as programmed cell death and differentiation. As will be discussed below, the developmental outcomes of the squid–*Vibrio* interaction exhibit marked similarities with the microbial impact on mammalian gut development.

Like the bacteria–mammalian gut interaction, the squid–*Vibrio* partnership is a mutually beneficial one. The bacteria inhabit the squid light organ and produce light that illuminates the lower surface of the animal. This provides a critical camouflage that allows the squid to evade detection by predators that detect dark shapes swimming above them (McFall-Ngai 2002). Both partners thus benefit from the association: The bacteria receive a protected, nutrient-rich niche, whereas the squid host receives protection from predation.

In establishing this partnership, *V. fischeri* induces a dramatic program of tissue remodeling and differentiation in the squid light organ. The light organs of young squid include ciliated epithelial cells that aid in harvesting bacterial symbionts from the surrounding seawater. Within hours, the bacteria irreversibly induce the loss of the ciliated epithelia (Montgomery and McFall-Ngai 1994), which involves the triggering of cell death by bacterial lipopolysaccharide (Foster and McFall-Ngai 1998; Foster et al. 2000). The bacteria also promote differentiation of light organ epithelial cells, causing profound changes in size and shape (Lamarcq and McFall-Ngai 1998). The result is a light organ morphology that fosters maintenance of a stable host-associated bacterial colony (McFall-Ngai 2002). Squid that are not exposed to *V. fischeri* remain in an arrested state of morphogenesis (Montgomery and McFall-Ngai 1994). Thus, *V. fischeri* actively shapes its niche in a way that promotes its association with its host. As will become clear below, this important theme is mirrored in the bacteria–mammalian gut interaction.

12.4. INDIGENOUS BACTERIA AS AN INDUCTIVE FORCE IN MAMMALIAN GUT DEVELOPMENT

In contrast to the squid light organ's simple, monospecific ecosystem, the mammalian intestinal ecosystem exhibits a staggering complexity. It has been estimated that at least 400 species of bacteria normally inhabit the gut (Savage 1977; Suau et al. 1999). Collectively, these microbes colonize to levels as high as 10^{11} organisms per milliliter of gut contents. Like squid, mammals acquire their microflora from the environment, as young mammals are sterile in utero and are inoculated during passage through the birth canal. The gut ecosystem thus develops during early postnatal life, increasing in complexity until an adult ecosystem is established (Savage 1977; Rotimi and Duerden 1981; Mackie et al. 1999).

Our understanding of mammalian gut microbial ecology is still limited, and is based primarily on culture-based studies that are biased toward

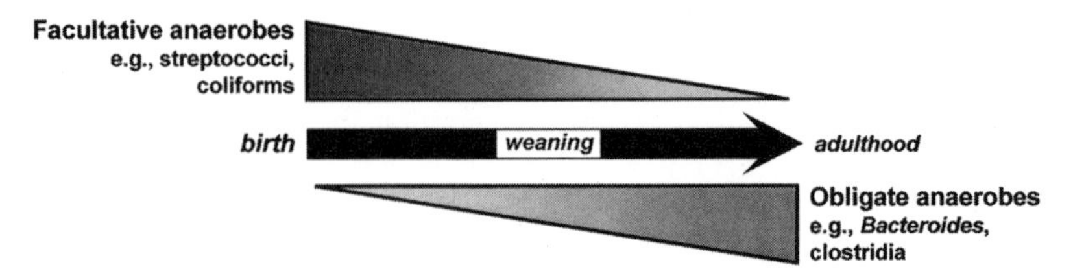

Figure 12.1. Ecology of the developing postnatal intestine in mammals.
Mammals are colonized at birth. Although the microbial ecology of the developing intestine is still poorly understood, in general, facultative anaerobes such as streptococci and coliforms (facultative rod-shaped anaerobes such as *Escherichia coli*) are the principal colonizers during early neonatal life. Following weaning, the gut ecosystem evolves into a stable climax community populated primarily by obligate anaerobes including *Bacteroides* and *Clostridium* species (Mackie et al. 1999; Moore and Holdeman 1974; Savage 1977).

organisms that can survive outside the gut. Nevertheless, almost all analyses to date indicate that profound changes occur in the intestinal ecosystem as the young are weaned from mother's milk onto solid food (Savage 1977; Mackie et al. 1999). The intestines of human neonates, for example, contain large numbers of facultative anaerobes, including *Escherichia coli* and streptococci (Fig. 12.1) (Mackie et al. 1999). These decline in number during weaning as obligate anaerobes establish a foothold in the ecosystem. Over time, the gut ecosystem evolves into a stable climax community consisting predominantly of obligate anaerobes including *Bacteroides* and *Clostridium* species (Savage 1977; Moore and Holdeman 1974; Mackie et al. 1999).

The weaning-associated shifts in microbial ecology are coincident with a marked functional and morphologic maturation of the gut. Dramatic alterations in the metabolic capacity of the small intestine facilitate the transition from a high-fat milk-based diet to a diet rich in carbohydrates. Important changes in immune function take place at the same time, alongside a withdrawal of the passive immunity conferred by maternal immunoglobulins present in milk. For example, increased numbers of B and T cells are found in the villus lamina propria and in the intraepithelial spaces following weaning (Steege et al. 1997). In addition, the polymeric immunoglobulin A receptor, which transports antibodies from the basolateral to the apical surfaces of gut epithelial cells, is markedly induced during this transition (Jenkins et al. 2003).

A key question is whether all intestinal developmental changes associated with weaning are genetically encoded, or whether at least some are triggered by the shifts in gut microbial ecology during this period. The technology of raising germ-free animals affords a powerful experimental approach for addressing this question. Such animals are raised under completely sterile conditions, and thus lack a resident gut microbiota. Mice raised germ-free from birth exhibit a number of phenotypic differences as compared to conventionally raised mice (which have a normal gut flora), including enlarged caeca (Wostmann 1981), altered kinetics of gut epithelial cell turnover (Savage et al. 1981), and an increased caloric intake (Wostmann et al. 1983). Moreover, as detailed below, studies in germ-free mice have revealed that bacteria provide essential inductive signals for key aspects of postnatal gut maturation.

12.4.1. Bacterial Modification of Glycoconjugate Expression on the Gut Epithelial Surface

Mammalian glycoconjugates (glycans) exhibit enormous diversity and structural complexity and are frequently regulated in a development- and tissue-specific manner (Varki 1993). Conventionally raised mice begin life by expressing intestinal epithelial glycans that terminate predominantly with sialic acid (Biol et al. 1992). During weaning (postnatal days 7 to 21 in mice), there is a pronounced shift toward expression of glycans terminating with the sugar fucose (Biol et al. 1992; Umesaki et al. 1995; Bry et al. 1996), and a concomitant change in the expression of glycosytransferases that direct synthesis of these terminal modifications (Umesaki et al. 1995; Bry et al. 1996). However, the shift to terminal fucose does not occur in germ-free mice, indicating that bacteria are required for the initiation and completion of this developmental program (Bry et al. 1996) (Fig. 12.2).

The arrest of the glycosylation program in germ-free mice is reversible upon bacterial colonization of the gut. Colonization of a germ-free mouse at any time during adulthood reinitiates the program of fucosylated glycan production. Moreover, inoculation of a single bacterial species, *Bacteroides thetaiotaomicron* (a normal bacterial inhabitant of the mouse and human gut), is sufficient to induce this developmental program (Bry et al. 1996; Hooper et al. 1999).

Genetic experiments with *B. thetaiotaomicron* have revealed that mutants lacking components of the fucose utilization operon lose the ability to induce the host fucosylation program. The link between bacterial fucose utilization and host fucosylation suggests that *B. thetaiotaomicron* depends on host-derived fucose as an energy source (Hooper et al. 1999). Like other Gram-negative obligate anaerobes, *B. thetaiotaomicron* normally colonizes the gut during weaning. Fucosylated host glycans may provide an inducible bacterial nutrient source in the early stages of colonization during this period, when Gram-negative anaerobes are carving out a niche in the competitive gut

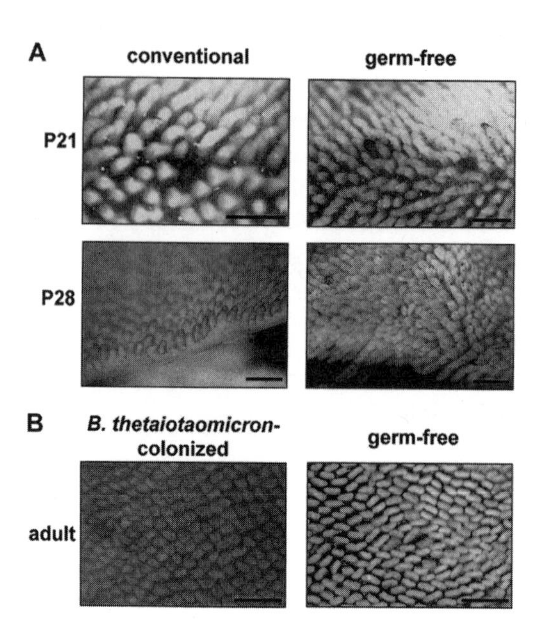

Figure 12.2. Bacterial modulation of intestinal glycan expression. Wholemounts of small intestines from postnatal day 21 (P21), P28, and adult mice were stained with peroxidase-conjugated *Ulex europeaeus* agglutinin. This lectin reacts with terminal α1,2-linked fucose produced by epithelial cells overlying finger-like villi, producing a brown staining pattern. Note that mice undergo weaning during P14–21. Scale bars = 0.5 mm. (**A**) Intestinal bacteria are required to initiate the developmentally regulated program of fucosylated glycan production, which is complete by P28. (**B**) Fucosylated glycans are induced in adult germ-free mice upon colonization with a single member of the indigenous gut flora, *B. thetaiotaomicron* (Bry et al. 1996; Hooper et al. 1999). (Figure originally published in *Glycobiology* [Oxford University Press] [Hooper and Gordon 2001]).

ecosystem. The host in turn is likely to benefit by actively promoting shifts in its gut ecology that allow it to adapt to a changing nutrient foundation. The *B. thetaiotaomicron*–gut interaction is thus conceptually similar to the *V. fischeri*–light organ association: By modulating host epithelial characteristics, both organisms shape their niche in a way that promotes and stabilizes the association.

12.4.2. Bacterial Induction of Gut Antimicrobial Protein Expression

The gut epithelium acts as a vital barrier against the vast microbial communities in the lumen. In addition to functioning as a physical obstacle to bacterial penetration of mucosal surfaces, gut epithelia also actively secrete antimicrobial proteins. Paneth cells are important participants in this type of innate mucosal defense in the small-intestine. Paneth cells are positioned at the base of small-intestinal crypts, just beneath the rapidly dividing multipotent stem cells that give rise to the differentiated mucosal epithelium. These specialized epithelial cells react to the presence of bacteria or bacterial lipopolysaccharides by discharging the microbicidal contents of their granules into the gut lumen (Ayabe et al. 2000). Their location at the crypt base suggests that they are strategically positioned to defend the stem cell niche. Such protection from microbial onslaught is likely to be important for maintaining the normal proliferative activity of epithelial progenitors and for promoting mucosal barrier function (Ayabe et al. 2000).

Angiogenin-4 (Ang4) is a prominent Paneth cell granule protein whose expression is developmentally regulated during early postnatal life (Hooper et al. 2003). Ang4 is a member of the angiogenin family, which includes proteins with proposed angiogenic functions (Strydom 1998). However, Ang4 (as well as other angiogenin family members) exhibits potent bactericidal activity that establishes its role in epithelial host defense (Hooper et al. 2003). Like fucosylated glycans, Ang4 expression in conventional mice rises dramatically during weaning and reaches adult levels soon thereafter (Fig. 12.3) (Hooper et al. 2003). In contrast, germ-free mice maintain low Ang4 expression levels (Fig. 12.3), suggesting that induction of Ang4 expression requires signals from gut bacteria. This idea was confirmed by showing that colonization of adult germ-free mice with a gut microbiota restores Ang4 expression to adult levels (Fig. 12.3). Furthermore, *B. thetaiotaomicron* alone is sufficient to stimulate conventional adult Ang4 expression levels (Hooper et al. 2003). Interestingly, bacterial induction of Ang4 mRNA expression is so far unique: Phospholipase A_2 and members of the α-defensin family do not exhibit a similar developmental pattern of expression, nor is their expression enhanced upon colonization of germ-free mice (Putsep et al. 2000; Hooper et al. 2003).

Bacterial interactions with Paneth cells thus contribute to shaping the intestinal antimicrobial arsenal during postnatal gut development. One view of the evolutionary rationale behind this interaction is that it represents an active host effort to maintain mucosal barrier integrity despite the ecological changes and the loss of passive immunity during this period. Alternatively, weaning colonizers such as *B. thetaiotaomicron* may actively "collaborate" with the host to produce bactericidal proteins that influence the composition of the evolving microbial community during postnatal gut development.

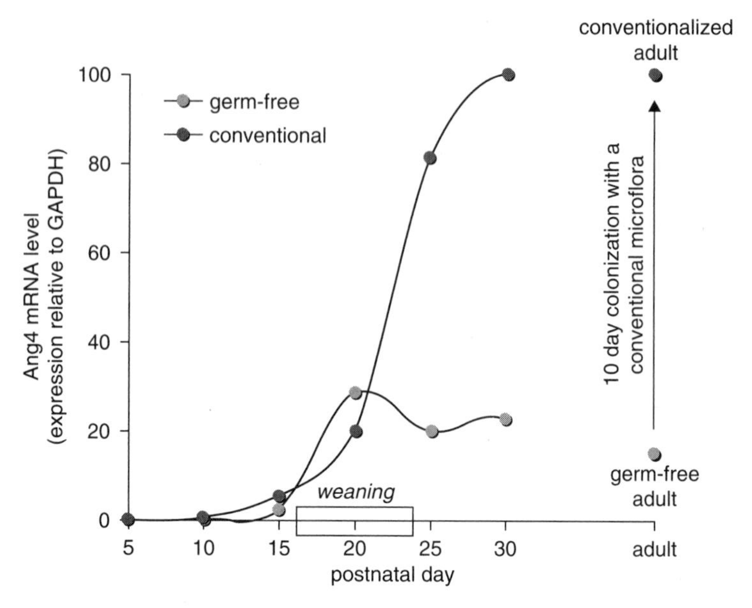

Figure 12.3. Bacterial induction of angiogenin-4 (Ang4) expression. Ang4 is a microbicidal protein expressed in small-intestinal Paneth cells (Hooper et al. 2003). Ang4 mRNA increases during weaning (P7–21) in the small intestines of developing conventionally raised mice. In contrast, its expression fails to reach adult levels in germ-free mice. Colonization of germ-free mice with a conventional gut microbiota at any time in adulthood induces Ang4 mRNA expression to conventional adult levels. Ang4 expression levels were determined by real-time quantitative reverse-transcriptase polymerase chain reaction on RNAs isolated from small intestine and are normalized to glyceraldehydes 3-phosphate dehydrogenase (GAPDH) mRNA levels (Hooper et al. 2003).

12.4.3. Bacterial Stimulation of Gut Angiogenesis

The squid–*Vibrio fischeri* interaction was among the first to reveal that symbiotic microbes can induce dramatic morphological changes during animal development. This naturally raises the question as to whether resident bacteria contribute to developmental morphogenesis in mammals. This question was recently answered when it was shown that indigenous gut bacteria provide critical inductive signals for intestinal blood vessel development.

One critical morphologic change that occurs in the gut during weaning is the development of a robust villus capillary network (Stappenbeck et al. 2002). In contrast to conventional mice, the villus capillaries of germ-free mice develop poorly during weaning and remain underdeveloped into adulthood, suggesting that gut bacteria are required for full intestinal blood vessel

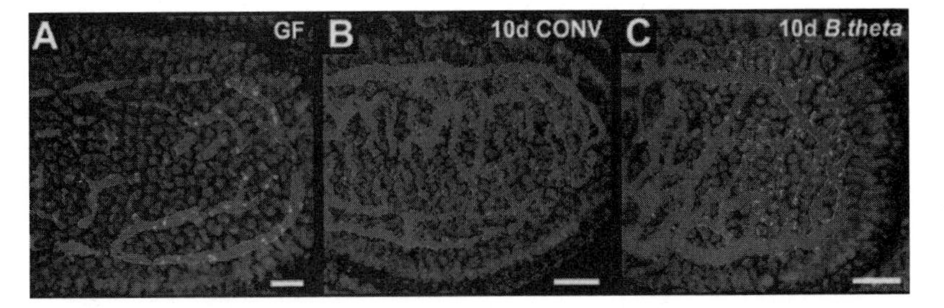

Figure 12.4. Bacterial induction of intestinal angiogenesis. Confocal scans of the capillary network in the upper third of small intestinal villi of (**A**) a germ-free mouse, (**B**) a mouse associated with a conventional (conv) microflora for 10 days, and (**C**) a mouse colonized for 10 days with *B. thetaiotaomicron* (*B. theta*). Blood vessels were detected by retro-orbital injection of FITC-conjugated dextran, followed by visualization of villi by confocal microscopy (Scale bars = 25 μm) (Stappenbeck et al. 2002). Note that the colonized mice show a higher-density capillary network. (Figure originally published in the *Proceedings of the National Academy of Sciences USA* [Stappenbeck et al. 2002]).

development. This idea was confirmed by showing that bacterial association of germ-free mice resulted in a rapid and dramatic reinduction of angiogenesis (Fig. 12.4). As with fucosylated glycans and Ang4, colonization with *B. thetaiotaomicron* alone stimulated villus capillary network development, revealing that a single bacterial species is sufficient to initiate the developmental program (Stappenbeck et al. 2002).

During villus capillary morphogenesis, Paneth cells reprise their role as mediators of a bacterially induced developmental transition. In addition to their contributions to innate immunity, Paneth cells are ideally situated to transduce bacterial signals to cells that underlie the mucosal epithelium, and have been shown to express intercellular signaling molecules such as tumor necrosis factor α (Porter et al. 2002; Lala et al. 2003). A line of transgenic mice in which Paneth cells have been genetically ablated (Garabedian et al. 1997) shows that these cells are required for full bacterial induction of villus angiogenesis (Stappenbeck et al. 2002). Thus, Paneth cells participate in at least two bacterially-induced developmental transitions: the postnatal expansion of the gut epithelial microbicidal arsenal and the development of a robust villus capillary network.

The molecular mechanisms whereby Paneth cells influence villus blood vessel development are still unclear. One possibility is that Paneth cells express factors that act directly on the vasculature. Although Ang4 is a member of the angiogenin family by homology (Hooper et al. 2003), it is not yet clear whether it is a molecular mediator of the bacterial effect on blood vessel development. Alternatively, Paneth cells may influence the expression of intercellular signaling factors by surrounding cell types. In addition, it will be important to determine whether Paneth cells are direct recipients of bacterial signals or whether they relay bacterial signals received by surrounding cells. In support of the direct interaction model, recent results indicate that Paneth cells express NOD2, a protein that is involved in bacterial recognition by mammalian cells (Ogura et al. 2003).

12.4.4. Common Themes Among Bacterial Contributions to Mammalian Gut Development

Several key similarities are exhibited by the developmental transitions represented by fucosylated glycan synthesis, expression of the microbicidal protein Ang4, and intestinal angiogenesis. First, all are normally induced during weaning in conventional mice but are arrested in germ-free mice. Second, the completion of all three developmental programs can be reinitiated in adult germ-free mice upon colonization with either a complex gut microflora or with a representative member of this flora, *B. thetaiotaomicron*. These similarities point to two general conclusions. The first is that the intestine is genetically preprogrammed to receive bacterial signals during weaning, and if these signals are lacking, gut development is arrested. The second is that resident bacteria are both necessary and sufficient to reverse this arrest in adulthood. The postnatal mammalian gut, like the squid light organ, thus appears to be poised for interaction with its microbial partners, which are essential for its normal development.

12.5. EDUCATING THE GUT IMMUNE SYSTEM

The intestinal immune system is separated from the vast luminal microflora by as little as a single epithelial layer. Gut lymphocytes furthermore undergo critical developmental transitions after encountering intestinal antigen in local lymphoid structures such as Peyer's patches (Mowat 2003). It thus seems likely that indigenous gut bacteria profoundly influence the development of the intestinal adaptive immune system.

Studies in germ-free mice have indicated that gut bacteria influence the maturation and function of several components of the mucosal immune system. Comparisons of germ-free and colonized mice have revealed that microbes drive production of mucosal immunoglobulin A (Shroff et al. 1995) through the involvement of dendritic cells (Macpherson and Uhr 2004). The microflora also contribute to the development of intraepithelial lymphocytes (IELs), as evidenced by the fact that numbers of $\alpha\beta$T cell receptor–bearing intestinal IELs ($\alpha\beta$IELs) are reduced in germ-free mice as compared with conventional mice (Umesaki et al. 1999; Umesaki et al. 1993). In addition, whereas $\gamma\delta$TCR-bearing IELs ($\gamma\delta$IELs) isolated from conventional mice are constitutively cytolytic, those isolated from germ-free mouse small intestine lack cytolytic activity (Lefrancois and Goodman 1989).

The epidemiology of inflammatory bowel diseases (IBDs) provides a circumstantial but compelling argument in favor of microbes as a driving force in mucosal immune system development. This group of diseases (which includes both ulcerative colitis and Crohn's disease) is characterized by a breakdown in immune tolerance toward resident gut bacteria (Shanahan 2002). The mechanisms underlying this dysregulation are still poorly understood. However, the incidence of IBDs has increased dramatically in the United States during the past fifty years, and has been linked to the presence of overly hygienic conditions in early childhood (Gent et al. 1994). These observations evoke the "hygiene hypothesis" (Wold 1998; Yazdanbakhsh et al. 2002): the idea that early childhood exposure to microbes can direct the maturing immune system to develop a tolerance to innocuous environmental antigens. The hygiene hypothesis is invoked most frequently to explain the rise in allergy incidence in industrialized countries, where there is increasing antibiotic usage and an emphasis on overly hygienic environments. However, IBDs and allergies are similar in that both are characterized by inappropriate immune responses to otherwise harmless environmental antigens. Thus, the hygiene hypothesis may also offer insight into the underlying causes of some IBDs.

The hygiene hypothesis is often placed in the context of T cell development. It has been suggested, for example, that lack of early microbial exposure directs T cells toward development of a proinflammatory phenotype (Th1 and Th2), whereas early exposure to key antigens promotes development of a regulatory phenotype (Tr), characterized by IL-10 and TGF-β secretion (Yazdanbakhsh et al. 2002). In fact, an extensive array of evidence shows that T cell development is heavily skewed toward inflammatory phenotypes in IBD models (De Winter et al. 1999). Furthermore, T cells with Tr phenotypes can ameliorate inflammatory responses in models of mucosal inflammation (Das et al. 2003). Thus, a key question in mucosal immune system development is whether exposure to key bacterial antigens during early postnatal life can direct the outcome of T cell development. If so, then it is likely that exposure to certain populations of gut microbes during early postnatal life is crucial for normal adaptive immune system development and the acquisition of tolerance to resident bacteria.

12.6. CONCLUSIONS

We are just beginning to understand the extent to which our indigenous microbes are intertwined in our biology. It is already becoming clear that our prokaryotic partners play an essential inductive role in normal gut maturation. However, many fundamental questions remain. What is the nature of the molecular conversation between normal resident bacteria and host intestinal cells? Which genes in each partner are required for this intergenomic dialogue? Obtaining answers to these questions will depend in large part on developing a comprehensive picture of the ecological changes that occur in the microbiota during gut development, and a more thorough definition of the microbial ecology of the adult host. This will require the application of modern molecular ecology techniques that are based on sequence comparisons of nucleic acids rather than culture-based techniques (Suau et al. 1999). In addition, ongoing studies of simple associations such as the squid–*Vibrio* interaction will certainly continue to yield important clues about the molecular foundations of animal–bacterial interactions. Finally, studies of host–microbial interactions in other vertebrate models, such as zebrafish, which are amenable to large-scale forward genetic screens, will undoubtedly generate many novel insights into how indigenous bacteria drive animal development. The future is thus likely to hold a dramatic shift in our thinking about our interactions with the prokaryotic world. Instead of something to be avoided at all costs, perhaps these relationships will eventually be viewed as an integral part of our biology.

ACKNOWLEDGMENTS

L.V.H. is supported by a Burroughs Wellcome Foundation Career Award in the Biomedical Sciences.

REFERENCES

Ayabe, T., Satchell, D. P., Wilson, C. L., Parks, W. C., Selsted, M. E., and Ouellette, A. J. (2000). Secretion of microbicidal alpha-defensins by intestinal Paneth cells in response to bacteria. *Nature Immunology* **1**, 113–118.

Biol, M. C., Martin, A., and Louisot, P. (1992). Nutritional and developmental regulation of glycosylation processes in digestive organs. *Biochimie* **74**, 13–24.

Bry, L., Falk, P. G., Midtvedt, T., and Gordon, J. I. (1996). A model of host-microbial interactions in an open mammalian ecosystem. *Science* **273**, 1380–1383.

Darveau, R. P., McFall-Ngai, M., Ruby, E., Miller, S., and Mangan, D. (2003). Host tissues may actively respond to beneficial microbes. *ASM News* **69**, 186–191.

Das, G., Augustine, M. M., Das, J., Bottomly, K., Ray, P., and Ray, A. (2003). An important regulatory role for CD4+CD8 alpha alpha T cells in the intestinal epithelial layer in the prevention of inflammatory bowel disease. *Proceedings of the National Academy of Sciences USA* **100**, 5324–5329.

De Winter, H., Cheroutre, H., and Kronenberg, M. (1999). Mucosal immunity and inflammation. II. The yin and yang of T cells in intestinal inflammation: pathogenic and protective roles in a mouse colitis model. *American Journal of Physiology* **276**, G1317–1321.

Dusheck, J. (2002). It's the ecology, stupid! *Nature* **418**, 578–579.

Foster, J. S., Apicella, M. A., and McFall-Ngai, M. J. (2000). *Vibrio fischeri* lipopolysaccharide induces developmental apoptosis, but not complete morphogenesis, of the *Euprymna scolopes* symbiotic light organ. *Developmental Biology* **226**, 242–254.

Foster, J. S. and McFall-Ngai, M. J. (1998). Induction of apoptosis by cooperative bacteria in the morphogenesis of host epithelial tissues. *Development Genes and Evolution* **208**, 295–303.

Garabedian, E. M., Roberts, L. J., McNevin, M. S., and Gordon, J. I. (1997). Examining the role of Paneth cells in the small intestine by lineage ablation in transgenic mice. *Journal of Biological Chemistry* **272**, 23729–23740.

Gent, A. E., Hellier, M. D., Grace, R. H., Swarbrick, E. T., and Coggon, D. (1994). Inflammatory bowel disease and domestic hygiene in infancy. *Lancet* **343**, 766–767.

Hooper, L. V. and Gordon, J. I. (2001). Glycans as legislators of host-microbial interactions: spanning the spectrum from symbiosis to pathogenicity. *Glycobiology* **11**, 1R–10R.

Hooper, L. V., Midtvedt, T., and Gordon, J. I. (2002). How host–microbial interactions shape the nutrient environment of the mammalian intestine. *Annual Review of Nutrition* **22**, 283–307.

Hooper, L. V., Stappenbeck, T. S., Hong, C. V., and Gordon, J. I. (2003). Angiogenins: a new class of microbicidal proteins involved in innate immunity. *Nature Immunology* **4**, 269–273.

Hooper, L. V., Xu, J., Falk, P. G., Midtvedt, T., and Gordon, J. I. (1999). A molecular sensor that allows a gut commensal to control its nutrient foundation in a competitive ecosystem. *Proceedings of the National Academy of Sciences USA* **96**, 9833–9838.

Jenkins, S. L., Wang, J., Vazir, M., Vela, J., Sahagun, O., Gabbay, P., Hoang, L., Diaz, R. L., Aranda, R., and Martin, M. G. (2003). Role of passive and adaptive immunity in influencing enterocyte-specific gene expression. *American Journal of Physiology: Gastrointestinal and Liver Physiology* **285**, G714–725.

Lala, S., Ogura, Y., Osborne, C., Hor, S. Y., Bromfield, A., Davies, S., Ogunbiyi, O., Nunez, G., and Keshav, S. (2003). Crohn's disease and the NOD2 gene: a role for paneth cells. *Gastroenterology* **125**, 47–57.

Lamarcq, L. H. and McFall-Ngai, M. J. (1998). Induction of a gradual, reversible morphogenesis of its host's epithelial brush border by *Vibrio fischeri*. *Infection and Immunity* **66**, 777–785.

Lefrancois, L. and Goodman, T. (1989). *In vivo* modulation of cytolytic activity and Thy-1 expression in TCR-gamma delta+ intraepithelial lymphocytes. *Science* **243**, 1716–1718.

Mackie, R. I., Sghir, A., and Gaskins, H. R. (1999). Developmental microbial ecology of the neonatal gastrointestinal tract. *American Journal of Clinical Nutrition* **69**, 1035S–1045S.

Macpherson, A. J. and Uhr, T. (2004). Induction of protective IgA by intestinal dendritic cells carrying commensal bacteria. *Science* **303**, 1662–1665.

McFall-Ngai, M. J. (2002). Unseen forces: the influence of bacteria on animal development. *Developmental Biology* **242**, 1–14.

Montgomery, M. K. and McFall-Ngai, M. (1994). Bacterial symbionts induce host organ morphogenesis during early postembryonic development of the squid *Euprymna scolopes*. *Development* **120**, 1719–1729.

Moore, W. E. and Holdeman, L. V. (1974). Human fecal flora: the normal flora of 20 Japanese-Hawaiians. *Applied Microbiology* **27**, 961–979.

Mowat, A. M. (2003). Anatomical basis of tolerance and immunity to intestinal antigens. *Nature Reviews in Immunology* **3**, 331–341.

Ogura, Y., Lala, S., Xin, W., Smith, E., Dowds, T. A., Chen, F. F., Zimmermann, E., Tretiakova, M., Cho, J. H., Hart, J., et al. (2003). Expression of NOD2 in Paneth cells: a possible link to Crohn's ileitis. *Gut* **52**, 1591–1597.

Porter, E. M., Bevins, C. L., Ghosh, D., and Ganz, T. (2002). The multifaceted Paneth cell. *Cellular and Molecular Life Sciences* **59**, 156–170.

Putsep, K., Axelsson, L. G., Boman, A., Midtvedt, T., Normark, S., Boman, H. G., and Andersson, M. (2000). Germ-free and colonized mice generate the same products from enteric prodefensins. *Journal of Biological Chemistry* **275**, 40478–40482.

Rotimi, V. O. and Duerden, B. I. (1981). The development of the bacterial flora in normal neonates. *Journal of Medical Microbiology* **14**, 51–62.

Savage, D. C. (1977). Microbial ecology of the gastrointestinal tract. *Annual Review of Microbiology* **31**, 107–133.

Savage, D. C. (1986). Gastrointestinal microflora in mammalian nutrition. *Annual Review of Nutrition* **6**, 155–178.

Savage, D. C., Siegel, J. E., Snellen, J. E., and Whitt, D. D. (1981). Transit time of epithelial cells in the small intestines of germfree mice and ex-germfree mice associated with indigenous microorganisms. *Applied and Environmental Microbiology* **42**, 996–1001.

Shanahan, F. (2002). Crohn's disease. *Lancet* **359**, 62–69.

Shroff, K. E., Meslin, K., and Cebra, J. J. (1995). Commensal enteric bacteria engender a self-limiting humoral mucosal immune response while permanently colonizing the gut. *Infection and Immunity* **63**, 3904–3913.

Stappenbeck, T. S., Hooper, L. V., and Gordon, J. I. (2002). Developmental regulation of intestinal angiogenesis by indigenous microbes via Paneth cells. *Proceedings of the National Academy of Sciences USA* **99**, 15451–15455.

Steege, J. C., Buurman, W. A., and Forget, P. P. (1997). The neonatal development of intraepithelial and lamina propria lymphocytes in the murine small intestine. *Developmental Immunology* **5**, 121–128.

Strydom, D. J. (1998). The angiogenins. *Cellular and Molecular Life Sciences* **54**, 811–824.

Suau, A., Bonnet, R., Sutren, M., Godon, J. J., Gibson, G. R., Collins, M. D., and Dore, J. (1999). Direct analysis of genes encoding 16S rRNA from complex communities reveals many novel molecular species within the human gut. *Applied and Environmental Microbiology* **65**, 4799–4807.

Umesaki, Y., Okada, Y., Matsumoto, S., Imaoka, A., and Setoyama, H. (1995). Segmented filamentous bacteria are indigenous intestinal bacteria that activate intraepithelial lymphocytes and induce MHC class II molecules and fucosyl asialo GM1 glycolipids on the small intestinal epithelial cells in the ex-germ-free mouse. *Microbiology and Immunology* **39**, 555–562.

Umesaki, Y., Setoyama, H., Matsumoto, S., Imaoka, A., and Itoh, K. (1999). Differential roles of segmented filamentous bacteria and clostridia in development of the intestinal immune system. *Infection and Immunity* **67**, 3504–3511.

Umesaki, Y., Setoyama, H., Matsumoto, S., and Okada, Y. (1993). Expansion of alpha beta T-cell receptor-bearing intestinal intraepithelial lymphocytes after microbial colonization in germ-free mice and its independence from thymus. *Immunology* **79**, 32–37.

Varki, A. (1993). Biological roles of oligosaccharides: all of the theories are correct. *Glycobiology* **3**, 97–130.

Wold, A. E. (1998). The hygiene hypothesis revised: is the rising frequency of allergy due to changes in the intestinal flora? *Allergy* **53**, 20–25.
Wostmann, B. S. (1981). The germfree animal in nutritional studies. *Annual Review of Nutrition* **1**, 257–279.
Wostmann, B. S., Larkin, C., Moriarty, A., and Bruckner-Kardoss, E. (1983). Dietary intake, energy metabolism, and excretory losses of adult male germfree Wistar rats. *Laboratory Animal Science* **33**, 46–50.
Yazdanbakhsh, M., Kremsner, P. G., and van Ree, R. (2002). Allergy, parasites, and the hygiene hypothesis. *Science* **296**, 490–494.

CHAPTER 13

Virulence or commensalism: Lessons from the urinary tract

Göran Bergsten, Björn Wullt, and Catharina Svanborg

13.1. INTRODUCTION

Chapter 11 focused on the interactions of Bordetellae with the respiratory tract. In this chapter, the subject is the dynamics of the interactions of Gram-negative bacteria with the urinary tract. The Koch-Henle postulates were formulated to prove causality between a microbial isolate and the disease in the patient from which it was isolated. To fulfil the postulates, the isolate should be cultured to homogeneity *in vitro*, and be shown to reproduce disease after re-infection of a susceptible host. This line of reasoning may also be applied to bacterial virulence factors. The molecular Koch postulates state that the role of a single virulence factor should be proven by discrete genetic manipulations that alter the expression of this virulence trait, and that host interactions or disease pathogenesis should be changed as a result of this manipulation (Falkow 1988).

The molecular Koch postulates focus on bacterial virulence and disease but do not consider the asymptomatic carrier state, even though this may be the most prevalent outcome of host–parasite interaction. Different host organisms offer highly defined ecological niches for bacterial colonisation, and strains that establish a critical mass at these sites have an evolutionary advantage as compared with the virulent strains that cause disease but are eliminated by the host defence. There is a need to explore the properties that regulate survival and persistence of the carrier strains *in vivo* and the molecular determinants that prevent the host response from eliminating those strains.

13.2. THE URINARY TRACT INFECTION MODEL

The urinary tract usually remains sterile, despite frequent microbial challenge. Yet, in susceptible individuals, bacteriuria is established. Urinary tract infections (UTIs) are quite common (Kunin 1987), as an estimated 150 million people are diagnosed annually with recurrent symptomatic UTI (Stamm and Norrby 2001) and about half of all adult women suffer from symptomatic UTI at least once (Kunin 1987; Stamm et al. 1991). These rough frequency estimates illustrate the magnitude of the problem but do not properly reflect the diversity of disease in the urinary tract. Urinary tract infections may be acute or chronic, symptomatic or varying in severity and localisation, sporadic or recurrent, and may cause end-stage renal disease (Fig. 13.1). Paradoxically, however, the most common form of UTI is asymptomatic bacteriuria (ABU) (Fig. 13.1). Patients with ABU may carry $>10^5$ cfu/ml of urine for months or years without developing symptoms or sequelae (Kass 1956; Lindberg 1975; Wullt et al. 2003). ABU is quite frequent, occurring in about one percent of girls, two to eleven percent of pregnant women and up to twenty percent of elderly men and women (Kunin 1987).

Urinary tract infections thus provide a highly relevant model in which to study the molecular basis of disease diversity and the interactions between bacteria and host that determine the transition from asymptomatic carriage to disease.

13.2.1. The "Two-Step" of Mucosal Activation and Disease Induction

The difference in disease severity reflects the virulence of the infecting strain and the propensity of the host to respond to infection (Svanborg et al. 2001a). In asymptomatic carriers, the mucosa remains inert despite large bacterial numbers in the lumen. Conversely, in acute pyelonephritis, both local and systemic inflammatory response pathways are activated.

We have identified molecular interactions that determine if bacteria in the lumen will break the inertia of the mucosal barrier and trigger the host response (step 1, Fig. 13.2). Without step 1, the host is not alerted to the presence of bacteria and the mucosal surfaces remain inert. Inertia may result either from a lack of bacterial virulence or from host unresponsiveness caused by aborted signalling. In both cases, the result is an asymptomatic carrier state, allowing more than 10^5 cfu/ml of bacteria to persist without evoking a host response (Wullt et al. 2003).

We have also identified important effector mechanisms involved in bacterial clearance (step 2, Fig. 13.2). Without step 2, infection causes an

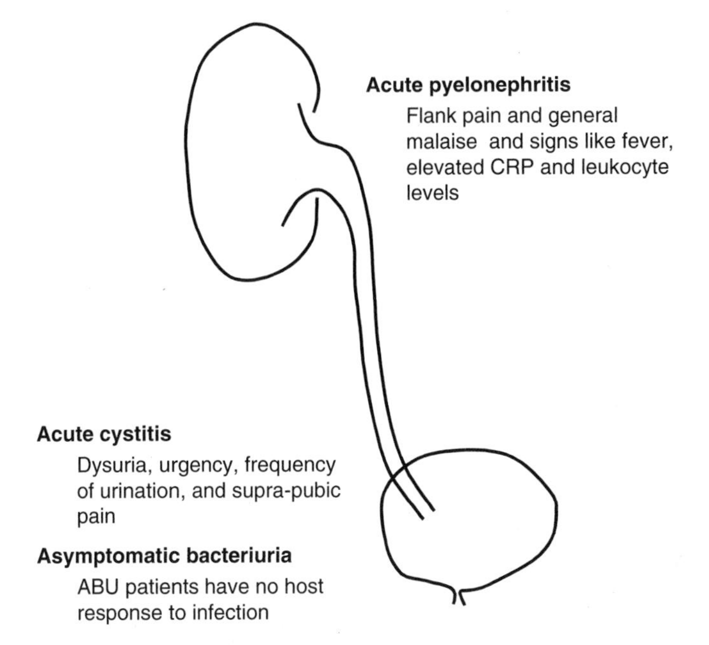

Figure 13.1. Symptomatic UTI and asymptomatic bacteriuria. Acute pyelonephritis is an infection of the renal pelvis and kidneys which may become systemic, with bacteremia. The symptoms are caused by local and generalised inflammation (de Man et al. 1991), giving rise to flank pain and general malaise and signs such as fever and elevated C-reactive protein (CRP) and leukocyte levels (Svanborg et al. 2001a). Before the use of antibiotics, acute pyelonephritis was lethal in about fifteen percent of the cases, and Gram-negative sepsis emanating from the urinary tract remains a major cause of death.

Acute cystitis is characterised by dysuria, urgency, and frequency of urination, and may be accompanied by supra-pubic pain (Kunin 1987). The symptoms reflect the inflammatory response in the lower urinary tract, and the patients have pyuria and sometimes hematuria. ABU is commonly detected at screening of defined populations or at follow-up visits after a first symptomatic infection (Kunin 1987). Classically, ABU patients have no host response to infection, but low-level cytokine responses and leukocytes in urine may occur (Lindberg 1975; Kass 1956; Wullt et al. 2003). Asymptomatic bacteriuria is quite frequent, occurring in about one percent of girls, two to eleven percent of pregnant women, and up to twenty percent of elderly men and women (Kunin 1987).

exaggerated inflammatory state with tissue destruction (Hang et al. 2000). In the mouse, step 2 depends on the expression of the mIL-8Rh receptor, and there is evidence that CXCR1 mutations may disarm the host defence and cause severe disease also in humans (Frendeus et al. 2000).

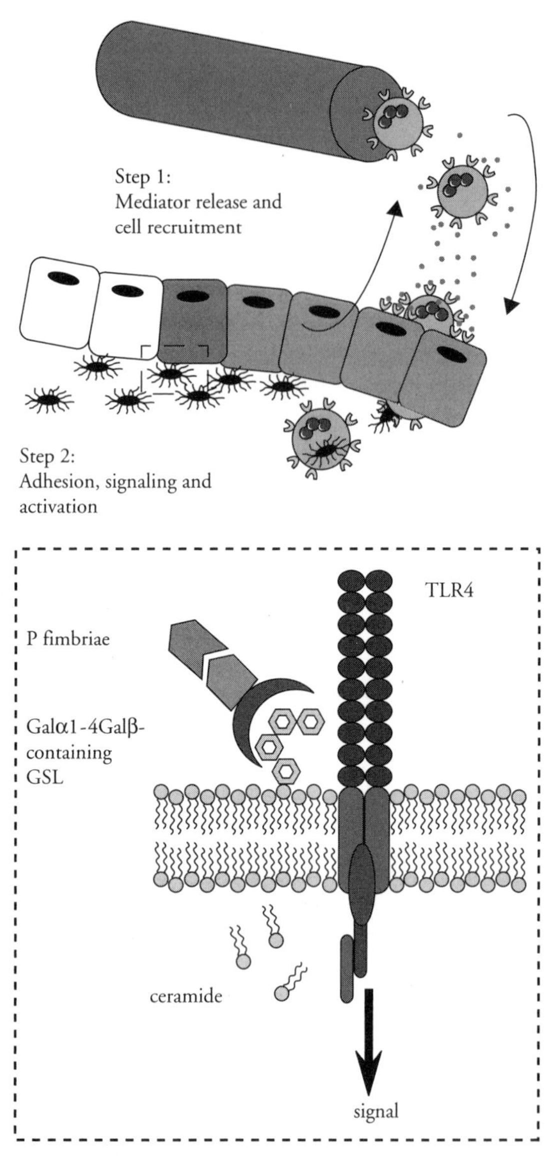

Figure 13.2. Host response induction by adhering bacteria. Bacterial attachment to epithelial cell receptors activates specific TLR4-dependent trans-membrane signalling pathways (step 1). Activated epithelial cells respond by secretion of cytokines and expression of receptors. Inflammatory cells are recruited to the mucosa and kill bacteria on their way across the mucosa (step 2). P fimbriae bind to the Galα1$\rightarrow$4Galβ receptor epitope in the globoseries of glycosphingolipids (GSLs) and activate epithelial cells through TLR4 in an LPS-independent, Ser/Thr protein kinase–dependent manner, and by the release of ceramide.

13.3. BACTERIAL DETERMINANTS OF CARRIAGE AND VIRULENCE

The uro-pathogens first establish a population outside the urinary tract, at a site that becomes the reservoir for infection. Virulence for the urinary tract may be "coincidental" as the properties which promote survival in the urinary tract also give the bacteria a colonisation advantage in the reservoir (Levin and Svanborg Eden 1990). For example, P-fimbriated bacteria colonise the large intestine, spread to the vaginal and peri-urethral areas, and ascend into the urinary tract more efficiently than strains lacking the *pap* (pyelonephritis-associated pili) operon (Plos et al. 1995).

The pathogenic *Escherichia coli* strains belong to a restricted set of clones which have the capacity to express: adherence factors, iron sequestering molecules, exo- and endo-toxins, capsules, and so on (Mabeck et al. 1971; Svanborg-Eden et al. 1976; Welch et al. 1983; Funfstuck et al. 1986; Stenqvist et al. 1987; Orskov et al. 1988; Johnson 1991) (Fig. 13.3). The virulence genes are located on chromosomal gene clusters called "pathogenicity islands" (Hacker et al. 1990 – see also Chapter 3 for a discussion of pathogenicity islands). The different virulence factors act in concert, and their expression can be regulated by the host and by environmental signals (Hacker et al. 1990, 1997).

The ABU strains, on the other hand, have been regarded as "avirulent" compared with the uro-pathogenic clones, and have been assumed to lack pathogenicity islands. Early studies showed that many carrier strains lacked

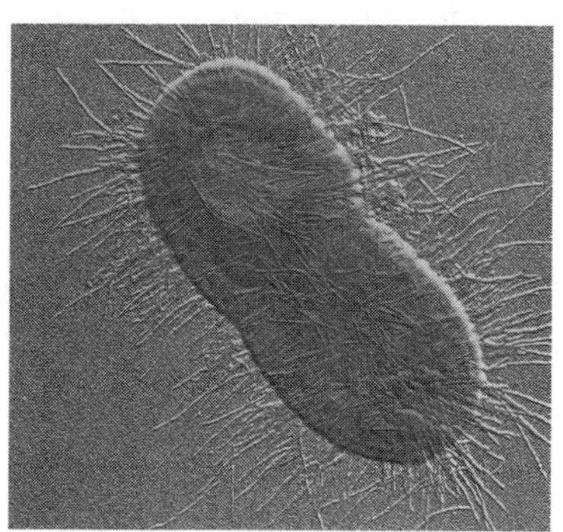

	Pyelonephritis	ABU
Adherence		
P fimbriae	76-100%	17-18%
type 1 fimbriae	91-100%	82-100%
Toxins		
LPS		
structure	complex	simple
amount	high	low
Hemolysin	60-75%	17-33%
Capsule	++	(+/-)

Figure 13.3. Bacterial virulence factors. Transmission electron micrograph of *E. coli* expressing P fimbriae after platinum shadowing (from Svanborg et al. 2002). Representation of virulence factors in uro-pathogenic *E. coli*. Proportion (%) of *E. coli* strains (pyelonephritis or ABU origin) expressing different virulence factors (data from Johnson 1991; Vaisanen-Rhen et al. 1984; O'Hanley et al. 1985a).

a defined O antigen (Lindberg et al. 1975), were un-encapsulated (Plos et al. 1990), and failed to adhere to uro-epithelial cells in vitro (Svanborg-Eden et al. 1976). More recently, the ABU strains have been shown to carry virulence genes even though they fail to express the phenotype (Plos et al. 1990, 1995). It is quite probable that the ABU strains express adherence factors and other critical adaptive properties while growing in the faecal reservoir and during the establishment of bacteriuria, but then down-regulate the phenotype to facilitate long-term persistence in the urinary tract. The bacterial properties that positively influence long-term persistence have not been investigated, but should be a focus of study.

13.3.1. Bacterial Adherence

Bacterial adherence initiates the tissue attack in UTI pathogenesis and determines if the innate host response will be activated. Early studies showed that the virulent clones bound more avidly to uro-epithelial cells than the carrier strains and identified adherence as a virulence factor in the human urinary tract (Svanborg-Eden et al. 1976). The uro-pathogenic clones are fimbriated (Svanborg-Eden et al. 1977) and may express several adhesive surface organelles, such as Dr, afa, S, P, and type 1 fimbriae (Leffler and Svanborg-Eden 1980; Walz et al. 1985; Virkola et al. 1988; Nowicki et al. 1989), but P fimbriae show the strongest association to acute disease severity, with at least ninety percent of acute pyelonephritis but less then twenty percent of ABU strains expressing this phenotype (Leffler and Svanborg-Eden 1981; Vaisanen et al. 1981; Plos et al. 1995). This is in contrast to type 1 fimbriae, which can be expressed by more than ninety percent of commensal and uro-pathogenic *E. coli* (Johnson 1991; Hagberg et al. 1981; Yamamoto et al. 1995).

13.3.2. P Fimbriae

P fimbriae are encoded by the *pap* gene cluster on the chromosome of uro-pathogenic *E. coli* (Hull et al. 1981) (Fig. 13.4). Six of the eleven genes encode structural proteins: PapA, PapH, PapK, PapE, PapF, and PapG (Lindberg et al. 1987; Hultgren et al. 1993; Kuehn et al. 1992). P fimbriae are composite fibres consisting of a thin tip fibrillum that is joined to the distal end of the pilus rod (Lindberg et al. 1987). The PapG adhesin is joined to the distal end of the fibrillum via the PapF adaptor, and the proximal end of the fibrillum is joined to the rod by the PapK adaptor (Jacob-Dubuisson et al. 1993; Dodson et al. 2001).

The uro-epithelial cells present Galα1→4Galβ receptor epitopes to the PapG adhesin (Virkola et al. 1988). These glycosphingolipid (GSL) receptors are abundant in the epithelial cells lining the urinary tract and the renal tubuli (Bock et al. 1985; Virkola et al. 1988). In addition, the receptor-GSLs are P blood group antigens (P_1, P_2, P^k). The expression varies between individuals of blood groups P_1 or P_2, and p individuals lack these receptor structures on epithelial cells and other cells (Marcus et al. 1976).

13.3.3. P Fimbriae Enhance Bacterial Virulence

1. By promoting the colonisation of the intestine and the spread to the urinary tract (Wold et al. 1992; Plos et al. 1995).
2. By promoting the early establishment of bacteriuria. The adherent state has been postulated to facilitate nutrient access, and bacterial growth (Zobell 1943).
3. By activating the innate host response. Adherence triggers mucosal inflammation by activation of specific transmembrane signalling pathways, causing cells in the urinary tract mucosa to secrete pro-inflammatory mediators. This step is critical to establish local inflammation and to recruit effector cells of the host defence system (Shahin et al. 1987; de Man et al. 1989; Hedges et al. 1991; Condron et al. 2003).
4. By resisting neutrophil killing (Svanborg-Eden et al. 1984; Tewari et al. 1994).

13.4. EVIDENCE FROM THE HUMAN UTI MODEL

The classical and molecular Koch postulates should ideally be proven in a host that is naturally susceptible to the disease that is being studied. Often, this is not possible, because of the potentially damaging effect of deliberate infection. Humans are the correct host for UTI, and all the disease epidemiology on which the virulence concept is based is from human patients. It is fortunate that the human urinary tract is amenable to inoculation and that there is a tradition of using local procedures to treat both infections and tumours. For example, deliberate inoculations with mycobacteria are routinely used in cancer therapy (Azuma and Seya 2001), and the carriage of *E. coli* in patients with ABU has been shown to protect against super-infection with fully virulent strains (Hansson et al. 1989a).

Epidemiologic studies have shown that ABU protects against symptomatic UTI. Children who were treated with antibiotics to remove bacteriuria

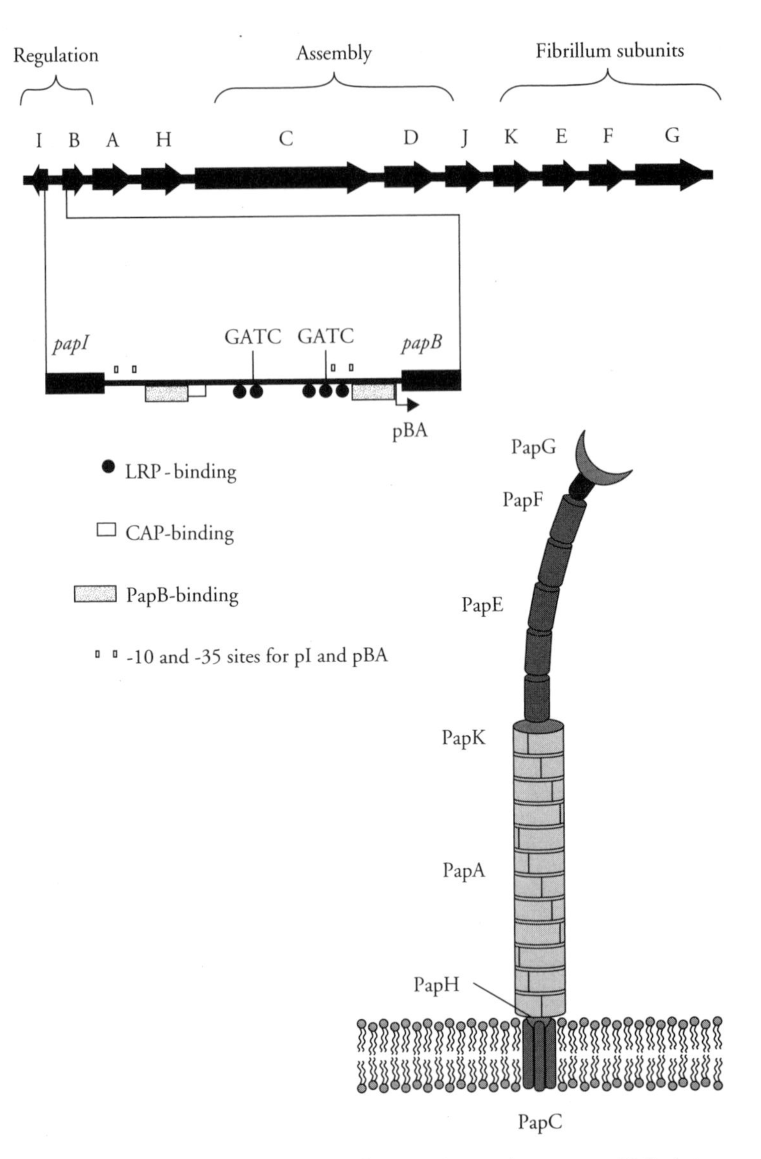

Figure 13.4. The *pap* gene cluster, *pap* regulatory region, and structure of P fimbriae. P fimbriae is a hetero-polymer of the structural protein, the membrane-anchoring proteins, and adhesive proteins. PapA is the major fimbrial subunit, PapH anchors the fimbrial rod to the cell surface, PapK is the adaptor of the fibrillum, PapE forms the major part of the fibrillum, and PapF connects PapE to the adhesin PapG. PapD is a periplasmic chaperon, aiding the transport of structural units to the usher, PapC. PapI and PapB are involved in the regulation of transcription of *pap*. The regulatory region of *pap* shows

were shown to develop symptomatic infections significantly more often than those who were left untreated and remained as asymptomatic carriers (Hansson et al. 1989a; Lindberg et al. 1978). This led to the hypothesis that ABU protects the urinary tract, by competitive exclusion of more virulent bacterial strains (Agace and Svanborg 1995). In a study by Hansson et al. (1989b), girls with ABU were left untreated and the bacterial strain was followed by multilocus enzyme electrophoresis and serotyping. The median duration of ABU was 2.5 years, and only twenty-four strain changes were observed in the total of 151 patient-years. This low change rate supports non-treatment of ABU as a safe way of avoiding symptomatic UTIs (Hansson et al. 1989b).

These findings have validated deliberate establishment of bacteriuria in UTI-prone patients who cannot be treated by conventional methods (Hagberg et al. 1989; Andersson et al. 1991). As a part of the deliberate inoculation protocol, it possible to vary the properties of the bacteria and to identify factors involved in colonisation and host response induction.

13.5. P FIMBRIAE ENHANCE BACTERIAL ESTABLISHMENT IN THE URINARY TRACT

We have tested if P fimbriae fulfil the molecular Koch postulates as independent virulence factors in the human urinary tract. The ABU strain *E. coli* 83972 was selected to represent the carrier strains that cause persistent bacteriuria without breaking the inertia of the mucosal barrier or causing disease. The strain is well adapted for survival in the urinary tract, and carries chromosomal DNA sequences homologous with fimbrial as well as some other virulence genes, but does not produce fimbriae. Thus, ABU strains carry incomplete virulence genes and fail to express the phenotype because of partial deletions in those genes. This mechanism may represent an efficient adaptation to survival in the urinary tract as the carrier strains both need to express adherence and adaptation factors, in order to establish a critical mass in the intestinal flora and persist in the urinary tract (Andersson et al. 1991; Hull et al. 1999).

The ABU host strain was transformed with plasmids carrying the *pap* cluster, and the P-fimbriated and non-fimbriated isogens were used to

Figure 13.4 (*cont.*) the location of Dam target sites, GATC, the binding site for Lrp, CAP, and PapB. Transcription initiates when the proximal GATC is methylated, and the distal GATC is un-methylated allowing the RNA polymerase to bind to the -10 and -35 sites for pBA. (Adapted from Blomfield and van der Woude 2002.)

inoculate human hosts. The P-fimbriated transformants established faster, and required fewer inoculations to reach 10^5 cfu/ml than *E. coli* 83972. Group-wise comparisons of all colonisation attempts showed that bacterial numbers were higher after inoculations with *E. coli* 83972 *pap+/prs+* than with *E. coli* 83972 ($P < 0.001$). Intra-individual analysis of patients colonised with both strains, but on different occasions, showed higher bacterial counts with the P-fimbriated transformants. If only one or two inoculations had been used, *E. coli* 83972 would have been established on fewer occasions than the P-fimbriated transformant ($p < 0.05$) (Wullt et al. 2000, 2001a).

To study the effect of adhesion *in vivo*, a *papG* deletion mutant was constructed, and the host response to the full-length and the non-adhesive fimbriae was compared. PapG is the tip adhesin of P fimbriae that mediates binding to Galα1$\rightarrow$4Galβ receptor epitopes in the globoseries of GSL on uro-epithelial cells. The coupling of PapG to this receptor activates the Toll-like receptor (TLR)4 dependent signalling cascades, leading to the secretion of pro-inflammatory cytokines (see Chapter 15 for further discussion of Toll-like receptors). In this study, *E. coli pap+* adhered to uro-epithelial cells *in vivo* and triggered a significant host response, but the papG deletion mutant failed to adhere and was unable to break the mucosal inertia. These findings confirm the importance of PapG-mediated host cell interactions for the host response (Bergsten et al. 2004).

The role of P fimbriae in bacterial persistence and virulence in the urinary tract has been debated. The early studies in the mouse UTI model showed that P fimbriae enhance bacterial persistence (Hagberg et al. 1983; O'Hanley et al. 1985b; Lanne et al. 1995), but later studies have produced contradictory results (Mobley et al. 1993; Bahrani-Mougeot et al. 2002). The importance of fimbrial expression in the human urinary tract environment has also been unclear (Andersson et al. 1991). We conclude that P fimbriation provides a colonisation advantage during the first few days of bacterial establishment, but that P fimbriae–mediated adherence may be less important for long-term bacterial carriage.

13.5.1. P Fimbriae Trigger Mucosal Responses in the Human Urinary Tract

The effect of P fimbriae on the innate host response was examined in the inoculated patients, by comparing inoculations with *E. coli* 83972 and the P-fimbriated transformants. Cytokine concentrations in urine were measured, and recruited inflammatory cells were counted.

The P-fimbriated strain triggered a local inflammatory response in all patients, but the responses to the ABU strain were low or absent. Intra-individual comparisons in patients who became carriers of the P-fimbriated transformants and *E. coli* 83972 on different occasions, invariably showed higher neutrophil numbers and IL-8 concentrations following inoculation with the P-fimbriated variant. The host response was also found to vary with the expression of P fimbriae, as urine samples that contained fimbriated bacteria also contained higher numbers of neutrophils and cytokines than did samples with the negative phenotype (Wullt et al. 2001a; Wullt et al. 2001b).

13.5.2. PapG Dependent Adhesion Breaks Mucosal Inertia and Triggers the Innate Host Response

The role of adhesion was addressed using isogenic variants of *E. coli* 83972 differing in *papG* genotype, and the in vivo expression of fimbriae was monitored, using a *gfp* reporter plasmid. The patients were subjected to intra-vesical inoculation with *E. coli pap+* or a *papG* deletion mutant and analysed for epithelial cell adherence by fluorescence microscopy and for in vivo regulation of fimbrial expression using the *gfp* reporter.

E. coli pap+ was shown to adhere to uro-epithelial cells in the human urinary tract, and to trigger the innate host response in each of the patients who was inoculated with this strain. No adherence or host responses were seen after inoculation of the same patients with the *papG* deletion mutant.

We conclude that carrier strains persist apparently without triggering the innate host response, and that P fimbriation converts a carrier strain to a host response inducer in the human urinary tract. This effect is adhesion dependent as P fimbriae with the PapG adhesin showed this effect but the PapG deletion mutant did not (Bergsten et al. 2004).

13.6. HOST DETERMINANTS OF THE MUCOSAL RESPONSE

13.6.1. Host Resistance Relies on Innate Immunity

The urinary tract is normally kept sterile despite frequent challenge from the environmental flora (Agace and Svanborg 1995; Cox and Hinman 1965). The anti-microbial defence of the urinary tract relies on innate immunity (Svanborg-Eden et al. 1985; Hagberg et al. 1985; Shahin et al. 1987). There is no pre-existing mucosal immune response in the uninfected urinary tract, yet following intra-vesical inoculation of *E. coli* into the bladder, bacteriuria is cleared within hours or days. A rapid initial reduction in bacterial counts during the first 2 hours is followed by a gradual decline until about 24 hours, and complete elimination by two to seven days, depending on virulence and host background. The reduction in bacterial numbers during the first two hours depends largely on the urine flow, but specific host defence molecules such as defensins may be involved (Parsons et al. 1977; Schulte-Wissermann et al. 1985; Ganz 2001). During this time, the epithelium is activated to produce inflammatory mediators (step 1) and neutrophils are recruited (Fig. 13.1). The subsequent reduction in bacterial numbers depends on these recruited cells, which may achieve complete clearance of infection (step 2, Fig. 13.1) (Hagberg et al. 1983; Shahin et al. 1987; Haraoka et al. 1999). Thus, in most cases, UTIs are cured before a specific immune response is activated. A specific immune response may occur later, in patients who develop severe symptomatic disease or chronic infection (Agace and Svanborg 1995).

We have studied the logistics of the innate host response and identified that epithelial cells are the first responders to the infecting bacterial strain, and directors of the subsequent mucosal response (Svanborg et al. 1999; Svanborg et al. 2001a). Uro-pathogenic *E. coli* stimulate cytokine production in human bladder and kidney cell lines and in freshly isolated uro-epithelial cells (Hedges et al. 1992; Agace et al. 1993a,b), with de novo synthesis of cytokine-specific mRNA (IL-1α, IL-1β, IL-6, and IL-8) (Svanborg et al. 1999; Hedges et al. 1994). Bacterial adherence enhances epithelial cell cytokine responses (Linder et al. 1991; Svensson et al. 1994), and P- and type 1–fimbriated *E. coli* strains elicit higher cytokine responses in vitro than do isogenic, non-fimbriated strains (Hedges et al. 1991; Svensson et al. 1994; Godaly et al. 1998); a similar effect was proven for P fimbriae in the human urinary tract (see above).

Although bacteria may vary host activation through the expression of virulence factors, the host controls the response through the expression of receptors. Two different host determinants have been identified.

1. The expression of recognition receptors for the P fimbriae varies between individuals.
2. There is genetic variation in the signalling pathways that control the innate response.

13.6.2. Variation in Recognition Receptors for P Fimbriae

Several members of the glycosphingolipid GSL receptor family provide binding sites for P fimbriae with variant iso-receptor recognition sites (Leffler and Svanborg-Eden 1980). The receptor GSLs are abundant in uro-epithelium and in kidney tissue, where the individual repertoire of receptor GSLs varies depending on the P blood group and secretor state (Lomberg et al. 1986). This variation provides a basis for the blood group–related difference in susceptibility to infection with P-fimbriated *E. coli*. Patients prone to UTI showed a higher density of epithelial cell receptors, and individuals of blood group P_1 run an increased risk of developing recurrent pyelonephritis (Lomberg et al. 1981, 1986). The receptor repertoire was also found to influence the P-fimbrial type on the infecting strain, as individuals of blood group A_1P_1, expressing the globoA receptor, become infected with bacteria recognising this receptor (Lindstedt et al. 1991). In theory, an individual lacking receptors would be resistant to UTI, but there are too few receptor-negative individuals to investigate this hypothesis.

Our studies predict that low receptor expression would impair the host response. This hypothesis was supported by pharmacological inhibition of epithelial receptor expression in vivo with the glucose analogue *N*-butyldeoxynojirimycin (*N*B-DNJ), which blocks the ceramide-specific glycosyl transferase involved in receptor biosynthesis (Svensson et al. 2001). The results suggested that the recognition receptor is one essential component in the host response, in vivo.

13.7. TOLL-LIKE RECEPTOR 4 AND EPITHELIAL CELL ACTIVATION

P-fimbriated *E. coli* use TLR4 as a co-receptor in trans-membrane signalling (Fig. 13.2) (Frendeus et al. 2001). The involvement of TLR4 was first shown in LPS non-responder mice, later identified as TLR4 mutant mice (Hagberg et al. 1984; Shahin et al. 1987; Svanborg-Eden et al. 1987; Poltorak et al. 1998). TLR4 deficient mouse strains of the C3H or C57Bl backgrounds do not respond to infection with recombinant P-fimbriated strains.

The cellular basis for TLR4 recruitment has been studied in human uro-epithelial cells (Samuelsson et al. 2004). The recognition receptors for P fimbriae are enriched in the cavaeoli and the membrane anchoring domain is ceramide, in the outer leaflet of the lipid bilayer. The coupling of P-fimbriated *E. coli* to their receptors triggers ceramide release, and the ceramide signalling pathway may be involved in TLR 4 activation by P-fimbriated bacteria (Hedlund et al. 1996, 1998).

In macrophages, TLR4 is activated by a complex of LPS, LPS-binding protein that binds CD14/MD-2 to the extra-cellular domain of TLR4 (Wright et al. 1990; Shimazu et al. 1999). The resulting dimerisation of TLR4 starts a cascade of intracellular signal events leading to the activation of nuclear factor (NF)-κB, resulting in the synthesis and release of pro-inflammatory mediators, including IL-6 and IL-8 (Schletter et al. 1995). Down-stream signalling involves a series of adapter proteins, including MyD88, TIRAP, and IRAK (Wesche et al. 1997; Muzio et al. 1997; Horng et al. 2001). More details of NF-κB signalling pathways are found in Chapters 15 and 17.

It still remains unclear whether P fimbriae trigger the response by delivering LPS to the tissues, and whether LPS participates in epithelial cell activation. Such an effect would be expected, as LPS is the classical activator of TLR4-dependent responses. Uro-epithelial cells are CD14 negative, however, and respond poorly to free LPS (Hedlund et al. 1996, 1999). The recognition receptor for P fimbriae may substitute for surface-bound CD14, allowing LPS to trigger the TLR4 pathway. To address this question, we inactivated the endotoxic lipid A portion of LPS by mutating the *msbB* gene that encodes an acyl transferase–coupling myristic acid to the lipid IV_A precursor (Somerville et al. 1996), and expressed P fimbriae in these two backgrounds. Interestingly, we found no evidence of LPS involvement with this approach (Hedlund et al. 1999). The P fimbriae thus appear to utilise an LPS-like mechanism to activate cells, which lack CD14 and are refractory to LPS itself.

In mice, the TLR4-signalling deficiency disrupts the inflammatory response, even though there are primary receptors for fimbriae on the mucosa, showing that the TLR4 co-receptor can override the primary receptor, as a regulator of cell activation. The TLR4-dependent unresponsiveness had two effects. The mice failed to develop symptoms and were unable to clear the infection. As a consequence, they developed asymptomatic bacteriuria, with high bacterial counts in the urine for several months.

This carrier state resembled ABU in patients who maintain high bacterial counts for months and years, if left untreated, and suggested that mutations in genes that control the TLR4 signalling pathway might underlie the unresponsiveness in ABU patients. Prospective clinical studies of the TLR4 genotype in patients with a history of primary ABU have supported this hypothesis.

13.8. GENETICS OF IL-8 RECEPTOR EXPRESSION IN PYELONEPHRITIS-PRONE CHILDREN

It has often been questioned if the susceptibility to symptomatic UTI is inherited. HLA typing has been performed on limited populations, but no relationship with disease has been found. We now propose that low IL-8 receptor expression may underlie differences in the susceptibility to acute pyelonephritis and renal scarring. IL-8 is a C–X–C chemokine involved in neutrophil recruitment and activation (Baggiolini et al. 1994). Local IL-8 concentrations in urine were elevated after deliberate colonisation of the urinary tract, and there was a strong correlation between urinary IL-8 levels and urinary neutrophil numbers, confirming that IL-8 is involved in the onset of pyuria (Agace et al. 1993a). Recent investigations in a mouse UTI model have shown that deletion of the IL-8 receptor severely impairs the neutrophil migration through the renal epithelial tissues and causes the entire syndrome of acute pyelonephritis and renal tissue damage (Hang et al. 2000; Frendeus et al. 2000; Godaly et al. 2000). The mIL-8Rh KO mice developed symptoms of acute pyelonephritis, and surviving mice later developed scarring of the kidneys (Hang et al. 2000; Frendeus et al. 2000). Neutrophils were trapped in the tissues, where their metabolites and proteolytic enzymes caused inflammation and scarring (Hang et al. 2000; Frendeus et al. 2000).

The progression from acute disease to renal scarring in the mIL-8Rh KO mice prompted us to test whether similar mechanisms might underlie the susceptibility to acute pyelonephritis in children. In a prospective clinical study, CXCR expression was compared between children with at least one episode of acute pyelonephritis and age-matched controls with no history of UTI. CXCR1 expression was dramatically lower in patients than in controls (Frendeus et al. 2000; Lundstedt et al. 2005). Sequence analyses revealed single-nucleotide polymorphisms (SNPs) in about forty percent of the patients, but not in the controls. The SNPs were observed in the intron, the coding region, and the 3'UTR region. No polymorphisms were found in the control group.

13.9. CONCLUSIONS

The molecular Koch postulates focus on bacterial virulence and disease induction, but do not take into consideration the asymptomatic carrier state, even though this may be the most important outcome, quantitatively. The human urinary tract offers a highly privileged site for mono-colonisation, and a strain establishing at this site has an evolutionary advantage as compared with the virulent strains that cause disease but are eliminated by host defence systems. There is a need to explore the properties that positively regulate survival and persistence of the carrier strains.

We have examined the mechanisms that regulate mucosal inertia and host responsiveness. We show that the acquisition of a single bacterial gene encoding P fimbriae converts a carrier strain to a host response inducer. In the host, glycolipid recognition receptors and TLR4 regulate the inertia of the mucosa and the initiation of a host response. The ensuing inflammatory response and defence against tissue damage depend on the other innate mechanisms controlled by the IL-8 receptor. The results suggest that the

Koch postulates should be rephrased to allow for a combined analysis of microbial adaptation factors and determinants of host susceptibility.

ACKNOWLEDGEMENTS

The work described was supported by the Swedish Medical Research Council (grants no. -07934, 14577, 14578) and The Royal Physiographic Society, The Medical Faculty, Lund University; Catharina Svanborg was a recipient of a Bristol-Myers Squibb unrestricted grant.

REFERENCES

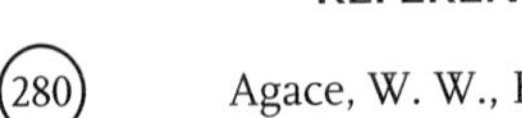

Agace, W. W., Hedges, S. R., Ceska, M., and Svanborg, C. (1993a). Interleukin-8 and the neutrophil response to mucosal gram-negative infection. *Journal of Clinical Investigation* **92**, 780–785.

Agace, W., Hedges, S., Andersson, U., Andersson, J., Ceska, M., and Svanborg, C. (1993b). Selective cytokine production by epithelial cells following exposure to *Escherichia coli*. *Infection and Immunity* **61**, 602–609.

Agace, W. and Svanborg, C. (1995). Mucosal immunity in the urinary system. In *Mucosal Vaccines*, ed. H. Kiyono, P. L. Ogra, and J. R. McGhee, pp. 389–402. San Diego: Academic Press.

Andersson, P., Engberg, I., Lidin-Janson, G., Lincoln, K., Hull, R., Hull, S., and Svanborg C. (1991). Persistence of *Escherichia coli* bacteriuria is not determined by bacterial adherence. *Infection and Immunity* **59**, 2915–2921.

Azuma, I. and Seya, T. (2001). Development of immunoadjuvants for immunotherapy of cancer. *International Immunopharmacology* **1**, 1249–1259.

Baggiolini, M., Dewald, B., and Moser, B. (1994). Interleukin-8 and related chemotactic cytokines–CXC and CC chemokines. *Advances in Immunology* **5**, 97–179.

Bahrani-Mougeot, F. K., Buckles, E. L., Lockatell, C. V., Hebel, J. R., Johnson, D. E., Tang, C. M., and Donnenberg, M. S. (2002). Type 1 fimbriae and extracellular polysaccharides are preeminent uropathogenic *Escherichia coli* virulence determinants in the murine urinary tract. *Molecular Microbiology* **45**, 1079–1093.

Bergsten, G., Samuelsson, M., Wullt, B., Leijonhufvud, I., Fischer, H., and Svanborg, C. (2004). PapG-dependent adherence breaks mucosal inertia and triggers the innate host response. *Journal of Infectious Disease* **189**, 1734–1742.

Blomfield, I. and van der Woude. M. (2002). Regulation and function of phase variation in *Escherichia coli*. In *Bacterial Adhesion to Host Tissues–Mechanisms and Consequences*, ed. M. Wilson, pp. 89–113. Cambridge: Cambridge University Press.

Bock, K., Breimer, M. E., Brignole, A., Hansson, G. C., Karlsson, K. A., Larson, G., Leffler, H., Samuelsson, B. E., Stromberg, N., Eden, C. S. et al. (1985). Specificity of binding of a strain of uropathogenic *Escherichia coli* to Gal alpha 1–4Gal-containing glycosphingolipids. *Journal of Biological Chemistry* **260**, 8545–8551.

Condron, C., Toomey, D., Casey, R. G., Shaffii, M., Creagh, T., and Bouchier-Hayes, D. (2003). Neutrophil bactericidal function is defective in patients with recurrent urinary tract infections. *Urology Research* **31**, 329–334.

Cox, C. E. and Hinman, F. (1965). Factors in resistance to infection in the bladder. 1. The eradication of bacteria by vesical emptying and intrinsic defense mechanisms. In *Progress in Pyelonephritis*, ed. E. H. Kass, p. 563. Philadelphia: F. A. Davis.

de Man, P., van Kooten, C., Aarden, L., Engberg, I., Linder, H., and Svanborg Eden, C. (1989). Interleukin-6 induced at mucosal surfaces by gram-negative bacterial infection. *Infection and Immunity* **57**:3383–3388.

de Man, P., Jodal, U., and Svanborg, C. (1991). Dependence among host response parameters used to diagnose urinary tract infection. *Journal of Infectious Diseases* **163**, 331–335.

Dodson, K. W., Pinkner, J. S., Rose, T., Magnusson, G., Hultgren, S. J., and Waksman, G. (2001). Structural basis of the interaction of the pyelonephritic *E. coli* adhesin to its human kidney receptor. *Cell* **105**, 733–743.

Falkow, S. (1988). Molecular Koch's postulates applied to microbial pathogenicity. *Reviews of Infectious Diseases* **10**, Suppl 2:S274–276.

Frendeus, B., Godaly, G., Hang, L., Karpman, D., Lundstedt, A. C., and Svanborg, C. (2000). Interleukin 8 receptor deficiency confers susceptibility to acute experimental pyelonephritis and may have a human counterpart. *Journal of Experimental Medicine* **192**, 881–890.

Frendeus, B., Wachtler, C., Hedlund, M., Fischer, H., Samuelsson, P., Svensson, M., and Svanborg, C. (2001). *Escherichia coli* P fimbriae utilize the Toll-like receptor 4 pathway for cell activation. *Molecular Microbiology* **40**, 37–51.

Funfstuck, R., Tschape, H., Stein, G., Kunath, H., Bergner, M., and Wessel, G. (1986). Virulence properties of *Escherichia coli* strains in patients with chronic pyelonephritis. *Infection* **14**, 145–150.

Ganz, T. (2001). Defensins in the urinary tract and other tissues. *Journal of Infectious Diseases* **183**, Suppl 1:S41–42.

Godaly, G., Frendeus, B., Proudfoot, A., Svensson, M., Klemm, P., and Svanborg, C. (1998). Role of fimbriae-mediated adherence for neutrophil migration across *Escherichia coli*-infected epithelial cell layers. *Molecular Microbiology* **30**, 725–735.

Godaly, G., Hang, L., Frendeus, B., and Svanborg, C. (2000). Transepithelial neutrophil migration is CXCR1 dependent in vitro and is defective in IL-8 receptor knockout mice. *Journal of Immunology* **165**, 5287–5294.

Hacker, J., Bender, L., Ott, M., Wingender, J., Lund, B., Marre, R., and Goebel, W. (1990). Deletions of chromosomal regions coding for fimbriae and hemolysins occur in vitro and in vivo in various extraintestinal *Escherichia coli* isolates. *Microbial Pathogogenesis* **8**, 213–225.

Hacker, J., Blum-Oehler, G., Muhldorfer, I., and Tschape, H. (1997). Pathogenicity islands of virulent bacteria: structure, function and impact on microbial evolution. *Molecular Microbiology* **23**, 1089–1097.

Hagberg, L., Jodal, U., Korhonen, T. K., Lidin-Janson, G., Lindberg, U., and Svanborg Eden, C. (1981). Adhesion, hemagglutination, and virulence of *Escherichia coli* causing urinary tract infections. *Infection and Immunity* **31**, 564–570.

Hagberg, L., Engberg, I., Freter, R., Lam, J., Olling, S., and Svanborg Eden, C. (1983). Ascending, unobstructed urinary tract infection in mice caused by pyelonephritogenic *Escherichia coli* of human origin. *Infection and Immunity* **40**, 273–283.

Hagberg, L., Hull, R., Hull, S., McGhee, J. R., Michalek, S. M., and Svanborg Eden, C. (1984). Difference in susceptibility to Gram-negative urinary tract infection between C3H/HeJ and C3H/HeN mice. *Infection and Immunity* **46**, 839–844.

Hagberg, L., Briles, D. E., and Eden, C. S. (1985). Evidence for separate genetic defects in C3H/HeJ and C3HeB/FeJ mice, that affect susceptibility to gram-negative infections. *Journal of Immunology* **134**, 4118–4122.

Hagberg, L., Price, A., Reid, G., Svanborg-Eden, C., Lincoln, K., and Lidin-Jansson, G. (1989). Colonization of the urinary tract with live bacteria from the normal faecal and urethral flora in patients with recurrent symptomatic urinary tract infections. In *Host Parasite Interactions in Urinary Tract Infections*, ed. E. H. Kass and C. Svanborg-Eden, pp. 194–197. Chicago: The University of Chicago Press.

Hang, L., Frendeus, B., Godaly, G., and Svanborg, C. (2000). Interleukin-8 receptor knockout mice have subepithelial neutrophil entrapment and renal scarring following acute pyelonephritis. *Journal of Infectious Diseases* **182**, 1738–1748.

Hansson, S., Caugant, D., Jodal, U., and Svanborg-Eden, C. (1989a). Untreated asymptomatic bacteriuria in girls: I Stability of urinary isolates. *British Medical Journal* **298**, 853–855.

Hansson, S., Jodal, U., Lincoln, K., and Svanborg-Eden, C. (1989b). Untreated asymptomatic bacteriuria in girls: II Effect of phenoxymethylpenicillin and erythromycin given for intercurrent infections. *British Medical Journal* **298**, 856–859.

Haraoka, M., Hang, L., Frendeus, B., Godaly, G., Burdick, M, Strieter, R., and Svanborg, C. (1999). Neutrophil recruitment and resistance to urinary tract infection. *Journal of Infectious Diseases* **180**, 1220–1229.

Hedges, S., Anderson, P., Lidin-Janson, G., de Man, P., and Svanborg, C. (1991). Interleukin-6 response to deliberate colonization of the human urinary tract with gram-negative bacteria. *Infection and Immunity* **59**, 421–427.

Hedges, S., Svensson, M., and Svanborg, C. (1992). Interleukin-6 response of epithelial cell lines to bacterial stimulation in vitro. *Infection and Immunity* **60**, 1295–1301.

Hedges, S., Agace, W., Svensson, M., Sjogren, A. C., Ceska, M., and Svanborg, C. (1994). Uroepithelial cells are part of a mucosal cytokine network. *Infection and Immunity* **62**, 2315–2321.

Hedlund, M., Svensson, M., Nilsson, A., Duan, R. D., and Svanborg, C. (1996). Role of the ceramide-signaling pathway in cytokine responses to P-fimbriated *Escherichia coli*. *Journal of Experimantal Medicine* **183**, 1037–1044.

Hedlund, M., Duan, R. D., Nilsson, A., and Svanborg, C. (1998). Sphingomyelin, glycosphingolipids and ceramide signalling in cells exposed to P-fimbriated *Escherichia coli*. *Molecular Microbiology* **29**, 1297–1306.

Hedlund, M., Wachtler, C., Johansson, E., Hang, L., Somerville, J. E., Darveau, R. P., and Svanborg, C. (1999). P fimbriae-dependent, lipopolysaccharide-independent activation of epithelial cytokine responses. *Molecular Microbiology* **33**, 693–703.

Horng, T., Barton, G. M., and Medzhitov, R. (2001). TIRAP: an adapter molecule in the Toll signaling pathway. *Nature Immunology* **2**, 835–841.

Hull, R. A., Gill, R. E, Hsu, P., Minshew, B. H., and Falkow, S. (1981). Construction and expression of recombinant plasmids encoding type 1 or D-mannose-resistant pili from a urinary tract infection *Escherichia coli* isolate. *Infection and Immunity* **33**, 933–938.

Hull, R. A., Rudy, D. C., Donovan, W. H., Wieser, I. E., Stewart, C., and Darouiche, R. O. (1999). Virulence properties of *Escherichia coli* 83972, a prototype strain associated with asymptomatic bacteriuria. *Infection and Immunily* **67**, 429–432.

Hultgren, S. J., Abraham, S., Caparon, M., Falk, P., St Geme, J. W., and Normark, S. (1993). Pilus and nonpilus bacterial adhesins: assembly and function in cell recognition. *Cell* **73**, 887–901.

Jacob-Dubuisson, F., Heuser, J., Dodson, K., Normark, S., and Hultgren, S. (1993). Initiation of assembly and association of the structural elements of a bacterial pilus depend on two specialized tip proteins. *EMBO Journal* **12**, 837–847.

Johnson, J. R. (1991). Virulence factors in *Escherichia coli* urinary tract infection. *Clinical Microbiology Reviews* **4**, 80–128.

Kass, E. H. (1956). Asymptomatic infections of the urinary tract. *Transactions of the Association of American Physicians* **69**, 56–64.

Kuehn, M. J., Heuser, J., Normark, S., and Hultgren, S. J. (1992). P pili in uropathogenic *E. coli* are composite fibres with distinct fibrillar adhesive tips. *Nature* **356**, 252–255.

Kunin, C. (1987). *Urinary Tract Infections. Detection, Prevention, and Management.* Baltimore: Williams & Wilkins.

Lanne, B., Olsson, B. M., Jovall, P. A., Angstrom, J., Linder, H., Marklund, B. I., Bergstrom, J., and Karlsson, K. A. (1995). Glycoconjugate receptors for P-fimbriated *Escherichia coli* in the mouse. An animal model of urinary tract infection. *Journal of Biological Chemistry* **270**, 9017–9025.

Leffler, H. and Svanborg-Eden, C. (1980). Chemical identification of a glycosphingolipid receptor for *Escherichia coli* attaching to human urinary tract epithelial cells and agglutinating human erythrocytes. *FEMS Microbiology Letters* **24** Suppl. 127–134.

Leffler, H. and Svanborg-Eden, C. (1981). Glycolipid receptors for uropathogenic *Escherichia coli* on human erythrocytes and uroepithelial cells. *Infection and Immunity* **34**, 920–929.

Levin, B. R. and Svanborg Eden, C. (1990). Selection and evolution of virulence in bacteria: an ecumenical excursion and modest suggestion. *Parasitology* **100**, Suppl., S103–115.

Lindberg, U. (1975). Asymptomatic bacteriuria in school girls. V. The clinical course and response to treatment. *Acta Paediatrica Scandinavica* **64**, 718–724.

Lindberg, U., Hanson, L. A., Jodal, U., Lidin-Janson, G., Lincoln, K., and Olling, S. (1975). Asymptomatic bacteriuria in schoolgirls. II. Differences in *Escherichia coli* causing asymptomatic bacteriuria. *Acta Paediatrica Scandinavica* **64**, 432–436.

Lindberg, U., Claesson, I., Hanson, L. A, and Jodal, U. (1978). Asymptomatic bacteriuria in schoolgirls. VIII. Clinical course during a 3-year follow-up. *Journal of Pediatrics* **92**, 194–199.

Lindberg, F., Lund, B., Johansson, L., and Normark, S. (1987). Localization of the receptor-binding protein adhesin at the tip of the bacterial pilus. *Nature* **328**, 84–87.

Linder, H., Engberg, I., Hoschutzky, H., Mattsby-Baltzer, I., and Svanborg, C. (1991). Adhesion-dependent activation of mucosal interleukin-6 production. *Infection and Immunity* **59**, 4357–4362.

Lindstedt, R., Larson, G., Falk, P., Jodal, U., Leffler, H., and Svanborg, C. (1991). The receptor repertoire defines the host range for attaching *Escherichia coli* strains that recognize globo-A. *Infection and Immunity* **59**, 1086–1092.

Lomberg, H., Jodal, U., Eden, C. S., Leffler, H., and Samuelsson, B. (1981). P1 blood group and urinary tract infection. *Lancet* **1**, 551–552.

Lomberg, H., Cedergren, B., Leffler, H., Nilsson, B., Carlstrom, A. S., and Svanborg-Eden, C. (1986). Influence of blood group on the availability of receptors for attachment of uropathogenic Escherichia coli. *Infection and Immunity* **51**, 919–926.

Lundstedt, A. C., McCarthy, S., Godaly, G., Karpman, D., Leijonhufvud, I., Samuelsson, M., Svensson, M., Andersson, B., and Svanborg C. (2005). Human chemokine (CXCR1) receptor polymorphisms and susceptibility to acute pyelonephritis. (Submitted for publication).

Mabeck, C. E., Orskov, F., and Orskov, I. (1971). *Escherichia coli* serotypes and renal involvement in urinary-tract infection. *Lancet* **1**, 1312–1314.

Marcus, D. M., Naiki, M., and Kundu, S. K. (1976). Abnormalities in the glycosphingolipid content of human Pk and p erythrocytes. *Proceedings of the National Academy of Sciences USA* **73**, 3263–3267.

Mobley, H. L., Jarvis, K. G., Elwood, J. P., Whittle, D. I., Lockatell, C. V., Russell, R. G., Johnson, D. E., Donnenberg, M. S., and Warren, J. W. (1993). Isogenic P-fimbrial deletion mutants of pyelonephritogenic Escherichia coli: the role of alpha Gal(1–4) beta Gal binding in virulence of a wild-type strain. *Molecular Microbiology* **10**, 143–155.

Muzio, M., Ni, J., Feng, P., and Dixit, V. M. (1997). IRAK (Pelle) family member IRAK-2 and MyD88 as proximal mediators of IL-1 signaling. *Science* **278**, 1612–1615.

Nowicki, B., Svanborg-Eden, C., Hull, R., and Hull, S. (1989). Molecular analysis and epidemiology of the Dr hemagglutinin of uropathogenic *Escherichia coli. Infection and Immunity* **57**, 446–451.

O'Hanley, P., Low, D., Romero, I., Lark, D., Vosti, K., Falkow, S., and Schoolnik, G. (1985a). Gal-Gal binding and hemolysin phenotypes and genotypes associated with uropathogenic *Escherichia coli. New England Journal of Medicine* **313**, 414–420.

O'Hanley, P., Lark, D., Falkow, S., and Schoolnik, G. (1985b). Molecular basis of *Escherichia coli* colonization of the upper urinary tract in BALB/c mice. Gal-Gal pili immunization prevents *Escherichia coli* pyelonephritis in the BALB/c mouse model of human pyelonephritis. *Journal of Clinical Investigation* **75**, 347–360.

Orskov, I., Svanborg Eden, C., and Orskov, F. (1988). Aerobactin production of serotyped *Escherichia coli* from urinary tract infections. *Medical Microbiology and Immunology (Berlin)* **177**, 9–14.

Parsons, C. L., Greenspan, C., Moore, S. W., and Mulholland, S. G. (1977). Role of surface mucin in primary antibacterial defense of bladder. *Urology* **9**, 48–52.

Plos, K., Connell, H., Jodal, U., Marklund, B. I., Marild, S., Wettergren, B., and Svanborg, C. (1995). Intestinal carriage of P fimbriated *Escherichia coli* and the susceptibility to urinary tract infection in young children. *Journal of Infectious Diseases* **171**, 625–631.

Plos, K., Carter, T., Hull, S., Hull, R., and Svanborg Eden, C. (1990). Frequency and organization of pap homologous DNA in relation to clinical origin of uropathogenic *Escherichia coli*. *Journal of Infectious Diseases* **161**, 518–524.

Poltorak, A., He, X., Smirnova, I., Liu, M. Y., Huffel, C. V., Du, X., Birdwell, D., Alejos, E., Silva, M., Galanos, C., Freudenberg, M., Ricciardi-Castagnoli, P., Layton, B., and Beutler, B. (1998). Defective LPS signaling in C3H/HeJ and C57BL/10ScCr mice: mutations in Tlr4 gene. *Science* **282**, 2085–2088.

Samuelsson, P., Gustafsson, E., Fischer, H., Godaly, G., Hedlund, M., Laurson, J., Svensson, M., and Svanborg, C. (2005). Glycosphingolipids as ligand recognition receptors in Toll-like receptor 4 – dependent cell activation. Manuscript in preparation.

Schletter, J., Heine, H., Ulmer, A. J., and Rietschel, E. T. (1995). Molecular mechanisms of endotoxin activity. *Archives of Microbiology* **164**, 383–389.

Schulte-Wissermann, H., Mannhardt, W., Schwarz, J., Zepp, F., and Bitter-Suermann, D. (1985). Comparison of the antibacterial effect of uroepithelial cells from healthy donors and children with asymptomatic bacteriuria. *European Journal of Pediatrics* **144**, 230–233.

Shahin, R. D., Engberg, I., Hagberg, L., and Svanborg Eden, C. (1987). Neutrophil recruitment and bacterial clearance correlated with LPS responsiveness in local gram-negative infection. *Journal of Immunology* **138**, 3475–3480.

Shimazu, R., Akashi, S., Ogata, H., Nagai, Y., Fukudome, K., Miyake, K., and Kimoto, M. (1999). MD-2, a molecule that confers lipopolysaccharide responsiveness on Toll-like receptor 4. *Journal of Experimental Medicine* **189**, 1777–1782.

Somerville, J. E. J., Cassiano, L., Bainbridge, B., Cunningham, M. D., and Darveau, R. (1996). A novel *Escherichia coli* lipid A mutant that produces anti-inflammatory lipopolysaccharide. *Journal of Clinical Investigation* **997**, 359–365.

Stamm, W. E., McKevitt, M., Roberts, P. L., and White, N. J. (1991). Natural history of recurrent urinary tract infections in women. *Reviews of Infectious Diseases* **13**, 77–84.

Stamm, W. E. and Norrby, S. R. (2001). Urinary tract infections: disease panorama and challenges. *Journal of Infectious Diseases* **183** Suppl 1:S1–4.

Stenqvist, K., Sandberg, T., Lidin-Janson, G., Orskov, F., Orskov, I., and Svanborg-Eden, C. (1987). Virulence factors of *Escherichia coli* in urinary isolates from pregnant women. *Journal of Infectious Diseases* **156**, 870–877.

Svanborg, C., Godaly, G., and Hedlund, M. (1999). Cytokine responses during mucosal infections: role in disease pathogenesis and host defence. *Current Opinion in Microbiology* **2**, 99–105.

Svanborg, C., Bergsten, G., Fischer, H., Frendeus, B., Godaly, G., Gustafsson, E., Hang, L., Hedlund, M., Karpman, D., Lundstedt, A. C., Samuelsson, M., Samuelsson, P., Svensson, M., and Wullt, B. (2001a). The 'innate' host response protects and damages the infected urinary tract. *Annals of Medicine* **33**, 563–570.

Svanborg, C., Frendeus, B., Godaly, G., Hang, L., Hedlund, M., and Wachtler, C. (2001b). Toll-like receptor signaling and chemokine receptor expression influence the severity of urinary tract infection. *Journal of Infectious Diseases* **183**, Suppl 1:S61–65.

Svanborg, C., Bergsten, G., Fischer, H., Frendéus, B., Godaly, G., Gustafsson, E., Hang, L., Hedlund, M., Lundstedt, A. C., Samuelsson, M., Samuelsson, P., Svensson, M., and Wullt, B. (2002). Adhesion, signal transduction and mucosal inflammation. In *Bacterial Adhesion to Host Tissues–Mechanisms and Consequences*, ed. M. Wilson, pp. 223–246. Cambridge: Cambridge University Press.

Svanborg-Eden, C. S., Hanson, L. A., Jodal, U., Lindberg, U., and Akerlund, A. S. (1976). Variable adherence to normal human urinary-tract epithelial cells of *Escherichia coli* strains associated with various forms of urinary-tract infection. *Lancet* **1**, 490–492.

Svanborg-Eden, C. S., Eriksson, B., and Hanson, L. A. (1977). Adhesion of *Escherichia coli* to human uroepithelial cells in vitro. *Infection and Immunity* **18**, 767–774.

Svanborg-Eden, C., Bjursten, L. M., Hull, R., Hull, S., Magnusson, K. E., Moldovano, Z., and Leffler, H. (1984). Influence of adhesins on the interaction of *Escherichia coli* with human phagocytes. *Infection and Immunity* **44**, 672–680.

Svanborg-Eden, C., Hagberg, L., Briles, D., McGhee, J., and Michalek, S. (1985). Suspectibility of *Escherichia coli* urinary tract infection and LPS responsiveness. In *Genetic Control of Host Resistance to Infection and Malignancy*, ed. E. Skamene, pp. 385–391. New York: Alan R Liss Inc.

Svanborg-Eden, C., Hagberg, L., Hull, R., Hull, S., Magnusson, K. E., and Ohman, L. (1987). Bacterial virulence versus host resistance in the urinary tracts of mice. *Infection and Immunity* **55**, 1224–1232.

Svensson, M., Lindstedt, R., Radin, N. S., and Svanborg, C. (1994). Epithelial glucosphingolipid expression as a determinant of bacterial adherence and cytokine production. *Infection and Immunity* **62**, 4404–4410.

Svensson, M., Platt, F., Frendeus, B., Butters, T., Dwek, R., and Svanborg, C. (2001). Carbohydrate receptor depletion as an antimicrobial strategy for prevention of urinary tract infection. *Journal of Infectious Diseases* **183**, Suppl 1:S70–73.

Tewari, R., Ikeda, T., Malaviya, R., MacGregor, J. I., Little, J. R., Hultgren, S. J., and Abraham, S. N. (1994). The PapG tip adhesin of P fimbriae protects *Escherichia coli* from neutrophil bactericidal activity. *Infection and Immunity* **62**, 5296–5304.

Vaisanen, V., Elo, J., Tallgren, L. G., Siitonen, A., Makela, P. H., Svanborg-Eden, C., Kallenius, G., Svenson, S. B., Hultberg, H., and Korhonen, T. (1981). Mannose-resistant haemagglutination and P antigen recognition are characteristic of *Escherichia coli* causing primary pyelonephritis. *Lancet* **2**, 1366–1369.

Vaisanen-Rhen, V., Elo, J., Vaisanen, E., Siitonen, A., Orskov, I., Orskov, F., Svenson, S. B., Makela, P. H., and Korhonen, T. K. (1984). P-fimbriated clones among uropathogenic *Escherichia coli* strains. *Infection and Immunity* **43**, 149–155.

Virkola, R., Westerlund, B., Holthofer, H., Parkkinen, J., Kekomaki, M., and Korhonen, T. K. (1988). Binding characteristics of *Escherichia coli* adhesins in human urinary bladder. *Infection and Immunity* **56**, 2615–2622.

Walz, W., Schmidt, M. A., Labigne-Roussel, A. F., Falkow, S., and Schoolnik, G. (1985). AFA-I, a cloned afimbrial X-type adhesin from a human pyelonephritic *Escherichia coli* strain. Purification and chemical, functional and serologic characterization. *European Journal of Biochemistry* **152**, 315–321.

Welch, R. A., Hull, R., and Falkow, S. (1983). Molecular cloning and physical characterization of a chromosomal hemolysin from *Escherichia coli*. *Infection and Immunity* **42**, 178–186.

Wesche, H., Henzel, W. J., Shillinglaw, W., Li, S., and Cao, Z. (1997). MyD88: an adapter that recruits IRAK to the IL-1 receptor complex. *Immunity* **7**, 837–847.

Wold, A., Caugant, D., Lidin-Janson, G., de Man, P., and Svanborg, C. (1992). Resident colonic *Escherichia coli* strains frequently display uropathogenic characteristics. *Journal of Infectious Diseases* **165**, 46–52.

Wright, S. D., Ramos, R. A., Tobias, P. S., Ulevitch, R. J., and Mathison, J. C. (1990). CD14, a receptor for complexes of lipopolysaccharide (LPS) and LPS binding protein. *Science* **249**, 1431–1433.

Wullt, B., Bergsten, G., Connell, H., Rollano, P., Gebretsadik, N., Hull, R., and Svanborg, C. (2000). P fimbriae enhance the early establishment of *Escherichia coli* in the human urinary tract. *Molecular Microbiology* **38**, 456–464.

Wullt, B., Bergsten G., Samuelsson, M., Gebretsadik, N., Hull, R., and Svanborg C. (2001a). The role of P fimbriae for colonisation and host response induction in the urinary tract. *Journal of Infectious Diseases* **183**, S43–46.

Wullt, B., Bergsten, G., Connell, H., Rollano, H., Gebretsadik, N., Hang, L., and Svanborg, C. (2001b). P-fimbriae trigger mucosal responses to *Escherichia coli* in the human urinary tract. *Cell Microbiology* 3, 255–264.

Wullt, B., Bergsten, G., Fischer, H., Godaly, G., Karpman, D., Leijonhufvud, I., Lundstedt, A. C., Samuelsson, P., Samuelsson, M., Svensson, M. L., and Svanborg, C. (2003). The host response to urinary tract infection. *Infectious Disease Clinics of North America* **17**, 279–301.

Yamamoto, S., Tsukamoto, T., Terai, A., Kurazono, H., Takeda, Y., and Yoshida, O. (1995). Distribution of virulence factors in *Escherichia coli* isolated from urine of cystitis patients. *Microbiology and Immunology* **39**, 401–404.

Zobell, C. (1943). The effect of solid surfaces on bacterial activity. *Journal of Bacteriology* **46**, 39–56.

IV Bacterial interactions with the immune system

CHAPTER 14

Host responses to bacteria: Innate immunity in invertebrates

L. Courtney Smith

14.1. INTRODUCTION

In Chapter 10, a brief introduction was provided of our current understanding of the bacterial recognition and effector systems involved in insect immunity. This chapter continues this theme but introduces the reader to a less well-known, but equally important, arena of innate immunity, namely that of marine invertebrates. Enormous advances have been made over the last decade in deciphering how marine organisms, such as the sea urchin, which live within ecosystems containing numerous bacteria, have evolved to maintain an equilibrium with such microbes.

14.2. THE MARINE ENVIRONMENT

14.2.1. Plankton

The marine environment is home to all ranges of organisms. A major subdivision of the marine habitat is the water column versus the sediment (see also Chapters 1 and 2). At the surface of the water column is the complex ecosystem of the plankton. The planktonic organisms are a complex assemblage of single-celled creatures including both eubacteria and archaebacteria, and eukaryotes – both single-celled and multicellular phytoplankton and zooplankton. Many of the single-celled organisms are described as picoplankton, based on their small size, 0.2 to 2.0 μm, and are too small to collect or concentrate by filtering, and if collected, are typically impossible to culture. Because of the difficulties in analyzing these populations, other approaches to characterize the plankton have employed more classic methods to estimate biomass and productivity of plankton in ocean waters

(Pedros-Alio et al. 1999), but correct or accurate estimates have not been previously obtained. Overall, it has been difficult to ascertain the actual population sizes and the true biomass of marine plankton. However, rRNA isolation from marine waters, followed by reverse-transcriptase and polymerase chain reaction (RT-PCR) and fragment sequencing, have indicated that the surface waters between the California coastline and Santa Barbara Island of the Channel Islands have 2×10^6 bacteria/ml and 5×10^5 archaea/ml (Massana et al. 2002). At a depth of 100 meters, these methods have demonstrated that there are 4×10^5 bacteria/ml and 5×10^5 archaea/ml, with the archaea being different from those at the surface. These results have led to more detailed analysis of the organisms present by preparing bacterial artificial chromosome (BAC) libraries from DNA collected from marine environments (DeLong 2001). The BAC plasmid allows the ligation of large DNA inserts (50 to 150 kb) so that very large amounts of chromosomal DNA can be incorporated into a genomic library. The generation of such libraries is described in Chapter 8. The production of BAC libraries has also been applied to an acidophilic biofilm surviving on drainage from a mine (Tyson et al. 2004) and to the prokaryotic populations in the oligotrophic water of the Sargasso Sea (Venter et al. 2004). Both communities included nonculturable organisms and were assumed to be of low complexity, which aided in disentangling the mixed genomic samples during data acquisition and evaluation. The assembly of whole and partial genomes from these environmental systems identified many new species and large numbers of previously unidentified genes, and resulted in the characterization of metabolic and cooperative strategies used for survival in habitats where the major resource was either geochemical, in the absence of light, or low levels of marine nutrients combined with high levels of light. These methods appear to be an excellent initial approach for characterizing a complex microbial community that can be used to direct and optimize future analyses of environmental microbial samples.

In addition to the prokaryote fraction of the plankton, there are large numbers of phyto- and zooplankton present in the surface waters. Analysis of phytoplankton plastid DNA has been used to characterize that fraction of the plankton (Rappe et al. 2000). The zooplankton includes many invertebrate larval forms that remain in the plankton for days or weeks and employ this part of their life cycle as an effective dispersal mechanism for species that have sedentary adult forms. There are two general types of larvae found in the zooplankton, based on whether or not they feed. Lecithotrophic larvae derive all their required nutrients from the large nutrient-rich egg, which enables them to develop and metamorphose into a juvenile adult without feeding (Villinski et al. 2002). Planktotrophic larvae do not have nutrient

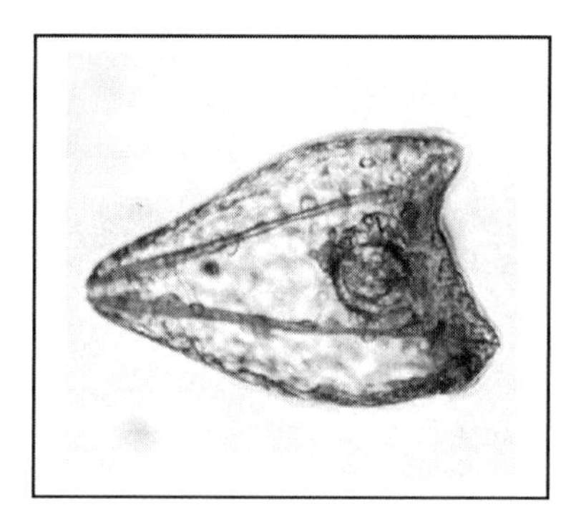

Figure 14.1. Early pluteus of the purple sea urchin, *Strongylocentrotus purpuratus*, also called the prism prior to the development of arms. The oral ectoderm to the right and the stomach is located just below. The image was provided by Ken Brown, George Washington University.

stores and feed on other members of the community to grow during their planktonic phase. Sea urchins exhibit both lecithotrophic and planktotrophic life histories, depending on species.

The sea urchin begins life as an egg that is fertilized in the water column and proceeds to undergo embryogenesis. After a few days, depending on water temperature and species, indirect development culminates in the production of a feeding pluteus (Fig. 14.1), which matures into a larva (Fig. 14.2). The pluteus and larvae are bilateral, with an internal skeleton constructed of calcium and protein that supports the organism, including several "arms" that extend out from the central body. The pluteus and larva are covered with a ciliated ectodermal layer and have a functional digestive tract with coelomic spaces between the ectoderm and the gut (Hyman 1955). The spaces are thinly populated with multipolar or stellate blastocoelar cells that reach long processes across the blastocoelar spaces, around the gut, and out into the arms (Tamboline and Burke 1992) (Fig. 14.3). The ciliated ectodermal cells have regions where the cells are columnar and are more tightly packed, and are located at the intersection between the oral and aboral regions. These cells and their high concentration of cilia are known as the ciliary bands or epaulettes (Strathmann 1975) and enable the larval sea urchin to swim and remain near the surface within the plankton and to simultaneously function to sweep food into the mouth (Strathmann 1971; Strathmann et al. 1972). Echinoderm larvae are omnivorous and will eat almost any particle that will fit down their esophagus, including all types of single-celled plankton and all types of multicellular organisms, including other members of the zooplankton.

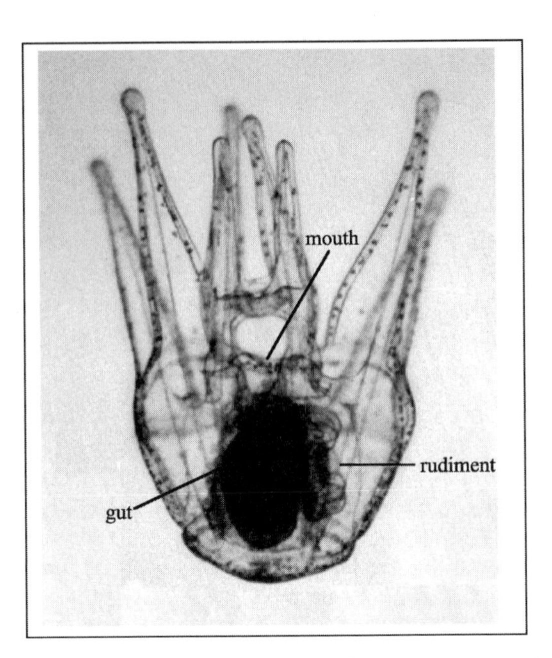

Figure 14.2. Two-week-old larva of the purple sea urchin, *S. purpuratus*. Laboratory-reared larva with eight arms and fully formed and functional gut, which is full of phytoplankton. The adult rudiment is located to the right of the gut. Pigment cells can be seen in the ectoderm. The image was generously contributed by R. Andrew Cameron, Caltech.

As the larva grows during its time in the plankton, it serves as a nutrient supplier and a protector for the adult rudiment (Fig. 14.2) (Hinegardner 1975). The rudiment initially appears on the left side of the gut and essentially forms the ventral half of the pentamerous adult sea urchin having five spines and five tube feet (Hinegardner 1975). Important cues to initiate metamorphosis of the larva into the juvenile of many sedentary invertebrates, including sea urchins, are low–molecular weight organic compounds derived from bacteria (Hinegardner 1969; Cameron and Hinegardner 1974). At the time of metamorphosis, the larvae leaves the plankton and sinks to the bottom. Given a receptive substrate and the proper cues, the rudiment is everted from the side of the larva with the remnant larval tissues remaining as a "turban" on the dorsal side (Hyman 1955). After metamorphosis, the reorganization of the larval turban tissues generates the remaining organ systems of the juvenile, including the dorsal half of the sea urchin, a new digestive tract, and other internal organs, including the gonads. Once the new gut becomes functional, the juvenile begins to feed on biofilms that form on rocks in the marine environment (Leahy et al. 1978), including surface-adhering diatoms

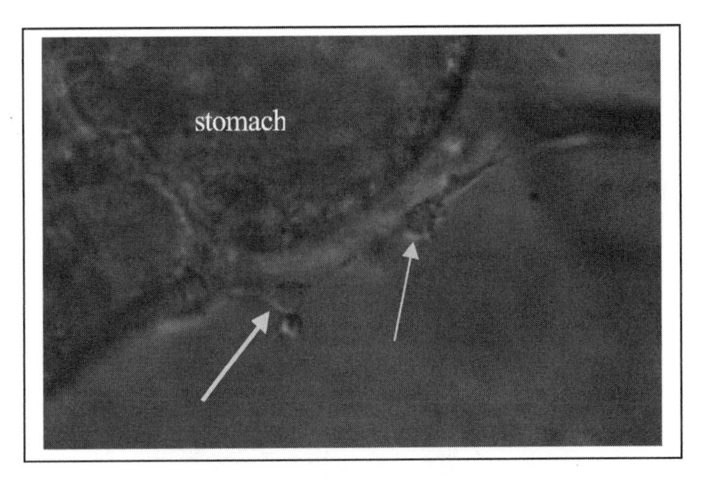

Figure 14.3. Blastocoelar cells. Cell bodies from two blastocoelar cells (arrows) are shown within the blastocoelar space near the wall of the stomach. Each cell is probably positioned near the basement membrane that lines the blastocoelar space, and each has long processes extending from the sides of the cell body. For more information, see Tamboline and Burke (1992). The image was generously contributed by Robert Burke, University of Victoria.

(Hinegardner 1975). As the juvenile sea urchin grows larger, it graduates to eating macroalgae such as giant kelp (Fig. 14.4).

14.3. MICROBIAL CONTACT WITH INVERTEBRATE SURFACES

14.3.1. Larvae

Marine invertebrates such as sea urchins live in environments rich in microbial assemblages. The hatched embryo and larvae contact multitudes of microbes in the water column. Adhesion and colonization of the surfaces of such invertebrates by microbes can lead to infection of the developing embryos, larvae, or adults unless microbial contact and proliferation are controlled by the surface layer of the host cells. Pigment cells located in the body wall of many sea urchin plutei and larvae, including *Strongylocentrotus purpuratus*, may be involved in surface protection. These cells arise from the secondary mesenchyme cells (Gustafson and Wolpert 1967; Gibson and Burke 1985; Kominami et al. 2001), which migrate across the basement membrane that lines the blastocoel, and enter the ectoderm. They extend two or three long processes in a stellate morphology, and can move about within the epithelium (Fig. 14.5) (Gibson and Burke 1985, 1987). These cells

Figure 14.4. Adult purple sea urchins. The image was taken near the coastline of Southern California by Dr. Susan Fuhs while on scuba.

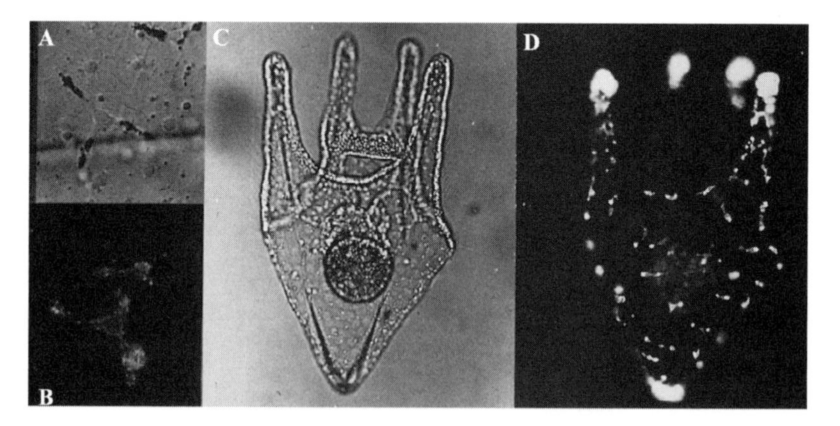

Figure 14.5. Pigment cells in a pluteus of *S. purpuratus*. **(A, B)** Close-ups of tripolar pigment cells showing granules containing echinochrome. **(C)** Pluteus with a gut containing food. The oral ectoderm and mouth are located at the base of the arms at the top of the body. The spherical structure within the body is the stomach. This bright field image is the matched pair to that shown in **D. (D)** The pluteus is incubated with a monoclonal antibody specific for pigment cells (Gibson and Burke 1985). The multipolar pigment cells are spread throughout the ectoderm, with concentrations at the tips of the arms and the tip of the aboral ectoderm at the bottom of the image. The plate was generously provided by Robert Burke, University of Victoria.

are conspicuously red because of the production of a naphthaquinone called echinochrome A, which is contained within spherical granules (Gibson and Burke 1987). Echinochrome is produced by a set of enzymes encoded by genes that are specifically expressed in pigment cells in the late embryo and pluteus (Calestani et al. 2003). Gene knockouts showed that echinochrome production could be blocked in the pigment cells, resulting in uncharacteristically clear embryos and plutei. Echinochrome A is a small double-ring compound with six hydroxyl groups and two carbonyl groups (Kuhn and Wallenfells 1940). Isolated echinochrome A functions as an effective antibacterial compound, particularly against vibrios, which are known to adversely affect sea urchins (Service and Wardlaw 1984, 1985). Although little is known of the functions of the pigment cells in the pluteus and larva, it is not too speculative to suggest that they have defense capabilities to protect the larval sea urchin from attachment and colonization of the ectoderm by marine microbes. This might include migration of the pigment cells to regions of microbial contact followed by degranulation of the pigment granules and release of echinochrome A.

14.3.2. Adults

Surface contact between microbes and the ectoderm of the adult sea urchin presents problems similar to those of the larva in the plankton. A number of marine bacterial species have been isolated and characterized from swabs of the peristomial membrane that surrounds the mouth of the eastern Atlantic sea urchin, *Echinus esculentus*. However, the bacterial composition resembled that of the sand and seawater in which the animal was living (Unkles 1977), suggesting that the microbes were not in permanent association with the host skin. The presence of large numbers of microbes in the sea urchin habitat may not be generally pathogenic, but may at least be opportunistic. Injuries or wear to the tips of the spines in addition to injuries to the skin from puncture wounds from other sea urchins sometimes results in the colonization at these sites by a variety of microbes (Johnson and Chapman 1970; Heatfield and Travis 1975a,b). Skin infections elicit responses by the red spherule cells (see description of coelomocytes below) that form red or black rings (depending on sea urchin species) around infected areas (Hinegardner 1975; Höbaus 1979). Commonly, skin infections result in loss of tissue and spines, but recovery has been observed with a regrowth of skin including new spines. Artificial injuries from surgical implants of allografted tissues also show an infiltration of red spherule cells to the edges of the graft bed, which dissipate after a few days

(Hinegardner 1975; Coffaro and Hinegardner 1977). The red spherule cells of the adult may have functions similar to those described for the pigment cells in the larva. They can extend short pseudopodia, are slightly mobile, and are filled with granules containing echinochrome A (Johnson 1969).

In addition to the activities of the red spherule cells and echinochrome A, extracts from coelomocytes and the body wall (plates, tube feet, spines, pedicellaria) from the Atlantic green sea urchin, *Strongylocentrotus droebachiensis*, were found to be active against *Vibrio anguilarum* and *Corynebacterium glutamicum* but not against *Escherichia coli* and *Staphylococcus aureus* (Haug et al. 2002). Extracts from cell-free coelomic fluid, gut, and eggs were not active against any of the bacteria. Extracts from other echinoderms including a sea star, *Asterias rubens*, and a sea cucumber, *Cucumaria frondosa*, showed antibacterial activities. The type of compounds responsible for these activities was suggested to be a complex mixture, some sensitive to heat or protease treatment, some not sensitive.

14.4. MICROBIAL CONTACT WITH THE GUT

Because the diet of the larva, juvenile, and adult sea urchin includes prokaryotes, either because they are small enough for the larva to eat or because they are present in or on the food ingested by the juvenile and adult, it can be assumed that when in the gut, microbes may become detrimental if not kept under control while passing through the gut. There is no information regarding microbes that have been isolated from the gut of the larva or the juvenile, but bacteria have been isolated and characterized from gut of the adult. All were Gram-negative bacteria, including vibrios, pseudomonads, aeromonads, and flavobacteria, with 2×10^7 culturable bacteria from three-centimeter gut sections (Unkles 1977). However, only fourteen percent of the sea urchins tested had culturable bacteria in their gut. It is not known whether the gut bacteria identified in these studies were just "tourists," microbes that were passing through and had originally been associated with the food eaten by the sea urchin, or if they were intestinal commensals or "residents." Because only a few of the sea urchins tested had culturable microbes, the conclusion is that they were tourists. On the other hand, the status of tourist versus resident may be based on the contributions to the host from the microbe. For example, several nitrogen-fixing *Vibrio* species were isolated from the gut of two sea urchin species (Guerinot and Patriquin, 1981), bacteria that might have been 'allowed' to colonise as they benefit the host.

14.5. DEFENSES IN BODY SPACES

14.5.1. Phagocytic Cells in Embryos, Plutei, and Larvae

In general, immune defense in embryonic and larval sea urchins have received very little attention. Silva (2000) showed that when yeast were injected into the blastocoel of the mid gastrula, they were readily phagocytosed by the secondary mesenchyme cells present in the blastocoel at that time of development. Similarly, when bacteria were injected into the blastocoel of four-day plutei, they were also phagocytosed, probably by blastocoelar cells (J. P. Rast, personal communication). It seems feasible that phagocytic activity in embryos is mediated by secondary mesenchyme cells and by the blastocoelar cells in plutei and larvae (Tamboline and Burke 1992). Both of these cell types may be of significant importance in host protection against microbes and other foreign contact in the internal body spaces of the embryo and larva.

14.5.2. Adult Coelomocytes

There are four morphologically distinct types of coelomocytes in the fluid of the coelomic cavity of the adult sea urchin (Johnson 1969; Gross et al. 2000), and are present in large numbers (1 to 7 $\times$ 10^6/ml) (Smith et al. 1992; Pancer et al. 1999). The most common type (sixty-six to seventy-five percent of all cells) is the phagocyte, which is potentially a complex set of several subtypes based on differences in both morphology and function (Edds 1993; Gross et al. 2000). These cells have an extensive cytoskeleton; are highly amoeboid, mobile, and phagocytic; and are thought to be the primary immune defense cells. The red spherule cells discussed above with respect to skin infections are usually found in the coelomic fluid (zero to twenty percent) and are considered to have important roles in immune functions based on their echinochrome A content. The spherical-shaped vibratile cells (approximately fifteen percent) are packed with granules and are not amoeboid, but have a single flagellum that is used for active swimming through the coelomic fluid. The colorless spherule cells (zero to five percent) are similar in morphology and mobility to the red spherule cells, but they do not produce echinochrome. Their functions and activities are unknown. The numbers of each type of coelomocyte in the coelomic fluid are estimates and may be quite variable. Significant changes in numbers of all coelomocytes (1.5-fold increase) have been noted in sea urchins after immune challenge (Clow et al. 2000). This result may be based mostly on the sevenfold increase in phagocytes containing the sea urchin complement homologue (see below). Coelomocytes have been known for some time to maintain sterility in the coelomic cavity of adult sea urchins. Injected bacteria and other foreign particles are quickly removed from the coelomic cavity by the coelomocytes (Unkles 1977; Yui and Bayne 1983; Plytycz and Seljelid 1993; for review, see Smith and Davidson 1994). Antibacterial activity is not only present in the red spherule cells based on the echinochrome activity (Messer and Wardlaw 1979), but coelomocytes also produce lysozyme, which is active against Gram-negative marine bacteria (Messer and Wardlaw 1979; Gerardi et al. 1990; Haug et al. 2002). In association with their active phagocytic activity, phagocytes have been shown to produce increased amounts of hydrogen peroxide with repeated foreign challenge (Ito et al. 1992).

14.5.3. Molecular Analysis of Immune Responses in Coelomocytes

The molecular analyses of echinoderm immunology initially employed genomics that capitalized on a previous characterization of profilin expression in coelomocytes. The gene encoding profilin, *SpCoel1*, was induced after injury or exposure to lipopolysaccharide (LPS) (Smith et al. 1992, 1995). It was assumed that profilin was involved in modulations of the cytoskeleton (dos Remedios et al. 2003) in the coelomocytes during immune activation based on the extensive cytoskeleton in the amoeboid phagocytes (Smith et al. 1992). Consequently, the *SpCoel1* expression pattern was used as a marker to identify sea urchins with activated coelomocytes to enable the identification of other genes activated by LPS. Activated coelomocytes were used in an expressed sequence tag (EST) study that resulted in the identification of a number of interesting immune response genes, including one encoding a C-type lectin and two others encoding complement components (Smith et al. 1996).

14.5.4. Lectins

Innate immunity is composed of numerous detection systems and effector responses directed toward classes or subclasses of microbes based on the macromolecular components or signatures of those microbes. Lectins compose a major type of effector molecule in multicellular organisms and function either as small, single-domain proteins or as one of many domains in a mosaic protein. They bind carbohydrates through their carbohydrate recognition domain (CRD) and are a complex set of proteins with at least seven different structural CRDs that bind a wide range of carbohydrates

(Drickamer and Fadden 2002). One domain type that has received a significant amount of attention is the C-type lectin, which is stabilized through calcium binding (Drickamer 1988, 1993, 1999). The functions of many lectins have been identified and characterized with regard to immune responses in eukaryotes including both vertebrates and invertebrates, and a number of them show enhanced expression in response to immune challenge (reviewed by Vasta 1992; Weis et al. 1998; Vasta et al. 1999; Feizi 2000; Kogelberg and Feizi 2001; Natori 2001; Vasta et al. 2001; Bianchet et al. 2002).

A C-type lectin identified originally as an EST in LPS-activated coelomocytes (Smith et al. 1996) showed significant similarities to echinoidin, a lectin from a different species of sea urchin, *Anthocidaris crassispina* (Giga et al. 1987). Echinoidin from *S. purpuratus* (called SpEchinoidin) is a single-domain C-type lectin with a CRD-binding motif composed of glutamine, proline, and asparagine, which is predictive for binding galactose and/or galactose derivatives in a Ca^{++}-dependent manner (Drickamer 1993). The expression pattern of SpEchinoidin was of particular interest with respect to immune responsiveness in the sea urchin because it was found exclusively in the phagocyte class of coelomocytes and only after activation by LPS (Smith et al., unpublished). Immune challenge of sea urchins with either LPS or heat-killed bacteria resulted in the appearance of a significant number of small lectins in the coelomic fluid between 12 and 48 hours after challenge. Subsets of the lectins isolated by differential affinity chromatography using galactose, mannose, N-acetyl-glucosamine, and fucose indicated that SpEchinoidin and the diverse array of inducible lectins is a dynamic response that may be essential for recognizing a variety of sugars on the surface of potential pathogens present in inappropriate places in the host body, such as the coelomic cavity.

14.5.5. Complement

Two other ESTs matched to homologues of complement components. *Sp064* encoded a homologue of vertebrate C3, called SpC3, and was the first complement component identified in an invertebrate (Smith et al. 1996; Al-Sharif et al. 1998). The deduced protein was 210 kd, with a conserved internal cleavage site that yields α and β chains that are disulfide linked. A conserved thioester site and an associated functional histidine were located in the α chain. Phylogenetic analysis of SpC3 compared with other members of the thioester protein family, which includes C3, C4, C5, and alpha 2 macroglobulin (α2M), indicated that SpC3 was the first divergent complement protein in the deuterostome lineage, falling at the base of the complement protein clade. Transcripts from *Sp064* are present in two subsets of phagocytes (Gross et al. 2000). Investigations of the SpC3 content in the coelomic fluid showed changes in response to both LPS and to injury, with faster increases in response to LPS (Clow et al. 2000). Expression of *Sp064* in coelomocytes also revealed slight increases with respect to immune challenge, and the number of coelomocytes containing SpC3 increased in response to challenge from LPS. These results indicated that responses to a perceived bacterial challenge by the sea urchin are similar in many ways to acute-phase reactants in higher vertebrates. More detailed analysis of the conserved thioester site of SpC3 indicated that it could probably mediate all the basic functions of C3 proteins that have been characterized in detail from mammals (Smith 2002). Analyses of opsonization functions showed that SpC3 may be a major opsonin in the coelomic fluid and significantly increased phagocytosis of SpC3-opsonized yeast (Smith 2001; Clow et al. 2004).

Almost no molecular investigations of immune function in embryos and larvae of marine invertebrates have been conducted; however, the presence of transcripts from *Sp064* throughout embryogenesis has been reported (Shah et al. 2003). An increase in expression was noted prior to and throughout gastrulation during normal embryogenesis, and exposure of early embryos to heat-killed marine bacteria significantly increased *Sp064* message content in late embryos and plutei. Given that SpC3 functions as a powerful opsonin (Smith 2001; Clow et al. 2004) and that the secondary mesenchyme and perhaps the blastocoelar cells are phagocytic (Silva 2000), embryos and larvae may employ a complement-mediated opsonization in their coelomic spaces to augment phagocytosis of foreign particles prior to their destruction.

The second sea urchin complement component was initially thought to be a complement receptor because of the short consensus repeats (SCRs) that were identified in the EST (Smith et al. 1996). However, sequence analysis of the full-length cDNA showed that it was a homologue of factor B (Bf) called SpBf (Smith et al. 1998). SpBf is a mosaic protein composed of five SCRs (vertebrate Bf proteins typically have three), a von Willebrand factor A domain, and a serine protease domain. Phylogenetic analysis of SpBf indicated that it was the most ancient member of the vertebrate Bf/C2 family. Additional phylogenetic analysis of the SCRs indicates that five SCRs in SpBf may be ancestral to three SCRs, which is the typical pattern in the vertebrate Bf/C2 proteins. The gene encoding SpBf, *Sp152*, is constitutively expressed in phagocyte coelomocytes and is not induced by LPS (Terwilliger et al. 2004). *Sp152* has three alternatively spliced messages in which either SCR1, SCR4, or both exons were deleted. Deletions of SCR1 introduced a frame shift. However, mRNAs with SCR4 deletions (SpBfΔ4) encoded a putatively functional protein with four SCRs rather than five. Comparisons among full-length SpBf, SpBfΔ4, and Bf/C2 proteins from other species, including carp (Nakao et al. 1998, 2002), suggested that the early evolution of this gene family may have involved gene duplications in addition to deletions of exons encoding SCRs.

14.5.6. Functions of the Complement System in Invertebrate Deuterostomes

Since the identification of a simpler complement system in the sea urchin (reviewed in Gross et al. 1999), an increasing number of complement homologues have been found in other invertebrates. Initially, only deuterostome invertebrates appeared to have complement homologues, including tunicates (Ji et al. 1997; Nonaka et al. 1999; Nair et al. 2000; Raftos et al. 2001; Sekine et al. 2001; Fujita 2002; Raftos et al. 2002; Marino et al. 2002; Endo et al. 2003; Azumi et al. 2003) and the cephalochordate *Branchiostoma* (Suzuki et al. 2002), suggesting that the system was specific to the deuterostome lineage. However, complement homologues have been identified in non-deuterostomes, including a gorgonian coral, *Swiftia exerta* (accession no. AAN86548), and ESTs that match to complement components have been found in the Hawaiian bobtail squid, *Euprymna scolopes*, a protostome and member of the Mollusca (M. McFall-Ngai, personal communication). Furthermore, complement-like proteins with conserved thioester sites and opsonin functions have been identified in insects (Lagueux et al. 2000; Levashina et al. 2001; Blandin and Levashina 2004). Consequently, it appears that complement and complement-like components are present throughout the animal kingdom, and therefore this system appears to be significantly more ancient than previously thought.

A simpler complement system composed of homologues of C3, factor B, and mannose binding–associated serine protease (MASP) was proposed to function based on mannose-binding lectin (MBL) recognition activity in association with the amplification feedback loop of the alternative pathway (Smith et al. 1999). Careful analysis of the deduced amino acid sequences of SpC3 and SpBf from the sea urchin revealed conserved cleavage sites for factor I and C3-convertase in SpC3 (Al-Sharif et al. 1998) and conserved sites in SpBf that were involved in both Mg^{++} binding and the activity of the serine protease domain as controlled by putative factor D protease cleavage at a conserved site (Smith et al. 1998). This was interpreted to predict the formation of a C3-convertase complex and feedback loop activation of the alternative pathway (Smith et al. 1999; Smith et al. 2001). The proposed selective advantage of an amplification feedback loop is that it would coat pathogens with complement proteins more quickly and more efficiently than a simple opsonin that binds upon contact as a result of simple diffusion (Smith et al. 2001). However, this prediction of invertebrate complement function may be too simple. Homologues of most of the components involved in the vertebrate complement pathways have been identified in the genome of the tunicate *Ciona intestinalis* (Azumi et al. 2003). This strongly suggests that at least the protochordates have a mostly complete complement system, which lacks only the ability to be activated by antibodies. It is currently not clear how many additional complement components may be found in other members of the animal kingdom.

Because complement proteins C3 and C4 in the higher vertebrates are known to form covalent thioester bonds with amines and hydroxyls on any molecule, they have the ability to bind to any surface, including self cells. The consequence of this chemical reactivity is the activation of opsonic functions in addition to the lytic functions of the terminal complement pathway that leads to cell lysis and inflammation. To control this activity, a complement regulatory system acts to protect self cells against autologous complement attack and thereby to direct the attack toward foreign pathogens, and it blocks depletion of complement components from uncontrolled activation (Liszewski et al. 1996). Because the sea urchin complement system resembles the alternative pathway and the tunicate system resembles most of the higher vertebrate system, both may require significant regulation. Two additional cDNAs from the sea urchin encode mosaic proteins with domains homologous to known complement regulatory proteins such as factor H and factor I (Multerer and Smith 2004). These domains include multiple SCRs, a fucolectin domain, ser/thr/pro-rich regions, a cys-rich region, and a factor I membrane attack complex (FIMAC) domain, which is also found on C6 and C7 of the terminal complement pathway. The genes appear to be members of a small gene family, are constitutively expressed in all tissues of the sea urchin, and are not induced in response to immune challenge. Although these are the first possible examples of invertebrate complement regulatory proteins, other proteins with similar function are expected to be identified in other invertebrates that employ thioester proteins in opsonization functions .

14.6. IMMUNE DIVERSITY IN INVERTEBRATES

Genes involved in invertebrate immunity are generally understood to be set in the genome as a result of evolutionary selection for protection against the pathogens with which the host comes into contact in its environment. The proteins encoded by these genes are designed to recognize or interact with microbial molecules that are essential for the viability of the microbes and are associated with broad classes of organisms. They have been described as pathogen-associated molecular patterns (PAMPs) by the late Charles Janeway (Janeway 1989) and include LPS (Gram-negative bacteria), flagellin (bacterial), double-stranded RNA (viral), unmethylated CpG motifs (bacterial), peptidoglycan (Gram-positive bacteria), techioic acid (Gram-positive), mannans (bacteria), and β-glucans (fungi). The major class of receptors that recognize the PAMPs, called pattern-recognition receptors (PRRs), are members of the Toll family that have been identified throughout both the animal and plant kingdoms (reviewed in Imler and Zheng 2004). These and other host–bacterial recognition receptors are reviewed in detail in Chapter 15. The best characterized set of Toll receptors are those in *Drosophila*, the organism in which Toll was originally identified as a gene involved in early dorsoventral polarity during embryogenesis (Lemaitre et al. 1997 – see Chapter 10). The PRRs in *Drosophila* initiate two signaling pathways, Toll and "immune deficiency" (*imd*), which result in the transcription of multiple genes encoding antimicrobial peptides (reviewed in Hoffmann 2003; Hetru et al. 2003; see Chapter 10). In general, the antimicrobial peptides are directed toward certain classes of microbes (Lemaitre et al. 1997; Hultmark 2003), including defensins, which act on Gram-positive bacteria; drosomycin, which is antifungal; and attacins, which are anti–Gram-negative. The immune response in *Drosophila* is very effective in protecting the host from all types of microbial attack and is designed for broad recognition and broad effector specificity.

Although the diversity of innate immune responsiveness in invertebrates has appeared to be limited, recent advances seem to suggest that the capabilities of generating significant immune diversity in invertebrates may be greater than previously known (Warr et al. 2003; Du Pasquier and Smith 2003). One example is the family of scavenger receptors with cysteine-rich (SRCR) domains that have been implicated in the development and regulation of the vertebrate immune system (Resnick et al. 1994; O'Keeffe et al. 1999). A family of perhaps 150 polymorphic SRCR genes that are constitutively expressed have been characterized in coelomocytes (Pancer et al. 1999; Pancer 2000, 2001). Comparisons among the encoded proteins showed that they are mosaic proteins with multiple SRCR domains plus a variety of other domains, including von Willebrand repeats, SCRs, transmembrane regions, extracellular matrix-like domains, RGD motifs, and epidermal growth factor repeats. Of note was that some SRCR genes were differentially expressed in

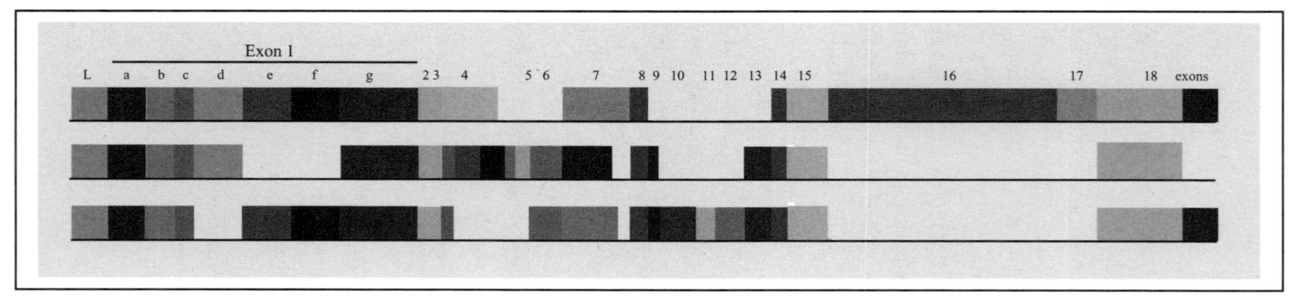

Figure 14.6. Alignment of three full-length *185/333* cDNA sequences. The amino acids deduced from the cDNA sequences were used to generate the alignment with Bioedit. Colored blocks represent exons, and missing blocks represent regions of deleted sequence due to alternative splicing. A leader and eighteen exons were identified from 81 full-length sequences. Exon 1 is shown as a set of subexons, "a" through "g," based on alternative splicing from within the exon as identified by alignments and by cryptic splice sites (see Fig. 14.7). Exons 4 and 7 are shown in multiple colors to indicate significant sequence variability. Exon 18 is shown in two colors of blue to denote a nucleotide variation that introduces an early stop codon.

coelomocytes from different sea urchins, whereas other SRCR genes were expressed with little variation among animals. Furthermore, the expression patterns of several SRCR genes indicated that they changed over time in response to immune challenge and injury, and revealed different expression patterns among the animals analyzed. The sea urchin SRCR gene family and its expression patterns are complex and dynamic and appear to mediate a diverse response to microbial challenge.

An unexpectedly higher level of immune diversity has recently been identified in another set of transcripts expressed in sea urchin coelomocytes in response to LPS or bacteria that are represented by two GenBank submissions, DD185 (accession no. AF228877: Rast et al. 2000) and EST333 (accession no. R62081; Smith et al. 1996), and provisionally called *185/333*. Preliminary analysis of *185/333* expression patterns showed that it was strikingly up-regulated in response to bacterial challenge and was not expressed in unchallenged coelomocytes (Rast et al. 2000), a result that was confirmed from screens of cDNA libraries constructed from bacterially activated versus nonactivated coelomocytes (Nair et al., unpublished). Hundreds of variants of these transcripts were identified by EST matches from an activated coelomocyte library using a subtracted probe specific for transcripts up-regulated by LPS. Comparisons among the EST sequences revealed two very surprising results. First, there was a significant level of alternative splicing of eighteen exons, which was confirmed from full-length cDNA sequences (Fig. 14.6). Complexity in some of the exons from different cDNA sequences was also identified. This is illustrated in exons 4 and 7, which are shaded with different colors and sizes (Fig. 14.6). These regions are shown in this way because the sequences did not align well for these exons and they may actually represent nonoverlapping regions rather than variations of the same exon. In addition, exon 1 is shown as subexons "a" through "g" rather than as seven different exons. This was based on alignments that showed alternative splicing possibly mediated by cryptic splice sites that were located in appropriate positions relative to the alignment (Fig. 14.7).

```
           |→exon spliced out in some clones←|
Gly Arg Arg Phe Asp~~~~~Gly Arg Arg Phe Asp
GGT AGG AGA TTC GAC~~~~~GGT AGG AGA TTC GAC
```

Figure 14.7. Cryptic splice sites within exon 1 of *185/333*. Splicing from within exon 1 was first identified from alignments and then from cryptic splice sites positioned at the edges of the subexons. An example of cryptic splice sites in exon 1 are indicated in bold and are located within the codons for arginine and phenylalanine. ~~, sequence omitted.

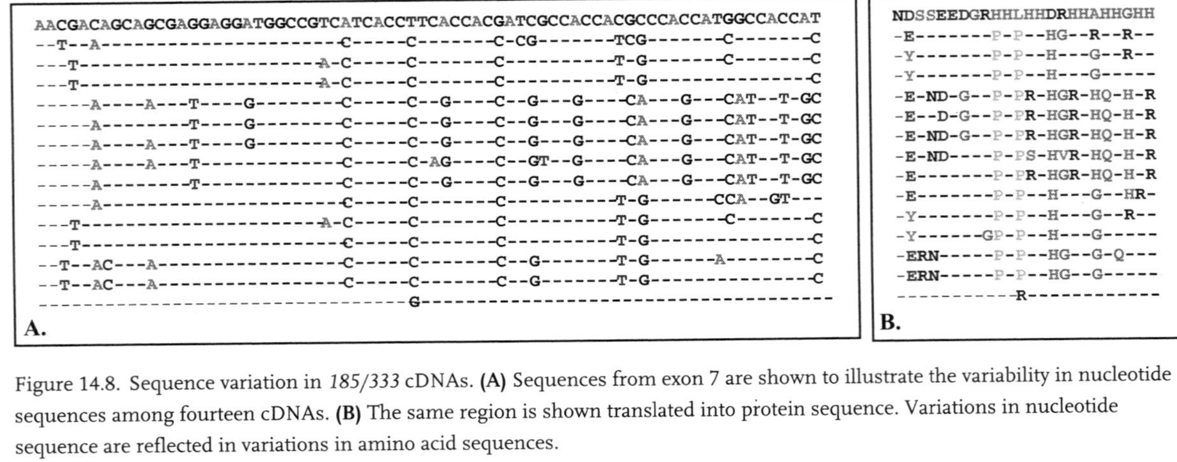

Figure 14.8. Sequence variation in *185/333* cDNAs. **(A)** Sequences from exon 7 are shown to illustrate the variability in nucleotide sequences among fourteen cDNAs. **(B)** The same region is shown translated into protein sequence. Variations in nucleotide sequence are reflected in variations in amino acid sequences.

The deduced 185/333 proteins were composed of an N-terminal glycine-rich region and a C-terminal histidine-rich region. All have a leader, a variety of repeats, acidic regions, histidine patches, and N-linked glycosylation sites only in the histidine-rich region, and most have an RGD motif. Proteins range in size from 26 kd to 55 kd, depending on splicing; are composed mostly of random coils, and do not contain cysteines.

The second unexpected result from the analysis of the *185/333* cDNAs was a significant level of nucleotide variations among the cDNA sequences. Part of exon 7 is shown and illustrates the nucleotide variation for fourteen sequences (Fig. 14.8A). This variation is reflected in the diversity of the corresponding amino acid sequences (Fig. 14.8B). The nucleotide variability was not random, was located throughout the cDNAs at specific positions, and was not clustered at exon boundaries. Based on a combination of nucleotide sequence variability plus alternative splicing, we identified sixty-four unique cDNAs of eighty-one full-length sequences. Estimates of the number of genes in the genome is presently unknown; however, preliminary analysis of the ESTs suggests that there may be hundreds. Overall, the *185/333* set of cDNAs from LPS-activated coelomocytes show an unexpectedly high diversity, perhaps because of the expression of many genes, a signficant amount of alternative splicing, and nucleotide variations within the same exon. If these transcripts are expressed, it is expected that they would generate significant diversity of a set of similar proteins. Based on the expression patterns in response to challenge from injected bacteria or LPS (Rast et al. 2000; Nair et al., unpublished), we hypothesize that the *185/333* genes encode antimicrobial proteins that are important in host defense.

14.7. CONCLUSIONS

Outside of insects, including *Drosophila* and mosquitoes, whose immune systems have received significant attention, little is known of the relevant genes, proteins, and immune mechanisms that function in most of the other invertebrates. Of interest is whether invertebrates, as a broad group, harbor resident gut-associated microbes and if there are residents present on the skin. Of course, the answer to this question may depend on the species under investigation. For example, Caribbean gorgonians have a resident microbial community found in association with the mucus covering on their surfaces (K. Kim, personal communication). In contrast, sea urchins may not (Unkles 1977). At present, little is known about this question and few investigators are pursuing it.

The rate of progress for understanding tidbits of invertebrate immunology has been slow. However, an exception is expected for investigations of the immune system of *Caenorhabditis elegans*. With the availability of a sequenced and assembled genome, a complete characterization of the developmental program, and multitudes of powerful genomic, proteomic, cellular, and molecular technologies that can be applied to this species, studies of the immune system is under way, and is expected to yield interesting and important information on how this primitive innate system functions (Millet and Ewbank 2004). Overall, the application of modern approaches to questions of immune responses to pathogens and interactions with bacterial assemblages that share habitats with invertebrates is expected to yield speedy advances in the future.

ACKNOWLEDGMENTS

The author wishes to acknowledge the generosity of Dr. Robert Burke, Dr. Andrew Cameron, and Dr. Susan Fuhs for supplying images for this chapter. I would like to thank Dr. Jonathan Rast and Dr. Kiho Kim for their communication of preliminary results. Writing of this manuscript was supported by funding from the National Science Foundation to LCS.

REFERENCES

Al-Sharif, W. Z., Sunyer, J. O., Lambris, J. D., and Smith, L. C. (1998). A homologue of the complement component C3 is specifically expressed in sea urchin coelomocytes. *Journal of Immunology* **160**, 2983–2997.

Azumi, K., De Santis, R., De Tomaso, A., Rigoutsos, I., Yoshizaki, F., Pinto, M. R., Marino, R., Shida, K., Ikeda, M., Ikeda, M., Arai, M., Inoue, Y., Shimizu, T., Satoh, N., Rokhsar, D. S., Du Pasquier, L., Kasahara, M., Satake, M., and Nonaka, M. (2003). Genomic analysis of immunity in a Urochordate and the emergence of the vertebrate immune system: "waiting for Godot." *Immunogenetics* **55**, 570–581.

Bianchet, M. A., Odom, E. W., Vasta, G. R., and Amzel, L. M. (2002). A novel fucose recognition fold involved in innate immunity. *Nature Reviews Structural Biology* **9**, 628–634.

Blandin, S. and Levashina, E. A. (2004). Thioester-containing proteins and insect immunity. *Molecular Immunology* **40**, 903–908.

Calestani, C., Rast, J. P., and Davidson, E. H. (2003). Isolation of pigment cell specific genes in the sea urchin embryo by differential macroarray screening. *Development* **130**, 4587–4596.

Cameron, R. A. and Hinegardner, R. T. (1974). Initiation of metamorphosis in laboratory cultured sea urchin. *Biological Bulletin* **146**, 335–342.

Clow, L. A., Gross, P. S., Shih, C. S., and Smith, L. C. (2000). Expression of SpC3, the sea urchin complement component, in response to lipopolysaccharide. *Immunogenetics* **51**, 1021–1033.

Clow, L. A., Raftos, D. A., Gross, P. S., and Smith, L. C. (2004). The sea urchin complement homologue, SpC3, functions as an opsonin. *Journal of Experimental Biology* **207**, 2147–2155.

Coffaro, K. A. and Hinegardner, R. T. (1977). Immune response in the sea urchin, *Lytechinus pictus. Science* **197**, 1387–1390.

DeLong, E. F. (2001). Microbial seascapes revisited. *Current Opinion in Microbiology* **4**, 290–295.

dos Remedios, C. G., Chhabra, D., Kekic, M., Dedova, I. V., Tsubakihara, M., Berry, D. A., and Nosworthy, N. J. (2003). Actin binding proteins: regulation of cytoskeletal microfilaments. *Physiological Reviews* **83**, 433–473.

Drickamer, K. (1988). Two distinct classes of carbohydrate-recognition domains in animal lectins. *Journal of Biological Chemistry* **263**, 9557–9560.

Drickamer, K. (1993). Evolution of Ca^{2+}-dependent animal lectins. *Progress in Nucleic Acid Research* **45**, 207–232.

Drickamer, K. (1999). C-type lectin-like domains. *Current Opinion in Structural Biology* **9**, 585–590.

Drickamer, K. and Fadden, A. J. (2002). Genomic analysis of C-type lectins. *Biochemical Society Symposium* **69**, 59–72.

Du Pasquier, L. and Smith, L. C. (2003). Workshop report: evolutionary immunobiology: new approaches, new paradigms. *Developmental and Comparative Immunology* **27**, 263–271.

Edds, K. T. (1993). Cell biology of echinoid coelomocytes. I. Diversity and characterization of cell types. *Journal of Invertebrate Pathology* **61**, 173–178.

Endo, Y., Nonaka, M., Saiga, H., Kakinuma, Y., Matsushita, A., Takahashi, M., Matsushita, M., and Fujita, T. (2003). Origin of mannose-binding lectin-associated serine protease (MASP)-1 and MASP-3 involved in the lectin complement pathway traced back to the invertebrate, *Amphioxus. Journal of Immunolology* **170**, 4701–4707.

Feizi, T. (2000). Carbohydrate-mediated recognition systems in innate immunity. *Immunological Reviews* **173**, 79–88.

Fujita, T. (2002). Evolution of the lectin-complement pathway and its role in innate immunity. *Nature Reviews Immunology* **2**, 346–353.

Gerardi, P., Lassegues, M., and Canicatti. C. (1990). Cellular distribution of sea urchin antibacterial activity. *Biolological Cell* **70**, 153–157.

Gibson, A. W. and Burke, R. D. (1985). The origin of pigment cells in embryos of the sea urchin *Strongylocentrotus purpuratus. Developmental Biology* **107**, 414–419.

Gibson, A. W. and Burke, R. D. (1987). Migratory and invasive behavior of pigment cells in normal and animalized sea urchin embryos. *Experimental Cell Research* **173**, 546–557.

Giga, Y., Ikai, A., and Takahashi, K. (1987). The complete amino acid sequence of echinoidin, a lectin from the coelomic fluid of the sea urchin *Anthocidaris crassispina*. *Journal of Biological Chemistry* **262**, 6197–6203.

Gross, P. S., Al-Sharif, W. Z., Clow, L. A., and Smith, L. C. (1999). Echinoderm immunity and the evolution of the complement system. *Developmental and Comparative Immunology*, **23**, 439–453.

Gross, P. S., Clow, L. A., and Smith, L. C. (2000). SpC3, the complement homologue from the purple sea urchin, *Strongylocentrotus purpuratus*, is expressed in two subpopulations of the phagocytic coelomocytes. *Immunogenetics* **51**, 1034–1044.

Guerinot, M. L. and Patriquin, D. G. (1981). N2-fixing vibrios isolated from the gastrointestinal tract of sea urchins. *Canadian Journal of Microbiology* **27**, 311–317.

Gustafson, T. and Wolpert, L. (1967). Cellular movement and contact in sea urchin morphogenesis. *Biological Review* **42**, 442–498.

Haug, T., Kjuuo, A. K., Styrvold, O. B., Sandsdalen, E., Olsen, O. M., and Stenvsag, K. (2002). Antibacterial activity in *Strongylocentrotus droebachiensis* (Echinoidea), *Cucumaria frondosa* (Holothruoidea), and *Asterias rubens* (Asteroidea). *Journal of Invertebrate Pathology* **81**, 94–102.

Heatfield, B. M. and Travis, D. F. (1975a). Ultrastructural studies of regenerating spines of the sea urchin *Strongylocentrotus purpuratus*. I. Cell types without spherules. *Journal of Morphology* **145**, 13–50.

Heatfield, B. M. and Travis, D. F. (1975b). Ultrastructural studies of regenerating splines of the sea urchin *Strongylocentrotus purpuratus*. II. Cell types with spherules. *Journal of Morphology* **145**, 51–72.

Hetru, C., Troxler, L., and Hoffmann, J. A. (2003). *Drosophila melanogaster* antimicrobial defense. *Journal of Infectious Disease* **187**, Suppl. 2, S327–S334.

Hinegardner, R. T. (1969). Growth and development of the laboratory cultured sea urchin. *Biological Bulletin* **137**, 465–475.

Hinegardner, R. T. (1975). Morphology and genetics of sea urchin development. *American Zoologist* **15**, 679–689.

Höbaus, E. (1979). Coelomocytes in normal and pathologically altered body walls of sea urchins. In *Proceedings of the European Colloquium on Echinoderms*, ed. M. Jangoux, pp. 247–249. Rotterdam, The Netherlands: A. A. Balkema.

Hoffmann, J. A. (2003). The immune response of *Drosophila*. *Nature* **426**, 33–38.

Hultmark, D. (2003). *Drosophila* immunity: paths and patterns. *Current Opinion in Immunology* **15**, 12–19.

Hyman, L. (1955). *The Invertebrates: Echinodermata*, Vol. IV. New York: McGraw-Hill Book Co., Inc.

Imler, J. L. and Zheng, L. (2004). Biology of Toll receptors: lessons from insects and mammals. *Journal of Leukocyte Biolology* **75**, 18–26.

Ito, T., Matsutani, T., Katsuyoshi, M., and Nomura, T. (1992). Phagocytosis and hydrogen peroxide production by phagocytes of the sea urchin *Strongylocentrotus nudus*. *Developmental and Comparative Immunology* **16**, 287–294.

Janeway, C. A. (1989). Approaching the asymptote? Evolution and revolution in immunology. *Cold Spring Harbor Symposium on Quantitative Biology* **54**, 1–13.

Ji, X., Azumi, K., Sasaki, M., and Nonaka, M. (1997). Ancient origin of the complement lectin pathway revealed by molecular cloning of mannan binding protein-associated serine protease from a urochordate, the Japanese ascidian, *Halocynthia roretzi*. *Proceeding of the National Academy of Sciences USA* **94**, 6340–6345.

Johnson, P. T. (1969). The coelomic elements of sea urchins (*Strongylocentrotus*). I. The normal coelomocytes: their morphology and dynamics in hanging drops. *Journal of Invertebrate Pathology* **13**, 25–41.

Johnson, P. T. and Chapman, F. A. (1970). Infection with diatoms and other icroorganisms in sea urchin spines (*Strongylocentrocus franciscanus*). *Journal of Invertebrate Pathololo**gy* **16**, 268–276.

Kogelberg, H. and Feizi, T. (2001). New structural insights into lectin-type proteins of the immune system. *Current Opinion of Structural Biology* **11**, 635–643.

Kominami, T., Takata, H., and Takaichi, M. (2001). Behavior of pigment cells in gastrula-stage embryos of *Hemicentrotus pulcherrimus* and *Scaphechinus mirabilis*. *Development Growth and Differentiation* **43**, 699–707.

Kuhn, R. and Wallenfells, K. (1940). Echinchrome als prosthetische gruppen hochmolekularern symplexe in den eirn von *Arbacia pustulosa*. *Ber.dt. Chem. Ges.* **73**, 458–464.

Lagueux, M., Perrodou, E., Levashina, E. A., Capovilla, M., and Hoffmann, J. A. (2000). Constitutive expression of a complement-like protein in toll and JAK gain-of-function mutants of *Drosophila*. *Proceedings of the National Academy of Sciences USA* **97**, 11427–11432.

Leahy, P. S., Tutschulte, T. C., Britten, R. J., and Davidson, E. H. (1978). A large-scale laboratory maintenance system for gravid purple sea urchins (*Strongylocentrotus purpuratus*). *Journal of Experimental Zoology* **204**, 369–380.

Lemaitre, B., Reichhart, J. M., and Hoffmann, J. A. (1997). *Drosophila* host defense: differential induction of antimicrobial peptide genes after infection by various classes of microorganisms. *Proceedings of the National Academy of Sciences USA* **94**, 14614–14619.

Levashina, E. A., Moita, L. F., Blandin, S., Vriend, G., Lagueux, M., and Kafatos, F. C. (2001). Conserved role of a complement-like protein in phagocytosis revealed by dsRNA knockout in cultured cells of the mosquito, *Anopheles gambiae*. *Cell* **104**, 709–718.

Liszewski, M. K., Farries, T. C., Lublin, D. M., Rooney, I. A., and Atkinson, J. P. (1996). Control of the complement system. *Advances in Immunology* **61**, 201–283.

Marino, R., Kimura, Y., De Santis, R., Lambris, J. D., and Pinto, M. R. (2002). Complement in urochordates: cloning and characterization of two C3-like genes in the ascidian *Ciona intestinalis*. *Immunogenetics* **53**, 1055–1064.

Massana, R., Guillou, L., Diez, B., and Pedros-Alio, C. (2002). Unveiling the organisms behind novel eukaryotic ribosomal DNA sequences from the ocean. *Applied Environmental Microbiology* **68**, 4554–4558.

Messer, L. G. and Wardlaw, A. C. (1979). Separation of the coelomocytes of *Echinus esculentus* by density gradient centrifugation. *Proceedings of the European Colloquium on Echionderms* Brussels, pp. 319–323.

Millet, A. C. and Ewbank, J. J. (2004). Immunity in *Caenorhabditis elegans*. *Current Opinion in Immunology* **16**, 4–9.

Multerer, K. A. and Smith, L. C. (2004). Two mosaic proteins in the purple sea urchin, *Strongylocentrotus purpuratus*, with multiple domains also found in factor H, factor I and complement components C6 and C7. *Immunogenetics*, **56**, 89–104.

Nair, S. V., Pearce, S., Green, P. L., Mahajan, D., Newton, R. A., and Raftos, D. A. (2000). A collectin-like protein from tunicates. *Comparative Biochemistry and Physiolology B, Biochemistry and Molecular Biology* **125**, 279–289.

Nakao, M., Fushitani, Y., Fujiki, K., Nonaka, M., and Yano, T. (1998). Two diverged complement factor B/C2-lik cDNA sequences from a teleost, the common carp (*Cyprinus carpio*). *Journal of Immunology* **161**, 4811–4818.

Nakao, M., Matsumoto, M., Nakazawa, M., Fujiki, K., and Yano, T. (2002). Diversity of complement factor B/C2 in the common carp (*Cyprinus carpio*): Three isotypes of B/C2-A expressed in different tissues. *Developmental and Comparative Immunolology* **26**, 533–541.

Natori, S. (2001). Insect lectins and innate immunity. *Advances in Experimental Medicine and Biology* **484**, 223–228.

Nonaka, M., Azumi, K., Ji, X., Namikawa-Yamada, C., Sasaki, M., Saiga, H., Dodds, A. W., Sekine, H., Homma, M. K., Matsushita, M., Endo, Y., and Fujita, T. (1999). Opsonic complement component C3 in the solitary ascidian, *Halocynthia roretzi*. *Journal of Immunology* **162**, 387–391.

O'Keeffe, M. A., Metcalfe, S. A., Cunningham, C. P., and Walker, I. D. (1999). Sheep CD4(+) alpha beta T cells express novel members of the T19 multigene family. *Immunogenetics* **49**, 45–55.

Pancer, Z. (2000). Dynamic expression of multiple scavenger receptor cysteine-rich genes in coelmocytes of the purple sea urchin. *Proceedings of the National Academy of Sciences USA* **97**, 13156–13161.

Pancer, Z. (2001). Individual-specific repertoires of immune cells SRCR receptors in the purple sea urchin (*S. purpuratus*). *Advances in Experimental Medicine and Biology* **484**, 31–40.

Pancer, Z., Rast, J. P., and Davidson, E. H. (1999). Origins of immunity: transcription factors and homologues of effector genes of the vertebrate immune system expressed in sea urchin coelomocytes. *Immunogenetics* **49**, 773–786.

Pedros-Alio, C., Calderon-Paz, J.-I., Guixa-Boixereu, N., Estrada, M., and Gasol, J. M. (1999). Bacterioplankton and phytoplankton biomass and production during summer stratification in the northwestern Mediterranean Sea. *Deep-Sea Research Part I Oceanographic Research Papers* **46**, 985–1019.

Plytycz, B. and Seljelid, R. (1993). Bacterial clearance by the sea urchin, *Strongylocentrotus droebachiensis*. *Developmental and Comparative Immunology* **17**, 283–289.

Raftos, D., Green, P., Mahajan, D., Newton, R., Pearce, S., Peters, R., Robbins, J., and Nair, S. (2001). Collagenous lectins in tunicates and the proteolytic activation of complement. *Advances in Experimental Medicine and Biology* **484**, 229–236.

Raftos, D. A., Nair, S. V., Robbins, J., Newton, R. A., and Peters, R. (2002). A complement component C3-like protein from the tunicate, *Styela plicata*. *Developmental and Comparative Immunology* **26**, 307–312.

Rappe, M. S., Suzuki, M. T., Vergin, K. L., and Giovannoni, S. J. (2000). Phylogenetic comparisons of a coastal bacterioplanton community with its counterparts in open ocean and freshwater systems. *FEMS Microbial Ecology* **33**, 219–232.

Rast, J. P., Pancer, Z., and Davidson, E. H. (2000). New approaches towards an understanding of deuterostome immunity. In *Origin and Evolution of the Vertebrate Immune System*, ed. L. Du Pasquier and G. W. Litman. *Current Topics in Microbiology and Immunology* **248**, 3–16.

Resnick, D., Pearson, A., and Krieger, M. (1994). The SRCR superfamily: a family reminiscent of the Ig superfamily. *Trends in Biochemical Sciences* **19**, 5–8.

Sekine, H., Kenjo, A., Azumi, K., Ohi, G., Takahashi, M., Kasukawa, R., Ichikawa, N., Nakata, M., Mizuochi, T., Matsushita, M., Endo, Y., and Fujita, T. (2001). An ancient lectin-dependent complement system in an ascidian: novel lectin isolated from the plasma of the solitary ascidian, *Halocynthia roretzi*. *Journal of Immunology* **167**, 4504–4510.

Service, M. and Wardlaw, A. C. (1984). Echinochrome-A as a bactericidal substance in the coelomic fluid of *Echinus esculentus* (L.). *Comparative Biochemistry and Physiology* **79B**, 161–165.

Service, M. and Wardlaw, A. C. (1985). Bactericidal activity of coelomic fluid of the sea urchin, *Echinus esculentus*, on different marine bacteria. *Journal of the Marine Biology Association of the United Kingdom* **65**, 133–139.

Shah, M., Brown, K. M., and Smith, L. C. (2003). The gene encoding the sea urchin complement proteins, SpC3, is expressed in embryos and can be upregulated by bacteria. *Developmental and Comparative Immunology* **27**, 529–538.

Silva, J. R. (2000). The onset of phagocytosis and identity in the embryo of *Lytechinus variegatus*. *Developmental and Comparative Immunology* **24**, 733–739.

Smith, L. C. (2001). The complement system in sea urchins. In *Phylogenetic Perspectives on the Vertebrate Immune Systems*, ed. G. Beck, M. Sugumaran, and E. Cooper. *Advances in Experimental Medicine and Biology* **484**, 363–372.

Smith, L. C. (2002). Thioester function is conserved in SpC3, the sea urchin homologue of the complement component C3. *Developmental and Comparative Immunology* **26**, 603–614.

Smith, L. C., Britten, R. J., and Davidson, E. H. (1992). SpCoel1, a sea urchin profilin gene expressed specifically in coelomocytes in response to injury. *Molecular Biology of the Cell* **3**, 403–414.

Smith, L. C. and Davidson, E. H. (1994). The echinoderm immune system: characters shared with vertebrate immune systems, and characters arising later in deuterostome phylogeny. In *Primordial Immunity: Foundations for the Vertebrate Immune System*, ed. G. Beck, E. L. Cooper, G. S. Habicht, and J. J. Marchalonis. *The New York Academy of Sciences* **712**, 213–226.

Smith, L. C., Britten, R. J., and Davidson, E. H. (1995). Lipopolysaccharide activates the sea urchin immune system. *Developmental and Comparative Immunology* **19**, 217–224.

Smith, L. C., Chang, L., Britten, R. J., and Davidson, E. H. (1996). Sea urchin genes expressed in activated coelomocytes are identified by expressed sequence tags (ESTs). Complement homologues and other putative immune response genes suggest immune system homology within the deuterostomes. *Journal of Immunology* **156**, 593–602.

Smith, L. C., Shih, C.-S. and Dachenhausen, S. (1998). Coelomocytes specifically express SpBf, a homologue of factor B, the second component in the sea urchin complement system. *Journal of Immunology* **161**, 6784–6793.

Smith, L. C., Azumi, K., and Nonaka, M. (1999). Complement systems in invertebrates. The ancient alternative and lectin pathways. *Immunopharmacology* **42**, 107–120.

Smith, L. C., Clow, L. A., and Terwilliger, D. P. (2001). The ancestral complement system in sea urchins. *Immunological Reviews* **180**, 16–34.

Strathmann, R. R. (1971). The feeding behavior of planktotrophic echinoderm larvae: mechanisms, regulation, and rates of suspension feeding. *Journal of Experimental Marine Biology and Ecology* **6**, 109–160.

Strathmann, R. R. (1975). Larval feeding in echinoderms. *American Zoolologist* **15**, 717–730.

Strathmann, R. R., Jahn, T. L., and Fonseca, J. R. C. (1972). Suspension feeding by marine invertebrate larvae: clearance of particles from suspension by ciliated bands of a rotifer, pluteus, and trochophore. *Biological Bulletin* **142**, 505–519.

Suzuki, M. M., Satoh, N., and Nonaka, M. (2002). C6-like and C3-like molecules from the cephalochordate, *Amphioxus*, suggest a cytolytic complement system in invertebrates. *Journal of Molecular Evolution* **54**, 671–679.

Tamboline, C. R. and Burke, R. D. (1992). Secondary mesenchyme of the sea urchin embryo: ontogeny of blastocoelar cells. *Journal of Experimental Zoology* **262**, 51–60.

Terwilliger, D. P., Clow, L. A., Gross, P. S., and Smith, L. C. (2004). Constitutive expression and alternative splicing of the SCR domains of *Sp152*, the sea urchin homologue of complement factor B. Implications on the evolution of the C2/BF gene family. *Immunogenetics* **56**, 531–543.

Tyson, G. W., Chapman, J., Hugenholtz, P., Allen, E. E., Ram, R. J., Richardson, P. M., Solovyev, V. V., Rubin, E. M., Rokhsar, D. S., and Banfield, J. F. (2004). Community structure and metabolism through reconstruction of microbial genomes from the environment. *Nature* **428**, 37–43.

Unkles, S. E. (1977). Bacterial flora of the sea urchin *Echinus esculentus*. *Applied and Environmental Microbiology* **34**, 347–350.

Vasta, G. R. (1992). Invertebrate lectins: distribution, synthesis, molecular biology, and function. In *Glycoconjugates: Composition, Structure, and Function*, ed. H. J. Ailan and E. C. Disallus, pp. 593–634. New York: Marcell Dekker, Inc.

Vasta, G. R., Quesenberry, M., Ahmed, H., and O'Leary, N. (1999). C-type lectins and galectins mediate innate and adaptive immune functions: their roles in the complement activatio pathway. *Developmental and Comparative Immunology* **23**, 401–420.

Vasta, G. R., Quesenberry, M. S., Ahmed, H., and O'Leary, N. (2001). Lectins from tunicates: structure-function relationships in innate immunity. *Advances in Experimental Medicine and Biology* **484**, 275–287.

Venter, J. C., Remington, D., Heidelberg, J. F., Halpern, A. L., Rusch, D., Eisen, J. A., Wu, D., Paulsen, I., Nelson, J. E., Nelson, W., Fouts, D. E., Levy, S., Knap, A. H., Lomas, M. W., Nealson, K., White, O., Peterson, J., Hoffman, F., Parsons, R., Baden-Tillson, H., Pfannkoch, C., Rogers, Y.-H., and Smith, H. O. (2004). Environmental genome shotgun sequencing of the Sargasso Sea. *Science* **304**, 66–74.

Villinski, J. T., Villinski, J. C., Byrne, M., and Raff, R. A. (2002). Convergent maternal provisioning and life-history evolution in echinoderms. *Evolution International Journal of Organic Evolution* **56**, 1764–1775.

Warr, G. W., Chapman, R. C., and Smith, L. C. (2003). Evolutionary immunobiology: new approaches, new paradigms. *Developmental and Comparative Immunology* **27**, 257–262.

Weis, W. I., Taylor, M. E., and Drickamer, D. (1998). The C-type lectin superfamily in the immune system. *Immunological Reviews* **163**, 19–34.

Yui, M. and Bayne, C. (1983). Echinoderm immunology: bacterial clearance by the sea urchin *Strongylocentrotus purpuratus*. *Biological Bulletin* **165**, 473–486.

CHAPTER 15

Bacterial recognition by mammalian cells

Clare E. Bryant and Sabine Tötemeyer

15.1. INTRODUCTION

Bacterial recognition by the host is a critical step in stimulating a protective immune response. The major paradigm explaining such recognition states that evolutionarily stable bacterial components which are termed pathogen-associated molecular patterns (PAMPs) are recognised by host receptors called pattern recognition receptors (PRRs). Toll-like receptors (TLRs) are one class of PRRs in plants, insects, birds, and mammals that recognise a broad spectrum of microbial pathogens, such as fungi and bacteria. In mammals, these receptors are particularly important in mediating the host recognition of pathogens such as bacteria and bacterial products including endotoxins, exotoxins, bacterial proteins, and DNA. Other receptors for bacterial PAMPs include the NOD proteins and other peptidoglycan recognition receptors. Here we will review the literature on the TLRs and other PAMP receptors to outline how they detect and respond to bacteria, with a focus comparing pathogenic and commensal species. Other chapters in this volume (e.g., 2, 10, 14, and 17) also review aspects of the bacterial recognition problem.

15.2. MAMMALIAN TOLL-LIKE RECEPTORS

The Toll-like receptors were first described as a family of *Drosophila* proteins that mediate the host response to fungal products. In 1997, Medzhitov and co-workers described a human homologue of the *Drosophila* Toll receptor that signalled via the nuclear factor kappa B (NF-κB) pathway to induce pro-inflammatory cytokines such as interleukin (IL)-1, IL-6, and IL-8 (Medzhitov et al. 1997 – see also Chapter 17 for further discussion of the NF-κB system).

Table 15.1. *Tissue distribution of Toll-like receptors*

	TLR1	TLR2	TLR3	TLR4	TLR5	TLR6	TLR7	TLR8	TLR9	TLR10
Heart	+	+	+	+	+	(−)	+	+	+	(−)
Brain	+	+	+	+	+	−	+	+	+	+
Placenta	+	+	+	+	+	+	+	+	+	+
Lung	+	+	+	+	+	+	+	+	+	+
Liver	+	+	+	+	+	(−)	(−)	+	+	(−)
Skeletal muscle	+	+	(−)	(−)	(−)	−	−	(−)	+	(−)
Kidney	+	+	+	+	+	(−)	+	(−)	+	(−)
Pancreas	+	+	+	+	+	+	+	+	+	+
Spleen	+	+	+	+	+	+	+	+	+	+
Lymph node							+	+	+	+
Thymus	+	+	+	+	+	+	+	+	+	+
Prostate	+	+	+	+	+	+	+	−	+	+
Testis	+	+	+	+	+	(−)	+	+	+	(−)
Ovaries	+	+	+	+	+	+	+	+	+	+
Tonsil							+	−	−	+
Bone marrow							−	+	+	−
Trachea							−	+		
Small intestine	+	+	+	+	+	−	+	+	+	+
Stomach							+	−		
Colon	+	+	+	+	+	+	+	+	+	+
Peripheral blood leucocytes	+	+	+	+	+	+	+	+	+	+

Results obtained by Northern blotting, reverse-transcriptase polymerase chain reaction (PCR), or TaqMan PCR; (−) detected only as very weak signal in TaqMan (less than 1/100 of signal for spleen sample); – not detected + = receptor is present, (−) = receptor not present.
(Chaudhary et al. 1998; Chuang and Ulevitch 2000, 2001; Du et al. 2000; Muzio et al. 2000; Rock et al. 1998; Takeuchi et al. 1999; Chuang et al. 2001; Matsumura et al. 2000; Zarember and Godowski 2002).

This TLR was found to be one of a large family of TLRs belonging to the IL1 receptor super-family (Rock et al. 1998). They share a cytoplasmic Toll/IL1 receptor (TIR) homology domain and hence activate similar intracellular signalling pathways (Kopp and Medzhitov 1999). Mutations Pro718His or Pro681His in the TIR of TLR4 and TLR2, respectively, turns the receptor into a dominant-negative mutant by disrupting the recruitment of MyD88 to the TLR (Xu et al. 2000). The extracellular domain of the TLR consists of a leucine-rich repeat (LRR) domain containing a number of cysteine-rich domains. The TLR ectodomain is also shared by the protein RP105, which modulates B cell responses (Ogata et al. 2000). TLRs are expressed in a variety of tissues and cell types (see Table 15.1). Currently ten different human TLRs have been sequenced (Rock et al. 1998; Schering Corp. patent DNAX Toll-like receptor 2–10; W09850547), and functional ligands have been convincingly described for TLR1, 2, 3, 4, 5, 7, and 9. TLRs may also act in synergy with each other (see Fig. 15.1), for example co-stimulation of TLR2 and TLR4 by a combination of their respective ligands results in a synergistic stimulation of host responses (Beutler et al. 2001).

15.2.1. Bacterial Ligands for TLRs

The early literature concerning ligands for TLRs focussed on these proteins as the receptor for *Escherichia coli* lipopolysaccharide (LPS). TLR2 was initially described as a receptor for LPS. This hypothesis was largely based upon transfection studies in which HEK cell lines, when over-expressed with TLR2, responded to high doses (1 to 10 μg/ml) of LPS (Yang et al. 1998; Kirschning et al. 1998). In contrast, genetic studies using the LPS-resistant mouse strains C3H/HeJ and C57BL/10ScCr showed that the former had a Pro712His dominant-negative mutation in TLR4 whereas the latter had a TLR4 gene deletion (Poltorak et al. 1998; Qureshi et al. 1999; Vogel et al. 1999). Over-expression of TLR4 in HEK cells conferred constitutive NF-κB activation on these cells, but not LPS responsiveness, whereas increased expression of TLR4 in macrophages enhanced their sensitivity to LPS (Yang et al. 1998; Kirschning et al. 1998; Du et al. 1999). In addition, Chinese hamster ovary (CHO) cells, which are known to be highly sensitive to LPS, showed a frame-shift mutation in the TLR2 sequence rendering the receptor non-functional (Heine et al. 1999). These experiments led to the conclusion that TLR4 rather than TLR2 is the receptor for enterobacterial LPS. This hypothesis has been confirmed in studies utilising knock-out mice in which TLR2$^{-/-}$ mice are able to respond to LPS whereas TLR4$^{-/-}$ animals are not (Takeuchi et al. 1999; Hoshino et al., 1999). Subsequent studies have demonstrated

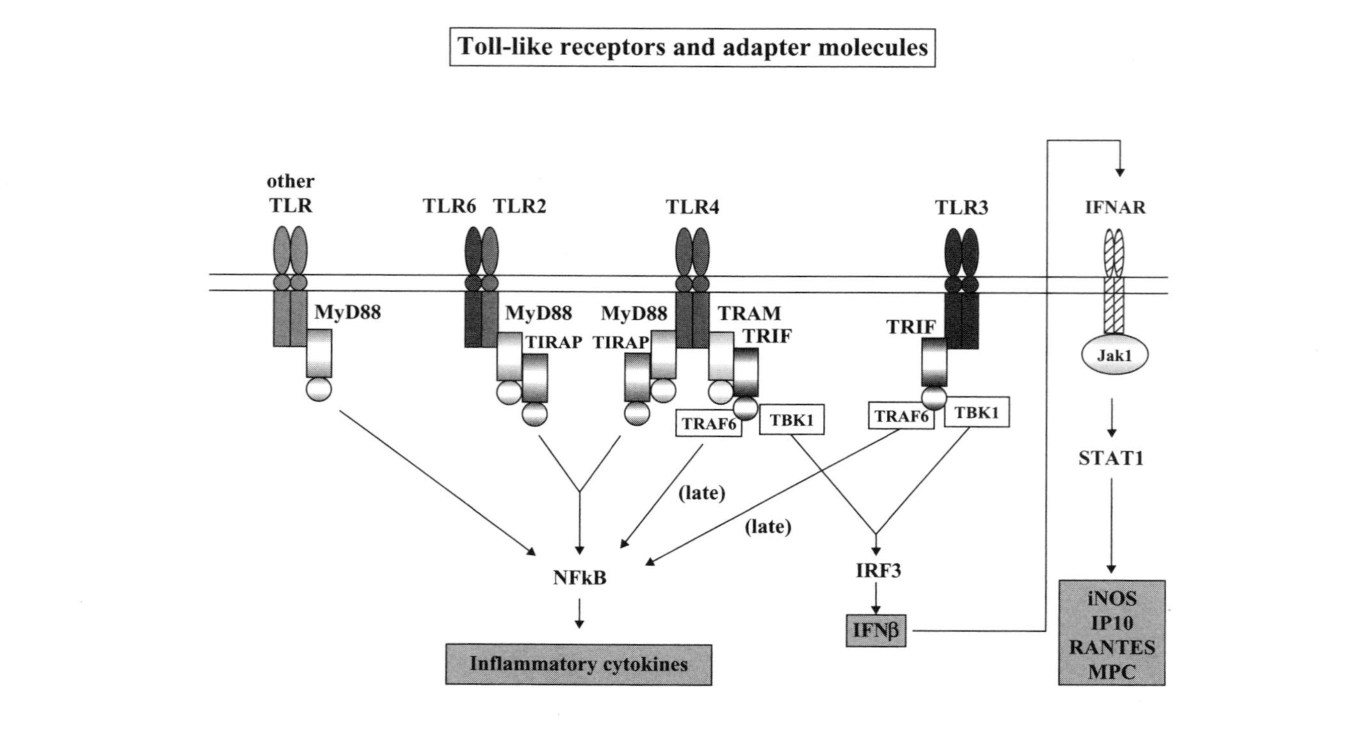
Toll-like receptors and adapter molecules
other TLR
TLR6
TLR2
TLR4
TLR3
IFNAR
MyD88
MyD88
TIRAP
MyD88
TIRAP
TRAM
TRIF
TRIF
Jak1
TRAF6
TBK1
TRAF6
TBK1
STAT1
(late)
(late)
NFkB
IRF3
IFNβ
Inflammatory cytokines
iNOS
IP10
RANTES
MPC

that human patients who are endotoxin resistant also have a mutation in TLR4 (Arbour et al. 2000). However, mice and people without TLR4 or with defective TLR4 function are more susceptible to Gram-negative bacterial infections (Hoshino et al. 1999; Agnese et al. 2002). The TLR4 receptor seems to respond to a number of ligands, but chiefly to LPS and lipid A. Other putative ligands that may activate this receptor include the anti-cancer drug taxol (Kawasaki et al. 2000), saturated fatty acids (Lee et al. 2001), a lipoteichoic acid (LTA)-related molecule from *Streptococcus pyogenes* (Okamoto et al. 2001), a heat-sensitive cell–associated factor from *Mycobacterium tuberculosis* (Means et al. 1999), and cross-linked polyclonal antibodies to the amino terminal of TLR4 (Nattermann et al. 2000). Heat shock protein 60 may also act as an endogenous TLR4 ligand (Ohashi et al. 2000). Currently, two antagonists (E5531, E5564) have been described that inhibit both murine and human TLR4 (Means et al. 2001) and their associated responses, including blocking NF-κB activation and cytokine production both in cell culture models and in vivo (Christ et al. 1995; Chow et al. 1999; Lien et al. 2001).

Recent studies have demonstrated that TLR2 can respond to Gram-positive bacterial products such as bacterial proteins (Brightbill et al. 1999; Aliprantis et al. 1999; Hirschfeld et al. 1999) and peptidoglycan (Yoshimura et al. 1999; Schwandner et al. 1999) from species such as *Listeria monocytogenes* (Flo et al. 2000) and *Staphylococcus aureus* (Takeuchi et al. 2000). It is possible that the early observations showing enterobacterial LPS stimulation of TLR2 were either the result of bacterial lipoprotein contamination (Kirikae et al. 1999; Hirschfeld et al. 2000) or because the TLR2 receptor is capable of responding to higher doses of LPS. Recent studies, however, have shown LPS derived from non-enterobacterial species such as *Leptospira interrogans* and *Porphyromonas gingivalis* is capable of activating TLR2 (Werts et al. 2001; Hirschfeld et al. 2001). Bacterial extracts such as lipoarabinomannan and

←

Figure 15.1 Toll-like receptors and adapter molecules. Downstream activation of signal transduction pathways in response to stimulation of toll-like receptors (TLRs) depends on which adapter molecules (MyD88, TRIF, TRAM, or TIRAP) are recruited to the receptor. Adapter molecules recruit downstream signalling proteins (such as TRAF-6 or TBK1) to activate transcription proteins such as nuclear factor κB (NF-κB) or interferon-regulatory factor-3 (IRF-3) to initiate gene transcription. This results either directly in the production of pro-inflammatory cytokines or in the activation of an IFNβ-dependent autocrine loop to stimulate the interferon-$\alpha\beta$ receptor (IFNAR) with the subsequent translocation of signal transducer and activator of transcription-1 (STAT-1) to the nucleus. STAT-1 activation results in the transcription of genes such as inducible nitric oxide synthase (iNOS) and the chemokines IP10, RANTES, and MPC.

other factors from *M. tuberculosis, Mycobacterium bovis,* and *Mycobacterium avium* can activate TLR2 (Underhill et al. 1999a; Means et al. 1999a,b). The Pro681His mutation of TLR2 blocks responses to *M. tuberculosis* (Underhill et al. 1999a). TLR2 is also activated by lipoproteins and glycolipids from *Borrelia burgdorferi* (Lien et al. 1999), mycoplasma products such as lipoproteins from *Mycoplasma fermentans,* and the R-isomer of myocoplasmal lipopeptide macrophage-activating lipopeptide-2 (Lien et al. 1999; Takeuchi et al. 2000b). The response of cells to the Gram-positive bacterial product LTA is, however, complex. Some studies suggest TLR2 is the LTA-responsive receptor but cells from TLR4$^{-/-}$ mice are non-responsive to this bacterial product (Schwandner et al. 1999; Takeuchi et al. 1999). It is possible that the LTA-driven TLR2-mediated response may reflect contamination of the LTA preparations with bacterial lipoproteins. A TLR2 mutation (Arg753Gly) has been detected in human patients who display less sensitivity to pathogenic peptides derived from *B. burgdorferi.* Like the TLR2$^{-/-}$ mice, these patients appear more sensitive to infection with Gram-positive organisms such as staphylococci (Lorenz et al. 2001; Takeuchi et al. 2000). These observations suggest that TLR2 plays a critical role in mediating host immunity to Gram-positive infections. The amino acids Ser40-Ile64 in the extracellular domain of TLR2 are important in recognition of Gram-positive bacterial ligands such as peptidoglycan (Mitsuzawa et al. 2001).

Functionally, TLR2 receptors can dimerise with other TLRs, such as TLR1 or TLR6. Association of TLR2 with TLR6 confers sensitivity of the receptor complex to modulin (a bacterial product secreted by *Staphylococcus epidermis*), peptidoglycan, the bacterial lipoprotein mycoplasmal macrophage-activating lipopeptide-2 zymosan, or Group B streptococci, whereas the TLR2-TLR1 dimer responds to a mycobacterial 19-kd lipoprotein (Ozinsky et al. 2000; Hajjar et al. 2001; Takeuchi et al. 2001; Henneke et al. 2001; Takeuchi et al. 2002). Therefore the ability of TLR2 to dimerise with other TLR receptor subtypes may explain how host cells are able to respond discretely to a broad range of pathogenic ligands specifically. A further complexity in the function of TLR2 is its ability to heterodimerise with receptors other than TLRs; for example, its interaction with the dectin receptor allows TLR2 to recognise fungal components, thus further complicating the potential ligand recognition by TLR2 (Gantner et al. 2003; Brown et al. 2003).

TLR9 and TLR5 also respond to bacterial ligands. TLR9 mediates the host inflammatory response to bacterial DNA (Hemmi et al. 2000; Bauer et al. 2001; Takeshita et al. 2001). There is no information available yet suggesting that animals or patients with a functionally defective TLR9 have altered susceptibility to bacterial infection. TLR5 is involved in the host response to

Salmonella infections by responding to bacterial flagellin protein (Sebastiani et al. 2000; Hayashi et al. 2001; Moors et al. 2001) at a binding site in the amino acids 386 to 407 region of the protein (Mizel et al. 2003). The binding site localises to a cluster of thirteen amino acid residues that participate in intermolecular interactions within flagellar protofilaments and that are required for bacterial motility (Smith et al. 2003). TLR5 is expressed on the surface of epithelial cells (Adamo et al. 2004) and is likely to be one of the primary detection mechanisms for enteric infections. TLR3, which responds to double-stranded RNA, mediates dendritic cell maturation, showing that another important patho-physiological response is mediated by these receptors (Alexopoulou et al. 2001). Small anti-viral compounds such as imiquimod activate immune cells via a TLR7-MyD88-dependent signalling pathway (Hemmi et al. 2002), but convincing roles and ligands for the other TLRs, such as TLR8 and TLR10 (present in humans but not mice), remain to be demonstrated. Bacterial ligands for TLRs are listed in Table 15.2.

15.2.2. Species Differences in Ligand Recognition by TLRs

A further level of complexity with Toll-receptor ligand recognition is that lipid A molecules from different bacteria differentially activate TLR4 from different mammalian species (see Table 15.3). For example, different acylation patterns of lipid A are recognised in humans, but not mice, at a hyper-variable region of TLR4 (Hajjar et al. 2002). The partial lipid A structure, lipid IVa, acts as an agonist in hamsters, mice, and horses, but as an antagonist in humans. LPS from *Rhodobacter sphaeroides* acts as an agonist for TLR4 in hamster and horse cells, but as an antagonist in human and murine cells (Lien et al. 1999; Lien et al. 2001). Mammalian species differences in ligand recognition have so far only been observed in TLR4 responses, but it is also possible that this phenomenon may be true for the other TLRs.

15.2.3. TLRs Exist in a Protein Complex: CD14, LBP, and MD-2

Prior to the discovery of the TLRs, the surface differentiation antigen CD14 was believed to be the receptor for bacterial LPS (Juan et al. 1995; Kusunoki et al. 1995; Wright 1995). CD14, either as a membrane-bound glycosylphosphatidylinositol (GPI)-linked protein, or in a soluble form, allows the binding of LPS and is therefore critically involved in mediating its signalling responses (Wright et al. 1990), although possibly not those to living bacteria (Moore et al. 2000). The LPS-induced signalling associated with CD14 is enhanced in the presence of a 60 kd LPS-binding protein

Table 15.2. *Ligands for Toll-like receptors*

Receptor(s)	Ligand	Reference
TLR4	*E. coli/Salmonella* LPS	Poltorak et al. 1998; Qureshi et al. 1999; Vogel et al. 1999
	LTA	Takeuchi et al. 1999
	S. pyogenes LTA	Okamoto et al. 2001
	M. tuberculosis heat-sensitive factor	Means et al. 1999 a and b
	Lipid A analogues	Lien et al. 1999, 2001
	Antagonist: E5531	Means et al. 2001
TLR2	*P. gingivalis* LPS	Hirschfeld et al. 2001
	Bacterial proteins	Brightbill et al. 1999; Aliprantis et al. 1999; Hirchfeld et al. 1999
	LTA	Schwandner et al. 1999
	Peptidoglycan	Yoshimura et al. 1999; Schwandner et al. 1999; Flo et al. 2000; Takeuchi et al. 2000
	Mycobacterium	Underhill et al. 1999b; Means et al. 1999
	Borrelia burgdorferi Mycoplasma fermentans	Lien et al. 1999
	Lipoproteins	Lien et al. 1999; Takeuchi et al. 2000
	Haemophilus influenzae	Shuto et al. 2001
TLR2–TLR6	*Staphylococcus epidermis* modulin Peptidoglycan	Hajjar et al. 2001; Ozinsky et al. 2000
	Group B streptococcus mycoplasmal	Henneke et al. 2001
	Diacylated mycoplasmal lipopeptide	Takeuchi et al. 2001; Takeuchi et al. 2002
TLR1–TLR2	Triacylated mycobacterial lipoprotein	Hajjar et al. 2001; Ozinsky et al. 2000; Takeuchi et al. 2001
TLR9	Bacterial DNA	Hemmi et al. 2000; Bauer et al. 2001; Takeshita et al. 2001

Table 15.3. *Mammalian species differences in response to Toll-like receptor 4 ligands*

Ligand	Mouse	Hamster	Human	Horse
E. coli LPS	+	+	+	+
Salmonella typhimurium LPS	+	+	+	+
Taxol	+	0	0	0
Lipid IVa	+	+	–	PA
Rhodobacter sphaeroides lipid A	–	+	–	+
E5531	–	–	–	0/–
Penta-acylated LPS (*Pseudomonas aeruginosa*)	+	?	0	?
Hexa-acylated LPS (*P. aeruginosa*)	+	?	+	?

+ agonist, – antagonist, 0 no effect. PA – partial agonist

(LBP; Fenton and Golenbock 1998). Cells from animal models in which either LBP or CD14 have been removed showed marked reductions in their responsiveness to LPS, although this is not necessarily true for their response to LPS in vivo (Wurfel et al. 1997). The CD14 protein is not transmembrane spanning and is therefore unable to transduce LPS-initiated signals without interaction with at least one additional cellular protein (Aderam and Ulevitch 2000) now known to be the TLRs. In studies in which TLR2 or TLR4 cDNA was transfected into cells that are non-responsive to LPS, such as HEK cells, both LBP and CD14 increased the sensitivity of the cellular response to LPS (Kirschning et al. 1998; Yang et al. 1998; Akashi et al. 2000). Thus, CD14 and LBP are indispensable to the recognition of bacterial products. Although these two proteins enhance the sensitivity of TLR4 to LPS, the receptor complex requires the presence of a further novel extra-cellular protein, MD-2 that targets TLR4 to be expressed in the cell membrane, to confer full LPS responsiveness (Shimazu et al. 1999; Nagai et al. 2002). Cross-linking studies show that LPS associates with TLR4 and MD-2 only when CD14 is co-expressed to recruit LPS-associated signalling molecules within lipid rafts (da Silva Correia et al. 2001; Triantafilou et al. 2002). Secreted MD-2 is also able to bind directly to LPS in the absence of both CD-14 and LBP (Viriyakosol et al. 2001; Visintin et al. 2001). N-linked glycosylations of MD-2 are required for full TLR4 activation and downstream signalling (Ohnishi et al. 2001). Specific amino acid residues in MD-2 are important for binding to TLR4, targeting the MD-2-TLR4 complex to the membrane (Mullen et al. 2003; Nagai et al. 2002; Re and Strominger 2003; Visintin et al. 2003) and for ligand recognition (Kawasaki et al. 2000, Kawasaki et al. 2001, Kawasaki et al. 2003, Visintin et al. 2003). MD-2 in a weaker association with TLR2 enhances its responses to Gram-positive bacterial products as well as enabling the receptor to respond to low doses of LPS and lipid A (Dziarski et al. 2001). Thus, the TLRs appear to function as part of a multi-protein complex that is involved in both presentation of and responses to the bacterial ligand.

CD14 also recognises, and is activated by, bacterial peptidoglycan (Dziarski 2003). It also interacts with LTA, lipoarabinomannan from *Mycoplasma*, various lipoproteins, synthetic lipopeptides, and a variety of other Gram-positive bacterial ligands (reviewed by Dziarski 2003). The interaction of CD14 with a broad spectrum of bacterial ligands is through a common sequence of amino acids (in the 51 to 64 region) within the protein as well as ligand-specific regions which differ for LPS and peptidoglycan (PGN), for example (Dziarski et al. 1998). CD14 is principally activated by high–molecular weight PGN and not by PGN fragments (Dziarski et al. 1998). CD14 alone is not able to activate macrophages in response to Gram-positive bacterial ligands, and it is likely to act as a co-receptor for proteins such as TLR2 in which the responses of this receptor to PGN are enhanced by CD14 (Dziarski. 2003)

15.2.4. Regulation of TLR Expression

Having established the cell type and tissue distribution for the various TLR family members, the question of how the expression of these receptors is regulated has arisen and currently is not fully understood as LPS tolerance is a commonly observed phenomenon (Ziegler-Heitbrock 2001). Early observations suggested that exposure of mammalian cells to LPS increases TLR4, but not TLR2, mRNA expression in monocytes whereas TLR2 expression is up-regulated in polymorphonuclear neutrophils (Muzio et al. 2000). More recent data, however, suggest TLR2, but not TLR4, expression can be up-regulated in macrophages and monocytes by LPS (through activation of TLR4), peptidoglycan, bacterial infection, and pro-inflammatory cytokines (IL1ß, IL2, IL10, IL15, GM-CSF) through an NF-κB-dependent mechanism (Musikacharoen et al. 2001; Matsuguchi et al. 2000; Zarember et al. 2002). TLR2 expression can be down-regulated by IL4. In contrast, the cytokines IFNγ and TNFα have been shown to both increase or decrease receptor expression (Musikacharoen et al. 2001; Liu et al. 2001; Flo et al. 2001). In endothelial cells, LPS and IFNγ increase both TLR2 and TLR4 expression (Faure et al. 2001). High concentrations of IFNγ also increase the expression of other TLRs, such as TLR9, in human monocytes (Takeshita et al. 2001). More importantly, TLR regulation occurs during *Salmonella* infection in vivo. TLR2 is up-regulated whilst TLR4 is down-regulated over the course of a 14-day infection, and the changes in TLR expression are TLR4 dependent (Tötemeyer et al. 2003). Analysis of the murine promoter for the TLR2 receptor shows binding sites for NF-κB, CCAAT/enhancer binding protein, cAMP response proteins, and Stat with promoter activation requiring NF-κB activation and the presence of Stat5 (Musikacharoen et al. 2001). In contrast, the human TLR2 promoter, which has no sequence homology to the murine promoter, contains putative binding sites for several transcription factors, including Sp1 and Ets family members, and is not influenced by inflammatory stimuli (Haehnel et al. 2002). Over-expression of MD-2 is an alternative means of enhancing both TLR2 and TLR4 expression (Dziarski et al. 2001). Bacterial pathogens and cytokines can regulate the expression of the various TLRs in different cell types, and it will be interesting to determine not only the absolute changes in receptor expression, but also changes in the sub-cellular localisation, such as has been shown for TLR2 (Underhill et al. 1999b), of these proteins during pathogenesis.

15.3. OTHER NON-TLR RECEPTORS

There are many observations in the literature suggesting that there are further pattern recognition receptors in addition to the TLRs. The majority of these receptors recognise bacterial peptidoglycan (PGN) or some of its constituents. In addition to TLR2, PGN is recognised by CD14, a family of PGN recognition proteins (PGRPs), NOD1, NOD2, and PGN-lytic enzymes (lysozyme, amidases). PGN consists of a polymer of B(1–4)-linked N-acetylglucosamine and N-acetylmuramic acid cross-linked by short peptides (for further details, see Dziarski 2003). The PGN recognition protein, PGRP-S, has been cloned in a variety of mammalian species, and in humans, three additional isoforms have been identified: PGRP-L, PGRP-Ia, and PGRP-Ib (Rehman et al. 2000; Liu et al. 2001, 2003; Tydell et al. 2002). The C-terminal domain of the protein is well conserved between species and contains an amidase in *Drosophila* which can hydrolyse the bond between muramic acid and the peptide bonds in PGN (Mellroth et al. 2003) but not in mice. PGRPs have a highly tissue-specific expression pattern, and as such, the different isoforms are likely to perform unique functions depending on their location. PGRP-S is located in neutrophils where it is bacteriostatic and plays a role in intra-cellular bacterial killing such that PGRP-S$^{-/-}$ mice are more susceptible to infection with low-virulence strains of Gram-positive bacteria (such as *Bacillus subtilis*), but, interestingly, not to high-virulence bacterial species (such as *S. aureus* or *E. coli*) (Dziarski et al. 2003). PGRP-S does not interact with either TLR2 or CD14.

NOD proteins contain leucine-rich repeats, a nucleotide-binding oligomerisation domain (NOD), and an N-terminal caspase recruitment domain (CARD). They structurally resemble plant resistance proteins. There are several structural analogues of these proteins in mammals, but most is known about NOD-1 and NOD-2. NOD-1 is widely expressed, whereas NOD-2 is restricted to monocytes. Activation of these receptors results in NF-κB activation, and mutations in NOD-2 are associated with increased susceptibility to Crohn's disease and Blau syndrome (Hugot et al. 2001; Ogura et al. 2001; Miceli-Richard et al. 2001). NODs are activated by low–but not high–, molecular weight PGN fragments such as muramyl dipeptide (of the L-D amino acid configuration) and muramyl tripeptide (Girardin et al. 2003). The synthetic versions of these peptides do not naturally occur, but N-acetylglucosamine-N-acetylmuramic acid isolated from Gram-negative bacteria such as *E. coli* is able to activate NOD-1 in mammalian epithelial cells (Girardin et al. 2003). The current literature suggests that the recognition of bacterial peptidoglycan may involve up to three different receptor types. This seems a complex solution to cellular recognition of bacteria, and further work needs to be done to clarify the role of these proteins in the cellular response to living bacteria.

There are a number of other molecules recognising LPS. The RP105 protein in B cells and dendritic cells, possibly in combination with a protein called MD-1, is believed to be important in mediating B-cell proliferation, and expression of co-stimulatory molecules in response to LPS (Ogata et al. 2000). Heat shock proteins 70 and 90 in combination with chemokine receptor 4 and growth differentiation factor 5 have been shown to be important CD14-independent mediators of LPS activation in cells, although this study did not rule out a potential association of this protein complex with TLR4 (Triantafilou et al. 2001). Therefore there are potentially many other supplementary receptors that may act to recognise pathogens and signal their presence to the host.

15.4. ROLE OF ADAPTER MOLECULES IN TLR SIGNALLING

Signalling of TLR receptor activation recruits scaffolding proteins (adapter proteins) that facilitate the activation of signalling kinases. These adapter molecules contain a TIR domain that interacts with the TIR domain of the TLRs. Five adapter molecules have been identified so far: MyD88, TIRAP/Mal, TRIF/TICAM, TRAM, and SARM (see Fig. 15.1). The function of the first four adapters has been established, whereas SARM has been identified only by sequence homology. As combinations of TLRs may mediate mammalian responsiveness to bacterial pathogens, their formation of both homo- and hetero-dimers may generate a mechanism for diversifying the repertoire of TLR-mediated responses. Diversification of the signalling pathways appears to happen at the level of the adapter molecules and their recruitment and interaction with IRAK isoforms 1 to 4, IRAK recruitment of TRAF6, and interaction with multiple MAP3Ks as well as intermediate points of bifurcation within this sequence of events generates the potential for further diversity in the propagation of intracellular signalling, activation of effectors, and ultimately diverse transcriptional and related functional responses.

MyD88 was the first adapter molecule identified and has been most fully characterised. It is involved in the signalling pathways of all the TLRs except TLR3 and plays a major role in the signalling in response to LPS (reviewed in O'Neill 2002). MyD88$^{-/-}$ mice are resistant to LPS-induced toxic shock (Kawai et al. 1999). Although MyD88 is the principal adapter molecule involved in the LPS-induced TLR4 signalling, there is also a MyD88 independent pathway leading to delayed NF-κB activation (Kawai et al. 2001). In addition, TLR4- and TLR3-dependent activation of IRF-3 and IFNβ induced gene expression is MyD88-independent (Kawai et al. 2001). TIRAP/Mal was identified as a second adapter molecule and was initially thought to be involved in MyD88-independent signalling (Fitzgerald et al. 2001; Horng et al. 2001). However, Mal/TIRAP$^{-/-}$ mice show the same phenotype in response to LPS as MyD88$^{-/-}$ mice, suggesting that Mal/TIRAP is not involved in MyD88-independent signalling but acts in cooperation with MyD88 (Horng et al. 2002; Yamamoto et al. 2002).

The adapter molecule TRIF/TICAM coordinates MyD88-independent signalling (Yamamoto et al. 2003a; Hoebe et al. 2003). Over-expression of TRIF/TICAM preferentially activates the IFNβ promoter but also shows weak activation of the NF-κB promoter (Yamamoto et al. 2003a; Oshiumi et al. 2003). In TRIF$^{-/-}$ KO macrophages, polyIC- and LPS-induced activation of IRF3 is lost. Additionally, inflammatory cytokine production in response to TLR4 ligands, but not to other TLR ligands, is severely impaired (Yamamoto et al. 2003a). The N-terminal portion of TRIF is required for the activation of IRF3, and the TIR domain of TRIF alone acts as a dominant negative inhibitor to TLR4-dependent NF-κB activation (Yamamoto et al. 2003a). Mice deficient in both MyD88 and TRIF showed complete loss of NF-κB activation in response to LPS activation of TLR4 (Yamamoto et al. 2003a). Additionally, the N-terminal of TRIF associates with TRAF6 and is required for NF-κB activation but not for IFNβ activation. TBK1 associates with TRIF at a different region, and that association is required for the activation of IFNβ-dependent gene expression (Sato et al. 2003). Hence TRIF acts as a scaffold to assemble the signalling proteins TRAF6 and TBK1, regulating two distinct pathways leading to the activation of the NF-κB and IFNβ promoter, respectively. TRAM has been identified as another adapter molecule involved in TLR4-mediated signalling. In TRAM$^{-/-}$ mice, TRL4- but not TLR3-mediated IFNβ and IFNβ-inducible gene expression is severely impaired (Yamamoto et al. 2003b).

Initially, cellular studies have focussed upon the activation of NF-κB as the functional readout for TLR activation. It is clear that TLR2, 4, 5, and 9 all activate the NF-κB transcription factor pathway. TLR2, 4, and 9 activate the MAP kinases of p38, JNK, and probably ERK. In addition, stimulation of TLR2 by bacterial lipoproteins initiates cellular apoptosis (Aliprantis et al. 2000). Furthermore, it must be recognised that activation of TLRs does not induce the expression of identical genes. For example, *E. coli* K235 LPS stimulation of TLR4 induces IL12p40 and MCP-5, whereas *P. gingivalis* stimulation of TLR2 does not (Hirschfeld et al. 2001). Overall the balance between the intracellular signalling pathways activated determines the profile of transcription factors activated and therefore gene transcription via interactions with heterogeneous promoter/enhancer sequences.

15.5. TLR RECOGNITION OF COMMENSAL AND PATHOGENIC BACTERIA

One of the key unresolved issues in TLR biology is why commensal bacteria, which express a wide range of PAMPs (or more appropriately, microbial-associated molecular patterns – MAMPs – see Chapter 16), activate different immune responses to pathogenic organisms. Obviously there is a whole range of bacterial proteins that modify the host immune response (reviewed by Galan 2001). A simple difference in PAMP recognition between commensal and pathogen is unlikely to be wholly responsible for disease pathogenesis, although it is likely that differences in TLR responses will be involved. Lipid A, for example, plays a critical role in the host response to *Salmonella typhimurium* (Khan et al. 1998; Takeuchi et al. 1999; Rosenberger et al. 2002; Royle et al. 2003). Work on commensal organisms has largely focussed on studies investigating gut-colonising organisms. Discrimination between commensals and pathogens by phagocytes is probably less important than for the epithelial cells lining the gut mucosa because breaching of the mucosal barrier by bacteria implies that an invasive insult has occurred, thus initiation of a robust host immune response is necessary. The gut is one of the major sites of commensal colonisation, and therefore the epithelial cells need to be able to distinguish the relatively rare pathogen from the massive commensal population to initiate a protective response (reviewed by Gewirtz 2003). There are at least two proposed mechanisms by which the host may avoid detection of commensal bacteria: low expression of PRRs and basolateral, rather than apical, expression of such receptors. TLR studies provide some evidence to support each of these hypotheses. Melmed et al. (2003) showed low expression of TLR2, 6, 1, and Tollip in gut epithelial cells. TLR5 is expressed on the basolateral surface of polarised epithelial cells (Gewirtz et al. 2001). In the field of PAMP biology, comparison between the cellular responses initiated between pathogenic and commensal bacteria is at a fairly early stage, but in the near future, it is likely to reveal a wide range of important information which should at least partially answer how the commensal organism avoids the host immune response.

15.6. CONCLUSIONS

In less than a decade we have gone from a situation in which there was a clear deficit in the number of host proteins able to recognise microbial pathogens to the situation today, in which there are dozens of cell surface and intracellular receptors able to recognise a large range of bacterial components and induce selective cell signalling. It is thus rapidly emerging that innate immunity is amply supplied with the apparatus needed to discriminate among the different microbial pathogens that cause infections. This, of course, raises the paradox of how these receptors cope with the massive numbers of cooperative bacteria that populate mammals. It is likely that the very large number of microbial receptors and their ability to indulge in cooperative signalling may be the answer to this question. After all, having a recognition system implies the ability to recognise different bacteria and that may be precisely what the integrated system created by the bacterial recognition receptors does.

ACKNOWLEDGEMENTS

The authors would like to thank Professor Duncan Maskell for his continued support. We also wish to acknowledge that limited space necessitates omission of a number of important reports in this field of research. Work in the authors' laboratory is supported by the Wellcome Trust, British Heart Foundation, and the Biotechnology and Biological Sciences Research Council.

REFERENCES

Adamo, R., Sokol, S., et al. (2004). *P. aeruginosa* flagella activate airway epithelial cells through asialoGM1 and TLR2 as well as TLR5. *American Journal of Respiratory Cellular and Molecular Biology*, in press.

Aderem, A. and Ulevitch, R. J. (2000). Toll-like receptors in the induction of the innate immune response. *Nature* **406**, 782–787.

Agnese, D. M., Calvano, J. E., et al. (2002). Human toll-like receptor 4 mutations but not CD14 polymorphisms are associated with an increased risk of gram-negative infections. *Journal of Infective Diseases* **186**, 1522–1525.

Akashi, S., Ogata, H., et al. (2000). Regulatory roles for CD14 and phosphatidylinositol in the signaling via toll-like receptor 4-MD-2. *Biochemical Biophysical Research Communications* **268**, 172–177.

Alexopoulou, L., Holt, A. C., et al. (2001). Recognition of double-stranded RNA and activation of NF-kappaB by Toll-like receptor 3. *Nature* **413**, 732–738.

Aliprantis, A. O., Yang, R. B., et al. (1999). Cell activation and apoptosis by bacterial lipoproteins through toll-like receptor-2. *Science* **285**, 736–739.

Aliprantis, A. O., Yang, R. B., et al. (2000). The apoptotic signaling pathway activated by Toll-like receptor-2. *EMBO Journal* **19**, 3325–3336.

Arbour, N. C., Lorenz, E., et al. (2000). TLR4 mutations are associated with endotoxin hyporesponsiveness in humans. *Nature Genetics* **25**, 187–191.

Bauer, S., Kirschning, C. J., et al. (2001). Human TLR9 confers responsiveness to bacterial DNA via species-specific CpG motif recognition. *Proceedings of the National Academy of Sciences USA* **98**, 9237–9242.

Beutler, E., Gelbart, T., et al. (2001). Synergy between TLR2 and TLR4: a safety mechanism. *Blood Cells Molecular Diseases* **27**, 728–730.

Brightbill, H. D., Libraty, D. H., et al. (1999). Host defense mechanisms triggered by microbial lipoproteins through toll-like receptors. *Science* **285**, 732–736.

Brown, G. D., Herre, J., Williams, D. L., Willment, J. A., Marshall, A. S., and Gordon, S. (2003). Dectin-1 mediates the biological effects of {beta}-glucans. *Journal of Experimental Medicine* **197**, 1119–1124.

Chaudhary, P. M., Ferguson, C., et al. (1998). Cloning and characterization of two Toll/Interleukin-1 receptor-like genes TIL3 and TIL4: evidence for a multigene receptor family in humans. *Blood* **91**, 4020–4027.

Chow, J. C., Young, D. W., et al. (1999). Toll-like receptor-4 mediates lipopolysaccharide-induced signal transduction. *Journal of Biological Chemistry* **274**, 10689–10692.

Christ, W. J., Asano, O., et al. (1995). E5531, a pure endotoxin antagonist of high potency. *Science* **268**, 80–83.

Chuang, T. H. and Ulevitch R. J. (2000). Cloning and characterization of a subfamily of human toll-like receptors: hTLR7, hTLR8 and hTLR9. *European Cytokine Network* **11**, 372–378.

Chuang, T. and Ulevitch, R. J. (2001). Identification of hTLR10: a novel human Toll-like receptor preferentially expressed in immune cells. *Biochimica Biophysica Acta* **1518**, 157–161.

da Silva Correia, J., Soldau, K., et al. (2001). Lipopolysaccharide is in close proximity to each of the proteins in its membrane receptor complex. transfer from CD14 to TLR4 and MD-2. *Journal of Biological Chemistry* **276**, 21129–21135.

Du, X., Poltorak, A., et al. (1999). Analysis of Tlr4-mediated LPS signal transduction in macrophages by mutational modification of the receptor. *Blood Cells and Molecular Diseases* **25**, 328–338.

Du, X., Poltorak, A., et al. (2000). Three novel mammalian toll-like receptors: gene structure, expression, and evolution. *European Cytokine Network* **11**, 362–371.

Dziarski, R. (2003). Recognition of bacterial peptidoglycan by the innate immune system. *Cellular and Molecular Life Sciences* **60**, 1793–1804.

Dziarski, R., Platt, K. A., et al. (2003). Defect in neutrophil killing and increased susceptibility to infection with nonpathogenic Gram-positive bacteria in peptidoglycan recognition protein-S (PGRP-S)-deficient mice. *Blood* **102**, 689–697.

Dziarski, R., Tapping, R., and Tobias, P. (1998). Binding of bacterial peptidoglycan to CD-14. *Journal of Biological Chemistry* **273**, 8680–8690.

Dziarski, R., Wang, Q., et al. (2001). MD-2 enables Toll-like receptor 2 (TLR2)-mediated responses to lipopolysaccharide and enhances TLR2-mediated responses to Gram-positive and Gram-negative bacteria and their cell wall components. *Journal of Immunology* **166**, 1938–1944.

Faure, E., Thomas, L., et al. (2001). Bacterial lipopolysaccharide and IFN-gamma induce Toll-like receptor 2 and Toll-like receptor 4 expression in human endothelial cells: role of NF-kappa B activation. *Journal of Immunology* **166**, 2018–2024.

Fenton, M. J. and Golenbock, D. T. (1998). LPS-binding proteins and receptors. *Journal of Leukocyte Biology* **64**, 25–31.

Fitzgerald, K. A., Palsson-McDermott, E. M., et al. (2001). Mal (MyD88-adapter-like) is required for Toll-like receptor-4 signal transduction. *Nature* **413**, 78–83.

Flo, T. H., Halaas, O., et al. (2000). Human toll-like receptor 2 mediates monocyte activation by *Listeria monocytogenes*, but not by group B streptococci or lipopolysaccharide. *Journal of Immunology* **164**, 2064–2069.

Flo, T. H., Halaas, O., et al. (2001). Differential expression of Toll-like receptor 2 in human cells. *Journal of Leukocyte Biology* **69**, 474–481.

Galan, J. E. (2001). Salmonella interactions with host cells: type III secretion at work. *Annual Review of Cell and Developmental Biology* **17**, 53–86.

Gantner, B. N., Simmons, R. M., et al. (2003). Collaborative induction of inflammatory responses by dectin-1 and Toll-like receptor 2. *Journal of Experimental Medicine* **197**, 1107–1117.

Gewirtz, A. T. (2003). Intestinal epithelial toll-like receptors: to protect. And serve? *Current Pharmaceutical Design* **9**, 1–5.

Gewirtz, A. T., Navas, T. A., et al. (2001). Cutting edge: bacterial flagellin activates basolaterally expressed TLR5 to induce epithelial proinflammatory gene expression. *Journal of Immunology* **167**, 1882–1885.

Girardin, S. E., Boneca, I. G., et al. (2003). Nod1 detects a unique muropeptide from gram-negative bacterial peptidoglycan. *Science* **300**, 1584–1587.

Haehnel, V., Schwarzfischer, L., et al. (2002). Transcriptional regulation of the human toll-like receptor 2 gene in monocytes and macrophages. *Journal of Immunology* **168**, 5629–5637.

Hajjar, A. M., Ernst, R. K., et al. (2002). Human Toll-like receptor 4 recognizes host-specific LPS modifications. *Nature Immunology* **3**, 354–359.

Hajjar, A. M., O'Mahony, D. S., et al. (2001). Cutting edge: functional interactions between toll-like receptor (TLR) 2 and TLR1 or TLR6 in response to phenol-soluble modulin. *Journal of Immunology* **166**, 15–19.

Hayashi, F., Smith, K. D., et al. (2001). The innate immune response to bacterial flagellin is mediated by Toll-like receptor 5. *Nature* **410**, 1099–1103.

Heine, H., Kirschning, C. J., et al. (1999). Cutting edge: cells that carry A null allele for toll-like receptor 2 are capable of responding to endotoxin. *Journal of Immunology* **162**, 6971–6975.

Hemmi, H., Kaisho, T., et al. (2002). Small anti-viral compounds activate immune cells via the TLR7 MyD88-dependent signaling pathway. *Nature Immunology* **3**, 196–200.

Hemmi, H., Takeuchi, O., et al. (2000). A Toll-like receptor recognizes bacterial DNA. *Nature* **408**, 740–745.

Henneke, P., Takeuchi, O., et al. (2001). Novel engagement of CD14 and multiple toll-like receptors by group B streptococci. *Journal of Immunology* **167**, 7069–7076.

Hirschfeld, M., Kirschning, C. J., et al. (1999). Cutting edge: inflammatory signaling by *Borrelia burgdorferi* lipoproteins is mediated by toll-like receptor 2. *Journal of Immunology* **163**, 2382–2386.

Hirschfeld, M., Ma, Y., et al. (2000). Cutting edge: repurification of lipopolysaccharide eliminates signaling through both human and murine toll-like receptor 2. *Journal of Immunology* **165**, 618–622.

Hirschfeld, M., Weis, J. J., et al. (2001). Signaling by toll-like receptor 2 and 4 agonists results in differential gene expression in murine macrophages. *Infection and Immunity* **69**, 1477–1482.

Hoebe, K., Du, X., et al. (2003). Identification of Lps2 as a key transducer of MyD88-independent TIR signalling. *Nature* **424**, 743–748.

Horng, T., Barton, G. M., et al. (2001). TIRAP: an adapter molecule in the Toll signaling pathway. *Nature Immunology* **2**, 835–841.

Horng, T., Barton, G. M., et al. (2002). The adaptor molecule TIRAP provides signalling specificity for Toll-like receptors. *Nature* **420**, 329–333.

Hoshino, K., Takeuchi, O., et al. (1999). Cutting edge: Toll-like receptor 4 (TLR4)-deficient mice are hyporesponsive to lipopolysaccharide: evidence for TLR4 as the Lps gene product. *Journal of Immunology* **162**, 3749–3752.

Hugot, J. P., Chamaillard, M., Zoualio, H., Lesage, S., Cezard, J., Belaiche, J. (2001). Association of NOD2 leucine-rich repeat varients with susceptibility to Crohn's disease. *Nature* **411**, 599–603.

Juan, T. S., Hailman, E., et al. (1995). Identification of a domain in soluble CD14 essential for lipopolysaccharide (LPS) signaling but not LPS binding. *Journal of Biological Chemistry* **270**, 17237–17242.

Kawai, T., Adachi, O., et al. (1999). Unresponsiveness of MyD88-deficient mice to endotoxin. *Immunity* **11**, 115–122.

Kawai, T., Takeuchi, O., et al. (2001). Lipopolysaccharide stimulates the MyD88-independent pathway and results in activation of IFN-regulatory factor 3 and the expression of a subset of lipopolysaccharide-inducible genes. *Journal of Immunology* **167**, 5887–5894.

Kawasaki, K., Akashi, S., et al. (2000). Mouse toll-like receptor 4.MD-2 complex mediates lipopolysaccharide-mimetic signal transduction by Taxol. *Journal of Biological Chemistry* **275**, 2251–2254.

Kawasaki, K., Gomi, K., et al. (2001). Cutting edge: Gln22 of mouse MD-2 is essential for species-specific lipopolysaccharide mimetic action of taxol. *Journal of Immunology* **166**, 11–14.

Kawasaki, K., Nogawa, H., et al. (2003). Identification of mouse MD-2 residues important for forming the cell surface TLR4-MD-2 complex recognized by anti-TLR4-MD-2 antibodies, and for conferring LPS and taxol responsiveness on mouse TLR4 by alanine-scanning mutagenesis. *Journal of Immunology* **170**, 413–420.

Khan, S. A., Everest, P., et al. (1998). A lethal role for lipid A in *Salmonella* infections. *Molecular Microbiology* **29**, 571–579.

Kirikae, T., Nitta, T., et al. (1999). Lipopolysaccharides (LPS) of oral black-pigmented bacteria induce tumor necrosis factor production by LPS-refractory C3H/HeJ macrophages in a way different from that of *Salmonella* LPS. *Infection and Immunity* **67**, 1736–1742.

Kirschning, C. J., Wesche, H., et al. (1998). Human toll-like receptor 2 confers responsiveness to bacterial lipopolysaccharide. *Journal of Experimental Medicine* **188**, 2091–2097.

Kopp, E. B. and Medzhitov, R. (1999). The Toll-receptor family and control of innate immunity. *Current Opinion in Immunology* **11**, 13–18.

Kusunoki, T., Hailman, E., et al. (1995). Molecules from *Staphylococcus aureus* that bind CD14 and stimulate innate immune responses. *Journal of Experimental Medicine* **182**, 1673–1682.

Lee, J. Y., Sohn, K. H., et al. (2001). Saturated fatty acids, but not unsaturated fatty acids, induce the expression of cyclooxygenase-2 mediated through Toll-like receptor 4. *Journal of Biological Chemistry* **276**, 16683–16689.

Lien, E., Chow, J. C., et al. (2001). A novel synthetic acyclic lipid A-like agonist activates cells via the lipopolysaccharide/toll-like receptor 4 signaling pathway. *Journal of Biological Chemistry* **276**, 1873–1880.

Lien, E., Sellati, T. J., et al. (1999). Toll-like receptor 2 functions as a pattern recognition receptor for diverse bacterial products. *Journal of Biological Chemistry* **274**, 33419–33425.

Liu, C. X. Z., Gupta, D., Dziarski, R. (2003). Peptidoglycan recognition proteins: a novel family of four human innate immunity pattern recognition molecules. *Journal of Biological Chemistry* **276**, 34686–34694.

Liu, Y., Wang, Y., et al. (2001). Upregulation of toll-like receptor 2 gene expression in macrophage response to peptidoglycan and high concentration of lipopolysaccharide is involved in NF-kappa b activation. *Infection and Immunity* **69**, 2788–2796.

Lorenz, E., Jones, M., et al. (2001). Genes other than TLR4 are involved in the response to inhaled LPS. *American Journal of Physiology: Lung Cellular and Molecular Physiology* **281**, L1106–1114.

Matsuguchi, T., Musikacharoen, T., et al. (2000). Gene expressions of Toll-like receptor 2, but not Toll-like receptor 4, is induced by LPS and inflammatory cytokines in mouse macrophages. *Journal of Immunology* **165**, 5767–5772.

Matsumura, T., Ito, A. et al. (2000). Endotoxin and cytokine regulation of toil-like receptor (TLR)2 and TLR4 gene expression in marine liver and hepatocytes. *Journal of Interferon Cytokine Research* **20**, 915–921.

Means, T. K., Jones, B. W., et al. (2001). Differential effects of a Toll-like receptor antagonist on *Mycobacterium tuberculosis*–induced macrophage responses. *Journal of Immunology* **166**, 4074–4082.

Means, T. K., Lien, E., et al. (1999a). The CD14 ligands lipoarabinomannan and lipopolysaccharide differ in their requirement for Toll-like receptors. *Journal of Immunology* **163**, 6748–6755.

Means, T. K., Wang, S., et al. (1999b). Human toll-like receptors mediate cellular activation by *Mycobacterium tuberculosis. Journal of Immunology* **163**, 3920–3927.`

Medzhitov, R., Preston-Hurlburt, P., et al. (1997). A human homologue of the *Drosophila* Toll protein signals activation of adaptive immunity. *Nature* **388**, 394–397.

Mellroth, P., Karlsson, J., and Steiner, H. (2003). A scavenger function for a *Drosophila* peptidoglycan recognition protein. *Journal of Biological Chemistry* **278**, 7059–7064.

Melmed, G., Thomas, L. S., et al. (2003). Human intestinal epithelial cells are broadly unresponsive to Toll-like receptor 2-dependent bacterial ligands: implications for host-microbial interactions in the gut. *Journal of Immunology* **170**, 1406–1415.

Miceli-Richard, C., Lesage, S., et al. (2001). CARD15 mutations in Blau syndrome. *Nature Genetics* **29**, 19–20.

Mitsuzawa, H., Wada, I., et al. (2001). Extracellular Toll-like receptor 2 region containing Ser40-Ile64 but not Cys30-Ser39 is critical for the recognition of *Staphylococcus aureus* peptidoglycan. *Journal of Biological Chemistry* **276**, 41350–41356.

Mizel, S. B., West, A. P., et al. (2003). Identification of a sequence in human toll-like receptor 5 required for the binding of Gram-negative flagellin. *Journal of Biological Chemistry* **278**, 23624–23629.

Moore, K. J., Andersson, L. P., et al. (2000). Divergent response to LPS and bacteria in CD14-deficient murine macrophages. *Journal of Immunology* **165**, 4272–4280.

Moors, M. A., Li, L., et al. (2001). Activation of interleukin-1 receptor-associated kinase by gram-negative flagellin. *Infection and Immunity* **69**, 4424–4429.

Mullen, G. E., Kennedy, M. N., et al. (2003). The role of disulfide bonds in the assembly and formation of MD-2. *Proceedings of the National Academy of Sciences USA* **100**, 3919–3924.

Musikacharoen, T., Matsuguchi, T., et al. (2001). NF-kappa B and STAT5 play important roles in the regulation of mouse Toll-like receptor 2 gene expression. *Journal of Immunology* **166**, 4516–4524.

Muzio, M., Bosisio, D., et al. (2000). Differential expression and regulation of toll-like receptors (TLR) in human leukocytes: selective expression of TLR3 in dendritic cells. *Journal of Immunology* **164**, 5998–6004.

Nagai, Y., Akashi, S., et al. (2002). Essential role of MD-2 in LPS responsiveness and TLR4 distribution. *Nature Immunology* **3**, 667–672.

Nattermann, J., Du, X., et al. (2000). Endotoxin-mimetic effect of antibodies against Toll-like receptor 4. *Journal of Endotoxin Research* **6**, 257–264.

Ogata, H., Su, I., et al. (2000). The toll-like receptor protein RP105 regulates lipopolysaccharide signaling in B cells. *Journal of Experimental Medicine* **192**, 23–29.

Ogura, Y., Bonen, D. K., Inohara, N., Nicolae, D. L., Chen, F. F., Ramos, R. (2001). A frameshift mutation in NOD2 associated with susceptibility to Crohn's disease. *Nature* **411**: 603–606.

Ohashi, K., Burkart, V., et al. (2000). Cutting edge: heat shock protein 60 is a putative endogenous ligand of the toll-like receptor-4 complex. *Journal of Immunology* **164**, 558–561.

Ohnishi, T., Muroi, M., et al. (2001). N-linked glycosylations at Asn(26) and Asn(114) of human MD-2 are required for toll-like receptor 4-mediated activation of NF-kappaB by lipopolysaccharide. *Journal of Immunology* **167**, 3354–3359.

Okamoto, M., Oshikawa, T., et al. (2001). Severe impairment of anti-cancer effect of lipoteichoic acid-related molecule isolated from a penicillin-killed *Streptococcus pyogenes* in toll-like receptor 4-deficient mice. *International Immunopharmacology* **1**, 1789–1795.

O'Neill, L. A. (2002). Signal transduction pathways activated by the IL-1 receptor/toll-like receptor superfamily. *Current Topics in Microbiology and Immunology* **270**, 47–61.

Oshiumi, H., Sasal, M., et al (2003). TIR-containing adapter molecule (TICAM)-2, a bridging adapter recruiting toll-like receptor 4 TICAM-1 that induces interferon-beta *Journal of Biological Chemistry* **278**, 49751–49756.

Ozinsky, A., Underhill, D. M., et al. (2000). The repertoire for pattern recognition of pathogens by the innate immune system is defined by cooperation between toll-like receptors. *Proceedings of the National Academy of Sciences USA* **97**, 13766–13771.

Poltorak, A., Smirnova, I., et al. (1998). Genetic and physical mapping of the Lps locus: identification of the toll-4 receptor as a candidate gene in the critical region. *Blood Cells Molecular Diseases* **24**, 340–355.

Qureshi, S. T., Lariviere, L., et al. (1999). Endotoxin-tolerant mice have mutations in Toll-like receptor 4 (Tlr4). *Journal of Experimental Medicine* **189**, 615–625.

Re, F. and Strominger, J. L. (2003). Separate functional domains of human MD-2 mediate toll-like receptor 4-binding and lipopolysaccharide responsiveness. *Journal of Immunology* **171**, 5272–5276.

Rehman, A., Taishi, P., Fang, J., Majde, J. A., and Krueger, J. M. (2000). The cloning of a rat peptidoglycan recognition protein (PGRP) and its induction in the brain by sleep deprivation. *Cytokine* **13**, 8–17.

Rock, F. L., Hardiman, G., et al. (1998). A family of human receptors structurally related to *Drosophila* Toll. *Proceedings of the National Academy of Sciences USA* **95**, 588–593.

Rosenberger, S. A., Finlay B. B. (2002). Macrophages inhibit *Salmonella typhimurium* through MEK/ERK kinase and phagocyte MADPH oxidase activities. *Journal of Biological Chemistry* **277**, 18753–18762.

Royle, M. C., Totemeyer, S., et al. (2003). Stimulation of Toll-like receptor 4 by lipopolysaccharide during cellular invasion by live *Salmonella typhimurium* is a critical but not exclusive event leading to macrophage responses. *Journal of Immunology* **170**, 5445–5454.

Sato, S., Sugiyama, M., et al. (2003). Toll/IL-1 receptor domain-containing adaptor inducing IFN-beta (TRIF) associates with TNF receptor-associated factor 6 and TANK-binding kinase 1, and activates two distinct transcription factors, NF-kappaB and IFN-regulatory factor-3, in the Toll-like receptor signaling. *Journal of Immunology* **171**, 4304–4310.

Schwandner, R., Dziarski, R., et al. (1999). Peptidoglycan- and lipoteichoic acid-induced cell activation is mediated by toll-like receptor 2. *Journal of Biological Chemistry* **274**, 17406–17409.

Sebastiani, G., Leveque, G., et al. (2000). Cloning and characterization of the murine toll-like receptor 5 (Tlr5) gene: sequence and mRNA expression studies in *Salmonella*-susceptible MOLF/Ei mice. *Genomics* **64**, 230–240.

Shimazu, R., Akashi, S., et al. (1999). MD-2, a molecule that confers lipopolysaccharide responsiveness on Toll-like receptor 4. *Journal of Experimental Medicine* **189**, 1777–1782.

Shuto, T., Xu, H., et al. (2001). Activation of NF-kappaB by nontypeable *Haemophilus influenzae* is mediated by toll-like rector 2-TAK1-dependent NIK-IKK alpha/beta-I kappa B alpha and MKK3/6-p38 MAP kinase signalling pathways inepithelial cells. *Proceedings of the National Academy of Sciences USA* **98**, 8774–8779.

Smith, K. D., Andersen-Nissen, E., et al. (2003). Toll-like receptor 5 recognizes a conserved site on flagellin required for protofilament formation and bacterial motility. *Nature Immunology* **4**, 1247–1253.

Takeshita, F., Leifer, C. A., et al. (2001). Cutting edge: Role of Toll-like receptor 9 in CpG DNA-induced activation of human cells. *Journal of Immunology* **167**, 3555–3558.

Takeuchi, O., Hoshino, K., et al. (1999). Differential roles of TLR2 and TLR4 in recognition of gram-negative and gram-positive bacterial cell wall components. *Immunity* **11**, 443–451.

Takeuchi, O., Hoshino, K., et al. (2000a). Cutting edge: TLR2-deficient and MyD88-deficient mice are highly susceptible to *Staphylococcus aureus* infection. *Journal of Immunology* **165**, 5392–5396.

Takeuchi, O., Kaufmann, A., et al. (2000b). Cutting edge: preferentially the R-stereoisomer of the mycoplasmal lipopeptide macrophage-activating lipopeptide-2 activates immune cells through a toll-like receptor 2- and MyD88-dependent signaling pathway. *Journal of Immunology* **164**, 554–557.

Takeuchi, O., Kawai, T., et al. (2001). Discrimination of bacterial lipoproteins by Toll-like receptor 6. *International Immunology* **13**, 933–940.

Takeuchi, O., Sato, S., et al. (2002). Cutting edge: role of Toll-like receptor 1 in mediating immune response to microbial lipoproteins. *Journal of Immunology* **169**, 10–14.

Tötemeyer, S., Foster, N., et al. (2003). Toll-like receptor expression in C3H/HeN and C3H/HeJ mice during *Salmonella enterica* serovar *Typhimurium* infection. *Infection and Immunity* **71**, 6653–6657.

Triantafilou, M., Miyake, K., et al. (2002). Mediators of innate immune recognition of bacteria concentrate in lipid rafts and facilitate lipopolysaccharide-induced cell activation. *Journal of Cell Science* **115**, 2603–2611.

Triantafilou, K., Triantafilou, M., et al. (2001). A CD14-independent LPS receptor cluster. *Nature Immunology* **2**, 338–345.

Tydell, C., Yount, N., Tran, D., Yuan, J., Selsted, M. E. (2002). Isolation, characterisation and antimicrobial properties of bovine oligosaccharide-binding protein. *Journal of Biological Chemistry* **277**, 19658–19664.

Underhill, D. M., Ozinsky, A., et al. (1999a). The Toll-like receptor 2 is recruited to macrophage phagosomes and discriminates between pathogens. *Nature* **401**, 811–815.

Underhill, D. M., Ozinsky, A., et al. (1999b). Toll-like receptor-2 mediates mycobacteria-induced proinflammatory signaling in macrophages. *Proceedings of the National Academy of Sciences USA* **96**, 14459–14463.

Viriyakosol, S., Tobias, P. S., et al. (2001). MD-2 binds to bacterial lipopolysaccharide. *Journal of Biological Chemistry* **276**, 38044–38051.

Visintin, A., Latz, E., et al. (2003). Lysines 128 and 132 enable lipopolysaccharide binding to MD-2, leading to Toll-like receptor-4 aggregation and signal transduction. *Journal of Biological Chemistry* **278**, 48313–48320.

Visintin, A., Mazzoni, A., et al. (2001). Secreted MD-2 is a large polymeric protein that efficiently confers lipopolysaccharide sensitivity to Toll-like receptor 4. *Proceedings of the National Academy of Sciences USA* **98**, 12156–12161.

Vogel, S. N., Johnson, D., et al. (1999). Cutting edge: functional characterization of the effect of the C3H/HeJ defect in mice that lack an Lpsn gene: in

vivo evidence for a dominant negative mutation. *Journal of Immunology* **162**, 5666–5670.
Werts, C., Tapping, R. I., et al. (2001). Leptospiral lipopolysaccharide activates cells through a TLR2-dependent mechanism. *Nature Immunology* **2**, 346–352.
Wright, S. D. (1995). CD14 and innate recognition of bacteria. *Journal of Immunology* **155**(1): 6–8.
Wright, S. D., Ramos, R. A., et al. (1990). CD14, a receptor for complexes of lipopolysaccharide (LPS) and LPS binding protein. *Science* **249**, 1431–1433.
Wurfel, M. M., Monks, B. G., et al. (1997). Targeted deletion of the lipopolysaccharide (LPS)-binding protein gene leads to profound suppression of LPS responses ex vivo, whereas in vivo responses remain intact. *Journal of Experimental Medicine* **186**, 2051–2056.
Xu, Y., Tao, X., et al. (2000). Structural basis for signal transduction by the Toll/interleukin-1 receptor domains. *Nature* **408**, 111–115.
Yamamoto, M., Sato, S. et al. (2002). Essential role for TIRAP in activation of the signalling cascade shared by TLR2 and TLR4. *Nature* **420**, 324–329.
Yamamoto, M., Sato, S. et al. (2003a). Role of adaptor TRIF in the MyD88-independent toll-like receptor signaling pathway. *Science* **301**, 640–643.
Yamamoto, M., Sato, S. et al. (2003b). TRAM is specifically involved in the Toll-like receptor 4-mediated MyD88-independent signaling pathway. *Nature Immunology* **4**, 1144–1150.
Yang, R. B., Mark, M. R., et al. (1998). Toll-like receptor-2 mediates lipopolysaccharide-induced cellular signalling. *Nature* **395**, 284–288.
Yoshimura, A., Lien, E., et al. (1999). Cutting edge: recognition of Gram-positive bacterial cell wall components by the innate immune system occurs via Toll-like receptor 2. *Journal of Immunology* **163**, 1–5.
Zarember, K. A. and Godowski, P. J. (2002). Tissue expression of human Toll-like receptors and differential regulation of Toll-like receptor mRNAs in leukocytes in response to microbes, their products, and cytokines. *Journal of Immunology* **168**, 554–561.
Ziegler-Heitbrock, L. (2001). The p50-homodimer mechanism in tolerance to LPS. *Journal of Endotoxin Research* **7**, 219–222.

CHAPTER 16

Moonlighting in protein hyperspace: Shared moonlighting proteins and bacteria–host cross talk

Brian Henderson

16.1. INTRODUCTION

It is not clear which multicellular species will win the prize for having the largest collection of cooperative bacteria. *Homo sapiens* must be in the running, with an estimated 500 bacterial species in the gut (Suau et al. 1999) and more than 700 different species in the oral cavity (Kazor et al. 2003). These bacteria constitute the normal human microbiota. Alternative names are the normal microflora or commensal bacteria (although see other chapters for criticism of the use of the term *commensal*). Note that the human microbiota or microflora will contain, in addition to bacteria, single-celled eukaryotes and archaea. We know almost nothing about the latter, and this chapter, along with all others in this volume, focuses only on cooperative bacteria. If the normal microbiota of the skin and urogenital system is included, the average human would seem to share her/his body with between 1,000 and 2,000 different species of bacteria. Most of these bacterial species cause no health problems. This is a truly remarkable concept, as the general feeling about bacteria is disease and death. However, only around forty bacterial species (some of which are also commensals) routinely cause human disease (Wilson et al. 2002). In some unexplained way, we live harmoniously with 1,000 to 2,000 bacterial species throughout our long lives.

This creates a fascinating paradox as has been touched on in many of the other chapters in this volume. All the bacteria we coexist with produce a range of proinflammatory molecules (described in detail in Chapter 15) that should activate mucosal surfaces and the immune cells associated with such surfaces. In Gram-negative bacteria, the major proinflammatory signal is lipopolysaccharide (LPS) and a range of associated moieties that constitute endotoxin (Henderson et al. 1996a, 1998). In Gram-positive bacteria, peptidoglycan, peptidoglycan breakdown products, and lipo/teichoic acids are the best-studied inflammogens (Henderson et al. 1996a,b). In addition to these nonproteinaceous factors, a growing number of exported and cell surface–associated proteins, including established bacterial toxins, have been shown capable of inducing the synthesis of proinflammatory cytokines (Henderson et al. 1997, 1998; Henderson 2000; Henderson and Seymour 2003).

16.2. BACTERIA–HOST INTERACTIONS PREVENTING MUCOSAL INFLAMMATION

The microbiotae existing on the various mucosal and epithelial surfaces of the human body are examples of microbial ecosystems that exist in the vicinity of a range of eukaryotic "ecosystems." Now clearly, the mucosal surface is able to identify the presence of pathogenic bacteria. How then does it interact with those bacteria normally resident in close apposition to it? One possibility is that the epithelial cells in the mucosal surfaces are unable to respond to bacteria, and it has been claimed that intestinal epithelial cells are insensitive to LPS and peptidoglycan (Melmed et al. 2003) and that cervicovaginal epithelial cells have no TLR4 receptors – although they express TLR2 (Fichorova et al. 2002). However, it is unlikely that such epithelia can be refractory to all bacterial signals and, as an example of this, intestinal epithelial cells are responsive to flagellin (Eaves-Pyles et al. 2001). Moreover, epithelia have a large collection of associated myeloid and lymphoid cells that should recognize a range of bacterial inflammogens. It is more likely that the suppression of mucosal inflammatory responses to members of the normal microbiota is not the result of indifference but is a deliberate and active process. If this is accepted, then the next question is – which cells (bacteria or host) are controlling this process? The answer will probably turn out to involve both the host and its normal microbiota. In the context of cooperative bacteria, we are beginning to identify some of the mechanisms utilized by bacteria to modulate host inflammatory responses. Such molecules are extremely varied, ranging from the low molecular mass acyl homoserine lactones used in quorum sensing, to some of the most complex bacterial enterotoxins, on to multicomponent bacteria secretions systems such as the type III and type IV mechanisms. These molecules and molecular systems are produced only by bacteria (and not by the host) and obviously have evolved to modulate host cell function. Bacterial components with anti-inflammatory effects have recently been reviewed in a sister volume (Henderson and Oyston 2003), and the ability of bacteria to interfere with the NF-κB inflammatory signaling system is reviewed in Chapter 17. The remainder of this chapter is devoted to a perplexing "class" of protein, termed moonlighting proteins, which are shared both by the host and by the microbiota. There is increasing evidence that these moonlighting proteins form part of the continuum of communication between the host and bacteria, including commensal organisms, and may play roles in discriminating among bacterial species.

16.3. THE MICROBIOTA AND PROTEIN HYPERSPACE

Let us assume that the final count for the number of cooperative bacterial species colonizing *Homo sapiens* is 2,000. We can also argue that the average number of genes in each of these organisms will be around 2,500. This allows us to add an extra 5 million genes to the "human" genome, which currently seems to run only to 23,000 genes. If we further conservatively estimate that five percent of these bacterial genes are not shared among the microbiota then, we have, potentially, 250,000 totally novel proteins to deal with, many of which may play key roles in bacteria–host cooperation. Large as these numbers are, they fade into insignificance when we consider the possible number of proteins that could exist in our universe. There are twenty protein building blocks – the amino acids – encoded for by the four nucleotides in DNA. If we take an average-sized protein of 300-odd amino acids, this is encoded for by a DNA sequence of 1,000 nucleotides. We can calculate the total number of permutations of these 1,000 nucleotides. The answer is 4^{1000} or 10^{600} possible gene/protein sequences. It is for this number of possible proteins that Smith and Morowitz, in a review written in the early 1980s, introduced the term *protein hyperspace* (Smith and Morowitz 1982). It is from this "almost infinite" hyperspace that evolutionary processes can sample. In a recent fascinating, and almost certainly controversial, book on evolution, Simon Conway Morris addresses the role of protein hyperspace in "the game of life" and suggests that significant constraints are involved in generating the working proteins on which evolution is based (Conway Morris 2003). Even with such constraints, it is surprising to find that there are a growing number of proteins, the moonlighting proteins, that have more than one function. This chapter introduces the concept of protein moonlighting and hypothesizes a role for these evolutionarily shared or convergent proteins in bacterial–host communication.

16.3.1. Moonlighting in Protein Hyperspace

Moonlighting, in the colloquial sense, means having two paid occupations – one during the day, the other at night. When applied to proteins, the term, which was introduced in the mid to late 1990s, describes those having two (or more) different functions (Jeffrey 1999, 2003). These different functions may take place in different parts of the cell or within the inside, versus the outside, of cells. The term *moonlighting* encompasses enzymes with a diverse range of substrate specificities (reviewed by Copley 2003) to proteins such as phosphoglucose isomerase (PGI), which is an intracellular enzyme of glycolysis, and also three distinct cytokines (Faik et al. 1988; Xu et al. 1996; Watanabe et al. 1996) and an embryonic implantation factor (Schulz and Bahr 2003). These nonglycolytic actions occur outside the cell. Enzymes involved in carbohydrate metabolism/cell oxidation form a theme in this chapter. For instance, evidence is emerging that TCA cycle enzymes such as fumarate dehydratase and succinate dehydrogenase moonlight as tumor suppressors in conditions such as inherited uterine fibroids (Tomlinson et al. 2002) and in paragangliomas – benign vascular tumors in the head and neck (Baysal et al. 2000). There is a growing list of moonlighting proteins (Table 16.1).

The question that immediately occurs when confronted with the concept of moonlighting is – are all proteins capable of moonlighting? The answer cannot be given at present, but it is unlikely that all proteins will have multiple functions. Equally, it is likely that many proteins will be found to moonlight. If this is the case, will this have to make us rethink current paradigms about protein evolution in which most mutations are seen as neutral when measured against the "function" of the protein chosen (e.g., hemoglobin) (see Bromham and Penn 2003 for a recent review)? If proteins have two or more "active sites" to promote their normal and moonlighting functions, this must put limits on the amino acid sequence in which neutral mutations can exist. Could this give rise to two or more molecular clock rates for individual moonlighting proteins, and what would the consequence for this be on the protein evolutionary landscape?

Leaving aside the evolution of moonlighting proteins, this chapter now sets out to discuss two groups of moonlighting proteins that are shared by both bacteria and by hosts. The first of these is the enzymes of glycolysis. A growing number of the glycolytic pathway enzymes are known to be involved in both human disease and in the communication between bacteria and host cells. The second group is the molecular chaperones which have been found in the last decade to act as cell–cell signaling proteins in addition to their original functions as protein-folding catalysts. Again, evidence is emerging that molecular chaperones can act as signals between bacteria and host cells.

Table 16.1. *Moonlighting proteins*

Protein	Originally described function	Moonlighting function
Various enzymes[1]	Substrate specificity 1	Substrate specificity 2
Mitochondrial tyrosyl tRNA synthetase	Charging $tRNA^{Tyr}$	Folding group 1 introns[2]
Cytochrome c	Mitochondrial electron transport	Stimulates apoptosis[3]
Gephyrin	GABA receptor clustering	Molybdenum metabolism[4]
Clf1p	Factor in pre-mRNA splicing	Initiation of DNA replication[5]
Proteasome complex	Proteolysis	RNA polymerase III function[6]
Thioredoxin	Intracellular redox controller	T cell cytokine/ chemokine[7]
Elongation factor 1	Polypeptide elongation factor	Multiple other actions[8]
Phosphoglucoisomerase	Glycolytic enzyme	Multiple other actions (see text)
Caeruloplasmin	Copper-binding protein	Multiple other functions[9]
Fumarate hydratase	TCA cycle enzyme	Tumor suppressor gene[10]
Succinate dehydrogenase	TCA cycle enzyme	Tumor suppressor gene[11]
CD26	Peptidase	Receptor, costimulatory protein[12]
Autolysins	Amidases	Receptors for various host ligands[13]
Enolase	Glycolytic enzyme	
GAPDH	Glycolytic enzyme	
Molecular chaperones	Protein-folding proteins	

[1]Copley (2003); [2]Caprara et al. (1996); [3]van Gurp et al. (2003); [4]Reiss et al. (2001); [5]Zhu et al. (2002); [6]Gonzalez et al. (2002); [7]Hirota et al. (2002); [8]Ejiri (2002); [9]Bielli and Calabrese (2002); [10]Tomlinson et al. (2002); [11]Baysal et al. (2000); [12]Boonacker and Van Noorden (2003); [13]Hell et al. (1998)

16.3.2. Moonlighting by Glycolysis

Biochemists are still taught that the glycolytic pathway (Fig. 16.1) is, as one popular textbook describes, "a nearly universal pathway that converts glucose into pyruvate with the concomitant production of a relatively small amount of ATP." Most biological scientists will have a hazy memory of the enzymes of glycolysis and, if asked, would probably ascribe to the view that eukaryotic glycolytic enzymes evolved from ancient bacterial genes. Indeed, it had been proposed that the eukaryotic glycolytic enzymes evolved from α-proteobacterial genes that had been transferred to pre-eukaryotic cells from the ancestor of our current mitochondria (Martin and Muller 1998). Attractive as this hypothesis is, a recent phylogenetic analysis of the glycolytic genes of Bacteria, Archaeae, and Eukarya failed to support the hypothesis that these genes are evolutionarily related (Canback et al. 2002). The inference would be that glycolysis was independently evolved by bacteria and

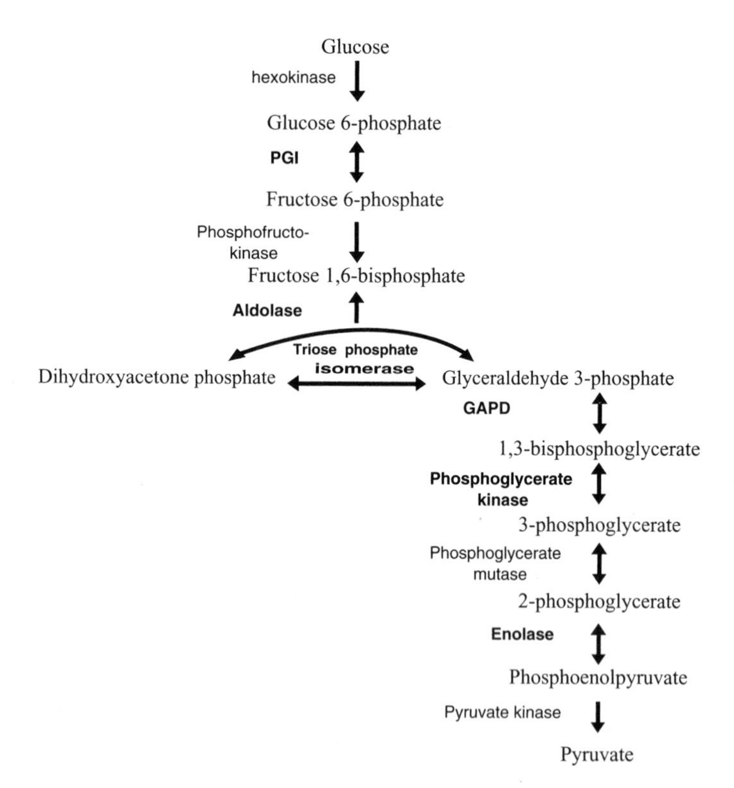

Figure 16.1. The glycolytic pathway of glucose oxidation. Enzymes of this pathway found on the surface of eukaryotic or prokaryotic cells are emboldened.

eukaryotes. If this is the case, then it makes the findings of the moonlighting actions of both bacterial and mammalian glycolytic enzymes all the more surprising.

The concept of the glycolytic pathway as a "complex" and able to interact with membranes has been around for some time (Gutowicz and Terlecki 2003). However, it has been only in the past few years that, aided by the development of proteomics, the number of glycolytic enzymes exported by cells, or associated with the cell surface, has begun to be determined. Currently, six glycolytic enzymes have been found on the surface of bacteria or eukaryotic cells (Table 16.2). Phosphoglucoisomerase (PGI) is the most fascinating of these proteins because, as described, it has four additional functions to its enzymic role. PGI contains a CXXC motif, characteristic of redox proteins such as thoredoxin (Table 16.1) and is possibly evolutionarily related to the CC and CXC motif of the chemokine gene family (Tsutsumi et al. 2003). Activity of exogenous PGI depends on binding to a 78 kd G–protein–coupled receptor (gp78/AMF-R) (Shimizu et al. 1999). Controversy exists over the role played by the enzyme active site in the moonlighting activities of PGI. Substrate analogues of glucose 6-phosphate block PGI binding to gp78 (Watanabe et al. 1996). Site-directed mutagenesis of the active site impaired the autocrine motility factor (AMF) activity of PGI (Tanaka et al. 2002). However, the same group has mutated the CXXC motif in PGI, which also blocks enzymic activity without, in this case, inhibiting the autocrine motility factor activity (Tsutsumi et al. 2003). The jury is currently out on this issue. Do bacterial PGIs have cytokine-like activity? The PGI from *Bacillus stearothermophilus*, a thermophile, has AMF activity (Sun et al. 1999). However, it is not known if the PGI from bacteria associated with humans has cytokine-like activity. The finding of PGI on the surface of *Streptococcus agalactiae* (Hughes et al. 2002) suggests that at least one bacterium may utilize this glycolytic protein. The consequence of bacterial PGIs having all the cytokine activities of the human enzyme could be significant.

Glyceraldehyde 3-phosphate dehydrogenase (GAPD) is another cause of constant surprise as its additional moonlighting actions are identified. Moonlighting functions of the mammalian enzyme include uracil DNA glycosylase activity, acting as an activator of transcription, binding to RNA, and an involvement in tubulin assembly. The most recently defined activity is as an initiator of apoptosis, and this seems particularly pertinent to neurological disease (Sirover 1999; Berry and Boulton 2000). It is not known if bacterial GAPD proteins will also exhibit these various activities.

In a number of Gram-positive bacteria, GAPD is either, or both, associated with the cell surface and secreted (Pancholi and Fischetti 1992; Nelson

Table 16.2. *Glycolytic enzymes associated with the cell surface or exported*

Enzyme	Organism	Other activity
Phosphoglucoisomerase	*Homo sapiens*	Neuroleukin, autocrine motility factor differentiation and maturation mediator implantation factor
	Streptococcus agalactiae	?
Aldolase	*Streptococcus oralis*	?
Triose phosphate isomerase	*Actinobacillus actinomycetemcomitans*	?
GAPD	Group A streptococci	Receptor for fibronectin, lysozyme, myosin, actin
	Group B streptococci	Receptor for lys and glu plasminogens
	Staphylococci	Receptor for transferrin
	Mycobacterium tuberculosis	Receptor for epidermal growth factor
Phosphoglycerate kinase	*Streptococcus agalactiae*	?
	Streptococcus oralis	?
Enolase	*Homo sapiens*	Plasminogen activation/binding
	Streptococcus pneumoniae	Plasmin(ogen) receptor
	Streptococcus agalactiae	?*
	Streptococcus oralis	?*
	Actinobacillus actinomycetemcomitans	?*
	Aeromonas hydrophila	?*

* It is presumed that these enolases also bind to plasmin(ogen).

et al. 2001; Hughes et al. 2002; Wilkins et al. 2003). This was first shown with Group A streptococci and the surface-associated GAPD, after purification, was reported to bind to fibronectin, lysozyme, myosin, and actin (Pancholi and Fischetti 1992) and to possess ADP-ribosylating activity (Pancholi and Fischetti 1993). Unexpectedly, this surface-associated GAPD could phosphorylate proteins on intact human epithelial cells and cause selected changes in the intracellular signaling of these eukaryotic cells (Pancholi and Fischetti 1997). The release of GAPD by an endocarditis strain of *Streptococcus gordonii* has been shown to be dependent on environmental pH. At pH 6.5, most of the enzyme was on the cell surface. In contrast, at pH 7.5, the enzyme was predominantly exported from the cell (Nelson et al. 2001). It is not clear how such patterns of secretion are controlled. The finding that surface-associated GAPD in streptococci binds to M and M-related fibrinogen-binding proteins (Costa et al. 2000) suggests this enzyme participates in immune evasion.

In addition to utilizing its enzymic activity for signaling, surface-associated GAPD in various bacteria, and in yeast, has been shown to act as a receptor for various ligands. In the yeast *Candida albicans*, surface-associated GAPD binds fibronectin and laminin (Gozalbo et al. 1998). In Group A (Lottenberg et al. 1992) and Group B (Seifert et al. 2003) streptococci, GAPD binds plasmin(ogen). The cell wall GAPD of *Staphylococcus aureus* has been claimed to bind transferrin (Modun and Williams 1999), although this has recently been questioned (Taylor and Heinrichs 2002). The causative agent of tuberculosis, *Mycobacterium tuberculosis*, also expresses a cell surface GAPD. The ligand for this particular GAPD protein is the human cytokine epidermal growth factor, and it has been shown that this protein can stimulate the growth of mycobacteria in culture (Bermudez et al. 1996; Parker and Bermudez 2000).

A fascinating example of functional convergence is found with the glycolytic enzyme enolase. Human alpha-enolase is exported to the cell surface, by an unknown mechanism, where it functions as a receptor for plasmin(ogen) and is responsible, in human cells, for the majority of plasminogen activation at the surfaces of leukocytes (Lopez-Alemany et al. 2003). In addition, mammalian enolase is an important autoantigen in autoimmune disease, it can function as a heat shock protein, and can bind to components of the cytoskeleton and chromatin (reviewed by Pancholi 2001). Bacterial alpha-enolase (specifically that from *Streptococcus pneumoniae*) has forty-seven percent sequence identity with the human enzyme, which seems to contradict the findings of Canback et al. (2002). The bacterial enolase is found on the surface of streptococci (Pancholi and Fischetti 1998; Ehinger et al. 2002) and the Gram-negative organisms *Actinobacillus actinomycetemcomitans* (Hara et al. 2000) and *Aeromonas hydrophila* (Sha et al. 2003). In the streptococci, the surface-bound alpha-enolase is a receptor for plasmin(ogen). The plasminogen binds to surface-bound alpha-enolase, and at this site, it is subsequently activated to the serine protease plasmin by plasminogen activator (tPA) or urokinase (uPA) emanating from the host. Two binding sites on *Strep. pneumoniae* enolase for plasminogen have been identified. Mutations in the internal binding motif, which causes loss of plasminogen binding, result in a decrease in virulence (Ehinger et al. 2004).

It therefore appears that a number of the members of the human commensal microflora (streptococci, staphylococci, and *A. actinomycetemcomitans*) can utilize the moonlighting function of two glycolytic enzymes to modulate the host fibrinolytic pathway. The consequence of this is presumably to enhance tissue penetration and colonization. A key question that needs to be addressed is how are these glycolytic proteins secreted?

Much more work needs to be done to address the role of secreted glycolytic proteins in bacteria–host interactions. In addition, two other glycolytic enzymes have been found on the surface of bacteria. These are aldolase on the surface of *Streptococcus oralis* (Wilkins et al. 2003) and triose phosphate isomerase on the surface of *A. actinomycetemcomitans* (Fletcher et al. 2001). It should be noted that the possession of these two enzymes plus GAPD and phosphoglycerate kinase could, potentially, enable bacteria to make ATP on their surface if supplied with an appropriate substrate. Such ATP synthesis could signal through the very large numbers of purinoceptors (la Sala et al. 2003) present on the surface of host cells.

It has been known for more than a century that streptococcal infections can result in neurological symptoms. It has recently emerged that certain human neurons also express many of the glycolytic enzymes on their cell surface. It is postulated that cross-reactivity between streptococcal glycolytic enzymes, such as enolase and cell surface neuronal enzymes, can produce the symptoms of a range of diseases including Tourette's syndrome and obsessive-compulsive disorder (Swedo 2001; Church et al. 2003). This finding raises all sorts of possibilities for the role of bacteria in neurological dysfunction.

16.4. THE STRESS OF MOONLIGHTING

Since 1962, it has been known that all cells respond to environmental stress by increasing the rate of transcription of a group of proteins, known variously as heat shock proteins, cell stress proteins, or molecular chaperones (Kregel 2002; Walter and Buchner 2002). Molecular chaperones are ubiquitous, often essential, often highly conserved, proteins with the common function of interacting with other proteins to induce them to fold, to refold, or to prevent misfolding (Walter and Buchner 2002). A growing number of proteins are now defined as molecular chaperones, and the major

Table 16.3. *The molecular chaperones*

Bacterial protein	Eukaryotic homologue
Chaperonin 10	Chaperonin 10 (mitochondrial)
Thoredoxin	Thioredoxin
Cyclophilin	Cyclophilin
Small HSPs (e.g., IbpA, 14 kd Mt* antigen)	Hsp27
E. coli DnaJ	Hsp40
E. coli GrpE	Mammalian mitochondrial GrpE
Chaperonin 60	Chaperonin 60 or Hsp60 (mitochondrial)
E. coli DnaK	Hsp70 (mitochondrial), HSc70, Bip
E. coli HtpG	Hsp90, gp96
Clp proteins	Hsp100

*Mt – *Mycobacterium tuberculosis.*

"classes" of these proteins shared between bacteria and mammals are shown in Table 16.3.

The discovery of molecular chaperones began in the 1960s, and by the 1990s, the mechanism of action of certain of these proteins was beginning to be defined. It was at this time that the first reports began to appear that some of the molecular chaperones from bacteria had the capacity to stimulate the proinflammatory actions of human or rodent myeloid cells (Friedland et al. 1993; Retzlaff et al. 1994). These reports were the first indication that another major class of moonlighting proteins – the molecular chaperones – existed (literature reviewed in Lewthwaite et al. 1998; Coates et al. 1999; Ranford et al. 2000; Maguire et al. 2002; Henderson 2003).

In the ten years since the discovery that the chaperonin (Cpn)60.2 protein of *M. tuberculosis* (a protein better known as hsp65) had the capacity to stimulate myeloid cells, a range of bacterial and mammalian (principally human) molecular chaperones have been tested for their ability to interact and modulate the function of a variety of cell populations. These studies are leading to the following conclusions: (i) a number of bacterial molecular chaperones can stimulate myeloid cells and vascular endothelial cells, and there is evidence that certain of these bacterial proteins can inhibit myeloid cell function; (ii) human molecular chaperone homologues can both stimulate, and *inhibit*, immune cell function; (iii) human molecular chaperones are found in the extracellular fluid of normal individuals. It is still very early days, but a simple hypothesis to explain these findings is that multicellular creatures utilize extracellular molecular chaperones as signals – perhaps to indicate levels of local cellular stress – and promote activation/inhibition of defense cells. Bacteria also produce highly homologous molecular chaperones that can interact with this host signaling system. It is not clear what the ultimate consequence is of the interaction of bacterial molecular chaperones with host cells.

One extremely interesting facet of the two groups of moonlighting proteins discussed in this chapter is their unusual immunogenicity. Predominant immune responses to a number of glycolytic enzymes such as GAPD and aldolase are found in patients with bacterial (Yamakami et al. 2000), protozoal (Goudot-Crozel et al. 1989), and filarial worm (McCarthy et al. 2002) infections. In addition, autoimmune responses to glycolytic enzymes are common and may be pathogenic. A fascinating example of this is the K/BxN mouse that develops a spontaneous disease resembling rheumatoid arthritis. The autoantigen in this disease is PGI (Mandik-Nayak et al. 2002). The significant immunogenicity of molecular chaperones such as Cpn60, Hsp70, and Hsp90 is well established, and their role in controlling immune responses is part of an ongoing debate (van Eden et al. 1998; Wallin et al. 2002). With both sets of moonlighting proteins, their immunogenicity is paradoxical in terms of current immunological paradigms of self versus nonself. Matzinger has proposed an alternative hypothesis for the function of immunity – namely the ability to perceive "danger signals." Among the proposed signals are the molecular chaperones (Matzinger 2002). However, it is not clear how such danger signals are perceived. One obvious way is if the danger *signals* act as agonists and bind and activate myeloid and lymphoid cells.

If we step back a pace and consider bacteria–host recognition in respect to bacterial infection, then a powerful explanatory paradigm is that developed by Janeway and collaborators, who proposed that the recognition of parasites by our first line of defense, innate immunity, relies on a limited number of germline-encoded receptors termed pattern-recognition receptors (PRRs). These receptors evolved to recognize conserved molecules produced by microbial pathogens *but not by the host*. These moieties have been termed pathogen-associated molecular patterns (PAMPs – Medzhitov and Janeway 1998) but should more correctly be termed microbial-associated molecular patterns (MAMPs) as they will also be produced by cooperative bacteria. We now know that immune cells recognize LPS, peptidoglycan, CpG DNA, flagella, and so on by employing a range of receptors such as the Toll-like receptors (TLRs), nucleotide oligomerization domain/caspase recruitment domain (NOD/CARD) receptors, and trigger receptor expressed on myeloid cells (TREM-1/2) (Girardin et al. 2002 – reviewed in Chapter 15). Now, bacterial molecular chaperones are highly conserved proteins and could be PAMPs. So are they recognized by any PRRs? The answer is – yes and no. Some bacterial molecular chaperones are recognized by TLR2/4 (Costa et al. 2002), others are apparently not (Lewthwaite et al. 2002b). However, what is striking is that human molecular chaperones such as Cpn60 (Hsp60) (Ohashi et al. 2000), Hsp70 (Vabulas et al. 2002a), and Hsp90 (Gp96) (Vabulas et al. 2002b) are also recognized by TLR2 and/or TLR4. Another identified receptor for bacterial and human Hsp70 is CD40, a member of the TNF receptor superfamily. What is interesting is that the bacterial and human proteins bind to different sites on CD40 (Wang et al. 2001; Becker et al. 2002). Another heat shock protein receptor recognizing Hsp90, a related Hsp90 protein termed gp96, and Hsp70 is CD91 (Basu et al. 2001). Indeed, gp96 has recently been shown not to be required for cell viability, but to be central to the surface expression of a limited number of receptors on the cell surface. Among these receptors are the TLRs, which are retained intracellularly in cells lacking gp96. Such cells are unresponsive to microbial PAMPs (Randow and Seed 2001).

Thus it emerges that both bacterial and host molecular chaperones are capable of interacting with the same host receptors important in the recognition of bacterial infection. This may be an artifactual situation – a result of sequence homology, and of no consequence if host molecular chaperones are not released from cells. However, it is rapidly becoming established that a number of human molecular chaperones are normally found in the human circulation (e.g., Walsh et al. 2001; Lewthwaite et al. 2002a; Delpino and Castelli 2002; Pockley 2003). Is it conceivable that some of the molecular chaperones have endocrine or cytokine functions?

Not all human molecular chaperones act to stimulate host immune responses. The Hsp70 family member BiP has anti-inflammatory actions both in vivo (Corrigall et al. 2001) and in vitro (Bodan-Smith et al. 2003; Corrigall et al. 2004). This protein is also an important autoantigen in patients with rheumatoid arthritis (Bodan-Smith et al. 2003; Purcell et al. 2003). The small heat shock protein Hsp27 stimulates human monocytes to preferentially synthesize and release the anti-inflammatory cytokine IL-10 (De et al. 2000). On the basis of these findings, it is possible to postulate that human molecular chaperones may form complex regulatory cell signaling networks acting, in part, as pro- and anti-inflammatory proteins whose functions are presumably related to the need to control cell stress and its consequences for

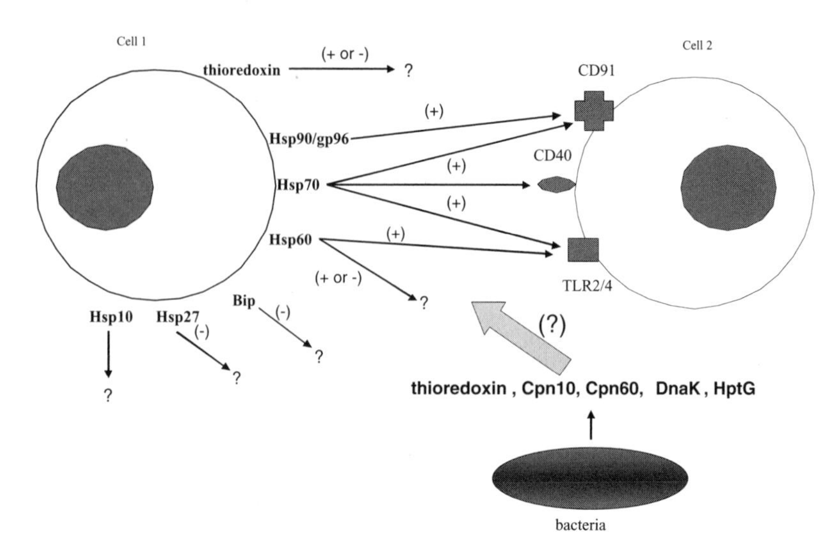

Figure 16.2. Schematic diagram outlining the eukaryotic cell stress proteins that have been shown to act as intercellular signals and the cellular receptors for these proteins identified to date, (TLRs, CD40, CD91). The question marks refer to unidentified host receptors for the various host molecular chaperones. To date, most host cell stress proteins cause cell activation (+). However, some of these proteins, such as Bip and Hsp27, appear to act as inhibitors of inflammation (−). A number of homologous molecular chaperones from bacteria have been reported to interact with host cells and induce alterations in cell behavior. Proteins such as DnaK and Cpn60 can interact with the same receptors as the host cell stress proteins. It is not yet clear if this functional "cross-reactivity" is part of a bacterial virulence mechanism or has evolved to aid in bacteria–host cooperation.

physiological and immunological homeostasis. Bacterial stress proteins may mimic or interfere with this host signaling process (Fig. 16.2).

A further complexity that has to be woven into this story of molecular chaperones and inflammatory signaling is the discovery that the mammalian receptor for LPS is a multimolecular complex that includes two molecular chaperones – Hsp70 and Hsp90 (Triantafilou et al. 2001). Thus host molecular chaperones can act as poacher and gamekeeper, being both agonists for cell surface receptors and cell surface receptors for agonists.

16.5. SUBTLETY IN PROTEIN HYPERSPACE

The authors' interest in molecular chaperones stemmed from early studies of the oral bacterium *A. actinomycetemcomitans*, a member of the oral microbiota and an organism associated with a severe form of periodontal disease (Henderson et al. 2003). In this disease, there is rapid bone destruction, and a potent osteolytic protein released by the bacterium was identified as Cpn60 (Kirby et al. 1995). The *cpn60* genes of bacteria and mitochondria are established to be conserved (Broccieri and Karlin 2000), and we expected all Cpn60 proteins to exhibit identical activity. The *Escherichia coli* Cpn60 protein, GroEL, turned out to be a potent inducer of bone resorption, but the Cpn60.2 protein of *Mycobacterium tuberculosis* and *M. leprae* were both inactive in the assay used to detect bone resorption (Kirby et al. 1995). The human (mitochondrial) Cpn60 protein is also an active bone-resorbing agonist (Meghji et al. 2003). It turns out that *M. tuberculosis* has two genes encoding Cpn60 proteins (Kong et al. 1993). The author and his colleagues have cloned and expressed both *M. tuberculosis cpn60* genes (*cpn60.1* and *cpn60.2*) and purified the recombinant proteins free of contaminating LPS (Lewthwaite et al. 2001; Maguire et al. 2003). In spite of more than seventy percent sequence identity, the two *M. tuberculosis* Cpn60 proteins differed significantly in potency and in their response to neutralizing antibodies to CD14. Thus the monocyte-stimulating activity of Cpn60.1 is completely blocked by anti-CD14 monoclonals that block the activity of LPS. In contrast, such monoclonals have no effect on Cpn60.2 (Lewthwaite et al. 2001). We have shown an even more marked difference with two of the three Cpn60 proteins of the plant symbiotic bacterium *Rhizobium leguminosarum*. We find that Cpn60.3 is able to activate human monocytes. In contrast, Cpn60.1 (which has eighty percent sequence identity) is completely inactive in this respect (Lewthwaite et al. 2002b). It would appear that Cpn60 proteins have sufficient sequence dissimilarity to be able to interact with different cellular receptors. This hypothesis has received experimental confirmation from the work of Hubert Kolb's group, which has shown that the commercially available mammalian and bacterial Cpn60 proteins do not compete with each other for binding to monocytes (Habich et al. 2003). Moreover, the human and murine Cpn60 proteins, which differ only by a handful of residues, differ in that the former binds to CD14 whereas the latter does not (Breloer et al. 2002) Further evidence for the marked plasticity of this protein has come from a study showing that the potent insect neurotoxin produced by the ant lion is in fact the Cpn60 protein of an oral symbiotic bacterium, *Enterobacter aerogenes*. This protein is almost identical to GroEL, and it was found that single-residue mutations in GroEL could turn this protein from an inactive into a potently active insect neurotoxin (Yoshida et al. 2001).

The two recombinant *M. tuberculosis* Cpn60 proteins have been prepared in large amounts for testing in various in vitro and in vivo model systems. The most notable effect to date is the ability of the Cpn60.1 protein to completely block the massive osteoclastic bone resorption found in rats with adjuvant arthritis (Winrow et al. 2002). This could have been the result of some interaction with activated T lymphocytes. However, it turns out that *M. tuberculosis* Cpn60.1 can directly inhibit agonist-induced bone resorption. More importantly, it does so by interfering with a newly discovered cytokine control pathway for driving the differentiation of the bone-resorbing cell population known as osteoclasts. In this pathway, the TNF-family member – receptor activator of NF-κB ligand (RANKL) – on the surface of the bone-forming osteoblast interacts with RANK on the surface of the preosteoclast to drive the process of cell fusion, cell differentiation, and cell activation to produce the fully active multinucleate osteoclast. This process can also be stimulated by activated T lymphocytes (Troen 2003). This control network is outlined in Figure 16.3. The *M. tuberculosis* Cpn60.1 is able to block the ability of RANKL to signal through RANK by either acting as a receptor antagonist or by interfering with intracellular signaling in the activated preosteoclast. The highly homologous Cpn60.2 is neither an activator nor inhibitor of bone resorption (Henderson, Meghji, Coates, Tormay, unpublished). To attempt to identify the structure–function relationship of these mycobacterial Cpn60 proteins, we have generated the three individual domains that constitute the monomeric proteins. These are the equatorial, intermediate, and apical domains. They have been produced by cloning and expression of the individual domains. For Cpn60.1, the majority of the biological activity of this protein, both as an inducer of human monocyte cytokine synthesis and as an inhibitor of bone resorption, resides in the equatorial domain (Tormay et al. 2005). This was also the conclusion from a structure–function study of human Cpn60 (Meghji et al. 2003).

The author and colleagues have also shown that *M. tuberculosis* Cpn60.1 can suppress experimental asthma in the mouse. The Cpn60.2 homologue was inactive (Riffo-Vasquez et al. 2004). Another group has tested the Cpn60 proteins from a number of bacteria: *M. tuberculosis* (Cpn60.2), *M. leprae*, bacillus Calmette-Guerin, *Strep. pneumoniae*, and *Helicobacter pylori*, in the same asthma model. Only the *M. leprae* Cpn60.2 protein was able to suppress lung inflammation and hyperresponsiveness (Rha et al. 2002). This strengthens the case that each Cpn60 protein has a different set of biological actions, presumably as a result of subtle variations in sequence or structure.

16.6. CONCLUSIONS

One of the major challenges of the twenty-first century will be to understand the interactions between cooperative bacteria and their hosts that allow

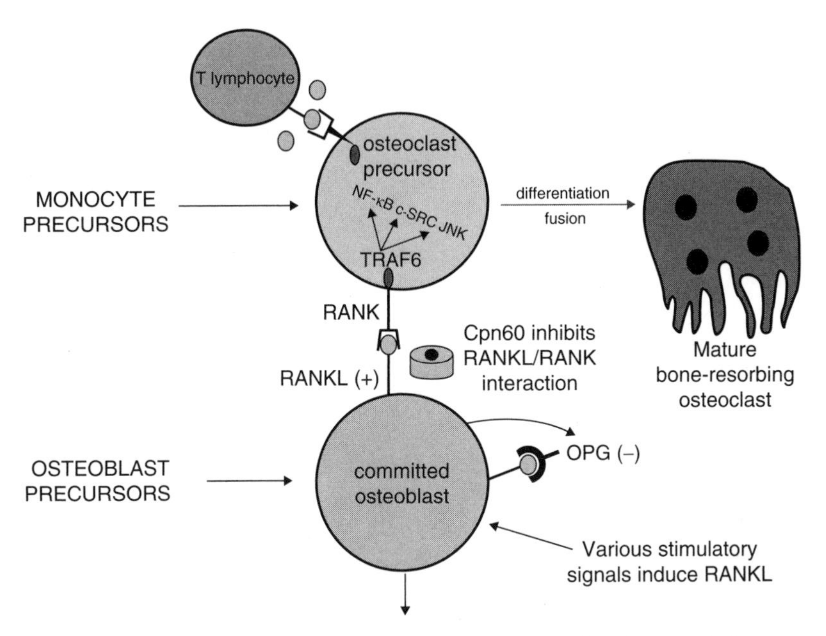

Figure 16.3. Schematic diagram of the cytokine network involved in the control of osteoclast generation. Osteoclast precursor cells arise from the circulating monocyte pool and express receptor activator of NF-κB (RANK) on their cell surfaces. The ligand for RANK is another NF-receptor (R) family member called RANK ligand (RANKL). This is expressed on stimulated osteoblasts (the cells that make bone matrix) or on activated T lymphocytes. RANKL binding to RANK drives the complex processes of cell fusion, differentiation, and activation that lead to the production of the bone-resorbing mature osteoclast. Another TNF-R, osteoprotegerin, acts as a decoy receptor for RANK and binds to RANKL, inhibiting osteoclast formation. Surprisingly, the Cpn60.1 protein from *M. tuberculosis* can also inhibit the activity of RANKL and can inhibit osteoclastic bone resorption both in vitro and in vivo. In contrast, the other Cpn60 protein of *M. tuberculosis* acts as neither a bone-resorbing agonist nor an antagonist.

multicellular creatures to coexist with large and diverse bacterial populations. The normal microbiota of *Homo sapiens* is enormously diverse, and the number of novel bacterial proteins that could be involved in interacting with the host may be in the order of tens to hundreds of thousands. With this potential population of signaling proteins, it is surprising to find that well-established groups of proteins, such as those involved in glycolysis and in the cell stress response, also participate in bacteria–host interactions. In the case of these two groups of moonlighting proteins, the host and bacterial proteins are both recognized by the host as signaling proteins. It is not known if the bacteria can respond to their own or host glycolytic proteins/stress proteins.

Our view of proteins is often colored by sequence similarities, and it is often accepted that proteins having more than fifty percent sequence identity are likely to have very similar functions. In terms of the moonlighting abilities of cell stress proteins, this general "rule" does not hold and proteins with very significant sequence identity can exhibit radically different cell–cell signaling activities. The influence of single-residue changes in GroEL has been described (Yoshida et al. 2001). It is now appreciated that there is substantial sequence variation in, for example, the *cpn60* genes of the enteric microbiota (Hill et al. 2002). The question then is how discriminating are the receptor systems that recognize cell stress proteins? Can the mucosa or immune cells within the mucosa recognize bacteria through sequence differences in the cell stress proteins, and is this one way of identifying bacterial invasion? This suggestion is unlikely to prove valid. However, it raises again the question of the ability of the host to recognize both its own and bacterial cell stress proteins. The recent finding that the cell stress protein, gp96, can act as a receptor for *E. coli* (Prasadarao et al. 2003) shows the potential complexity of the interactions between bacteria and host cell stress proteins. It is suggested that understanding the nature of the signaling induced by these cross-reactive signals will help us to understand how the balance is kept between ourselves and our normal microbiota.

ACKNOWLEDGMENTS

Financial support from the ARC in the form of program grant H0600 is gratefully acknowledged.

REFERENCES

Basu, S., Binder, R. J., Ramalingam, T., and Srivastava, P. K. (2001). CD91 is a common receptor for heat shock proteins gp96, hsp90, hsp70, and calreticulin. *Immunity* **14**, 303–313.

Baysal, B. E., Ferrell, R. E., Willett-Brozick, J. E. et al. (2000). Mutations in SDHD, a mitochondrial complex II gene, in hereditary paraganglioma. *Science* **287**, 848–851.

Becker, T., Hartl, F.-U., and Wieland, F. (2002). CD40, an extracellular receptor for binding and uptake of Hsp70-peptide complexes. *Journal of Cell Biology* **158**, 1277–1285.

Bermudez, L. E., Petrofsky, M., and Shelton, K. (1996). Epidermal growth factor-binding protein in *Mycobacterium avium* and *Mycobacterium tuberculosis*: a possible role in the mechanism of infection. *Infection and Immunity* **64**, 2917–2922.

Berry, M. D. and Boulton, A. A. (2000). Glyceraldehyde-3-phosphate dehydrogenase and apoptosis. *Journal of Neuroscience Research* **60**, 150–154.

Bielli, P. and Calabrese, L. (2002). Structure to function relationships in ceruloplasmin: a 'moonlighting' protein. *Cellular and Molecular Life Science* **59**, 1413–1427.

Bodan-Smith, M. D., Corrigall, V. M., Kemeny, D. M., and Panayi, G. S. (2003). BiP, a putative autoantigen in rheumatoid arthritis, stimulates IL-10-producing CD8-positive T cells from normal individuals. *Rheumatology* **42**, 637–644.

Boonacker, E. and Van Noorden, C. J. (2003). The multifunctional or moonlighting protein CD26/DPPIV. *European Journal of Cell Biology* **82**, 53–73.

Breloer, M., More, S. H., Osterloh, A., Stelter, F., Jack, R. S., and Bonin A. A. (2002). Macrophages as main inducers of IFN-gamma in T cells following administration of human and mouse heat shock protein 60. *International Immunology* **14**, 1247–1253.

Broccieri, L. and Karlin, S. (2000). Conservation among HSP60 sequences in relation to structure, function, and evolution. *Protein Science* **9**, 476–486.

Bromham, L. and Penny, D. (2003). The modern molecular clock. *Nature Reviews in Genetics* **4**, 216–224.

Canback, B., Andersson, S. G., and Kurland, C. G. (2002). The global phylogeny of glycolytic enzymes. *Proceedings of the National Academy of Sciences USA* **99**, 6097–6102.

Caprara, M. G., Lehnert, V., Lambowitz, A. M., and Westhof, E. (1996). A tyrosyl-tRNA synthetase recognizes a conserved tRNA-like structural motif in the Group 1 intron catalytic core. *Cell* **87**, 1135–1145.

Church, A. J., Dale, R. C., Lees, A. J., Giovannoni, G., and Robertson, M. M. (2003). Tourette's syndrome: a cross sectional study to examine the PANDAS hypothesis. *Journal of Neurology and Neurosurgery and Psychiatry* **74**, 602–607.

Coates, A. R. M., Henderson, B., and Mascagni, P. (1999). The unfolding story of the chaperonins. *Biotechnology Genetic Engineering Reviews* **16**, 393–405.

Conway Morris, S. (2003). *Life's Solution: Inevitable Humans in a Lonely Universe.* Cambridge: Cambridge University Press.

Copley, S. D. (2003). Enzymes with extra talents: moonlighting functions and catalytic promiscuity. *Current Opinion in Chemical Biology* **7**, 265–272.

Corrigall, V. M., Bodman-Smith, M. D., Fife, M. S., Canas, B., Myers, L. M., Wooley, P. H., Soh, C., Staines, N. A., Pappin, D. J. C., Berlo, S. E., van Eden, W., van der Zee, R., Lanchbury, J. S., and Panayi, G. S. (2001). The human endoplasmic reticulum molecular chaperone BiP is an autoantigen for rheumatoid arthritis and prevents the induction of experimental arthritis. *Journal of Immunology* **166**, 1492–1498.

Corrigall, V. M., Bodman-Smith, M. D., Brunst, M., Cornell, H., and Panayi, G. S. (2004). Inhibition of antigen-presenting cell function and stimulation of human peripheral blood mononuclear cells to express an anti-inflammatory cytokine profile by the stress protein BiP. *Arthritis and Rheumatism* **50**, 1164–1171.

Costa, C. P., Kirschning, C. J., Busch, D., Durr, S., Jennen, L., Heinzmann, U., Prebeck, S., Wagner, H., and Miethke, T. (2002). Role of chlamydial heat shock protein 60 in the stimulation of innate immune cells by *Chlamydia pneumoniae*. *European Journal of Immunology* **32**, 2460–2470.

De, A. K., Kodys, K. M., Yeh, B. S., and Miller-Graziano, C. (2000). Exaggerated human monocyte IL-10 concomitant to minimal TNF-alpha induction by heat-shock protein 27 (Hsp27) suggests Hsp27 is primarily an antiinflammatory stimulus. *Journal of Immunology* **165**, 3951–3958.

Delpino, A. and Castelli, M. (2002). The 78 kDa glucose-regulated protein (GRP78/BIP) is expressed on the cell membrane, is released into cell culture medium and is also present in human peripheral circulation. *Bioscience Reports* **22**, 407–420.

Eaves-Pyles, T., Murthy, K., Liaudet, L., Virag, L., Ross, G., Soriano, F. G., Szabo, C., Salzman, A. L. (2001). Flagellin, a novel mediator of *Salmonella*-induced epithelial activation and systemic inflammation: I kappa B alpha degradation, induction of nitric oxide synthase, induction of proinflammatory mediators, and cardiovascular dysfunction. *Journal of Immunology* **166**, 1248–1260.

Ehinger, S., Schubert, W. D., Bergmann, S., Hammerschmidt, S., Heinz, D.W. (2004). Plasmin (ogen)-binding alpha-enolase from *Streptococcus pneumoniae*: crystal structure and evaluation of plasmin (ogen) binding sites. *Journal of Molecular Biology* **343**, 997–1005.

Ejiri, S. (2002). Moonlighting functions of polypeptide elongation factor 1: from actin bundling to zinc finger protein R1-associated nuclear localization. *Bioscience Biotechnology Biochemistry* **66**, 1–21.

Faik, P., Walker, J. I., Redmill, A. A., and Morgan, M. J. (1988). Mouse glucose-6-phosphate isomerase and neuroleukin have identical 3′ sequences. *Nature* **332**, 455–457.

Fichorova, R. N., Cronin, A. O., Lien, E., Anderson, D. J., and Ingalls, R. R. (2002). Response to *Neisseria gonorrhoeae* by cervicovaginal epithelial cells occurs in the absence of toll-like receptor 4-mediated signaling. *Journal of Immunology* **168**, 2424–2432.

Fletcher, J. M., Nair, S. P., Ward, J. M., Henderson, B., and Wilson, M. (2001). Analysis of the effect of changing environmental conditions on the expression patterns of exported surface-associated proteins of the oral pathogen *Actinobacillus actinomycetemcomitans. Microbial Pathogenesis* **30**, 359–368.

Friedland, J. S., Shattock, R., Remick, D. G., and Griffin, G. E. (1993). Mycobacterial 65-kD heat shock protein induces release of proinflammatory cytokines from human monocytic cells. *Clinical and Experimental Immunology* **91**, 58–62.

Girardin, S. E., Sansonetti, P. J., and Philpott, D. J. (2002). Intracellular vs extracellular recognition of pathogens – common concepts in mammals and flies. *Trends in Microbiology* **10**, 193–199.

Gonzalez, F., Delahodde, A., Kodadek, T., and Johnston, S. A. (2002). Recruitment of a 19S proteasome subcomplex to an activated promoter. *Science* **296**, 548–550.

Goudot-Crozel, V., Caillol, D., Djabali, M., and Dessein, A. J. (1989). The major parasite surface antigen associated with human resistance to schistosomiasis is a 37-kD glyceraldehyde-3P-dehydrogenase. *Journal of Experimental Medicine* **170**, 2065–2080.

Gozalbo, D., Gil-Navarro, I., Azorin, I., Renau-Piqueras, J., Martinez, J. P., and Gil, M. L. (1998). The cell wall-associated glyceraldehyde-3-phosphate dehydrogenase of *Candida albicans* is also a fibronectin and laminin binding protein. *Infection and Immunity* **66**, 2052–2059.

Gutowicz, J. and Terlecki, G. (2003). The association of glycolytic enzymes with cellular and model membranes. *Cellular and Molecular Biology Letters* **8**, 667–680.

Habich, C., Kempe, K., van der Zee, R., Burkart, V., and Kolbe, H. (2003). Different heat shock protein 60 species share pro-inflammatory activity but not binding sites on macrophages. *FEBS Letters* **533**, 105–109.

Hara, H., Ohta, H., Inoue, T., Ohashi, T., Takashiba, S., Murayama, Y., and Fukui, K. (2000). Cell surface-associated enolase in *Actinobacillus actinomycetemcomitans. Microbiology and Immunology* **44**, 349–356.

Hell, W., Meyer, H. G. W., and Gatermann, S. G. (1998). Cloning of *aas*, a gene encoding a *Staphylococcus saprophyticus* surface protein with adhesive and autolytic properties. *Molecular Microbiology* **29**, 871–881.

Henderson, B., Poole, S., and Wilson, M. (1996a). Bacterial modulins: A novel class of virulence factor which causes host tissue pathology by inducing cytokine synthesis. *Microbiology Reviews* **60**, 316–341.

Henderson, B., Poole, S., and Wilson, M. (1996b). Bacterial/host interactions in health and disease: Who controls the cytokine network? *Immunopharmacology* **35**, 1–21.

Henderson, B., Wilson, M., and Wren, B. (1997), Are bacterial exotoxins cytokine network regulators? *Trends in Microbiology* **5**, 454–458.

Henderson, B., Poole, S., and Wilson, M. (1998). *Bacteria/Cytokine Interactions in Health and Disease.* London: Portland Press.

Henderson, B. (2000). Therapeutic control of cytokines: Lessons from microorganisms. In *Novel Cytokine Inhibitors,* eds. G. A. Higgs and B. Henderson, pp. 243–261. Basle: Birkhauser Verlag.

Henderson, B. (2003). Chaperonins: Chamelion proteins that influence myeloid cells. In *Progress in Inflammation Research,* ed. W. van Eden, pp. 175–191. Basle: Birkhauser Verlag.

Henderson, B. and Oyston, P. C. F. (eds) (2003). *Bacterial Evasion of Host Immune Responses.* Cambridge: Cambridge University Press.

Henderson, B. and Seymour, R. M. (2003). Microbial modulation of cytokine networks. In *Bacterial Evasion of Host Immune Responses,* eds. B. Henderson and P. C. Oyston, pp. 223–242. Cambridge: Cambridge University Press.

Hill, J. E., Seipp, R. P., Betts, M., Hawkins, L., Van Kessel, A. G., Crosby, W. L., and Hemmingsen, S. M. (2002). Extensive profiling of a complex microbial community by high-throughput sequencing. *Applied and Environmental Microbiology* **68**, 3055–3066.

Hirota, K., Nakamura, H., Masutani, H., and Yodoi, J. (2002). Thioredoxin superfamily and thioredoxin-inducing agents. *Annals of the New York Academy of Sciences* **957**, 189–199.

Hughes, M. J., Moore, J. C., Lane, J. D., Wilson, R., Pribul, P. K., Younes, Z. N., Dobson, R. J., Everest, P., Reason, A. J., Redfern, J. M., Greer, F. M., Paxton, T., Panico, M., Morris, H. R., Feldman, R. G., and Santangelo, J. D. (2002). Identification of major outer surface proteins of *Streptococcus agalactiae. Infection and Immunity* **70**, 1254–1259.

Jeffery, C. J. (1999). Moonlighting proteins. *Trends in Biochemistry* **24**, 8–11.

Jeffery, C. J. (2003). Moonlighting proteins: Old proteins learning new tricks. *Trends in Genetics* **19**, 415–417.

Kazor, C. E., Mitchell, P. M., Lee, A. M., Stokes, L. N., Loesche, W. J., Dewhirst, F. E., and Paster, B. J. (2003). Diversity of bacterial populations on the tongue dorsa of patients with halitosis and healthy patients. *Journal of Clinical Microbiology* **41**, 558–563.

Kirby, A. C., Meghji, S., Nair, S. P., White, P., Reddi, K., Nishihara, T., Nakashima, K., Willis, A. C., Sim, R., Wilson, M., and Henderson, B. (1995). The potent bone resorbing mediator of *Actinobacillus actinomycetemcomitans* is homologous to the molecular chaperone GroEL. *Journal of Clinical Investigation* **96**, 1185–1194.

Kong, T. H., Coates, A. R. M., Butcher, P. D., Hickman, C. J., and Shinnick, T. M. (1993). *Mycobacterium tuberculosis* expresses two chaperonin-60 homologs. *Proceedings of the National Academy of Sciences USA* **90**, 2608–2612.

Kregel, K. C. (2002). Heat shock proteins: modifying factors in physiological stress responses and acquired thermotolerance. *Journal of Applied Physiology* **92**, 2177–2186.

la Sala, A., Ferrari, D., Di Virgilio, F., Idzko, M., Norgauer, J., and Girolomoni, G. (2003). Alerting and tuning the immune response by extracellular nucleotides. *Journal of Leukocyte Biology* **73**, 339–343.

Lewthwaite, J., Skinner, A. C., Henderson, B. (1998). Are molecular chaperones microbial virulence factors? *Trends in Microbiology* **6**, 426–428.

Lewthwaite, J. C., Coates, A. R. M., Tormay, P., Singh, M., Mascagni, P., Poole, S., Roberts, M., Sharp, L., and Henderson, B. (2001). *Mycobacterium tuberculosis* chaperonin 60.1 is a more potent cytokine stimulator than chaperonin 60.2 (hsp 65) and contains a CD14-binding domain. *Infection and Immunity* **69**, 7349–7355.

Lewthwaite, J., Owen, N., Coates, A. R. M., Henderson, B., and Steptoe, A. D. (2002a). Circulating heat shock protein (Hsp)60 in the plasma of British civil servants: Relationship to physiological and psychosocial stress. *Circulation* **106**, 196–201.

Lewthwaite, J. C., George, R., Lund, P. A., Poole, S., Tormay, P., Sharp, L., Coates, A. R. M., and Henderson, B. (2002b). *Rhizobium leguminosarum* chaperonin 60.3, but not chaperonin 60.1, induces cytokine production by human monocytes: activity is dependent on interaction with cell surface CD14. *Cell Stress and Chaperones* **7**, 130–136.

Lopez-Alemany, R., Longstaff, C., Hawley, S., Mirshahi, M., Fabregas, P., Jardi, M., Merton, E., Miles, L. A., and Felez, J. (2003). Inhibition of cell surface mediated plasminogen activation by an antibody against alpha-enolase. *American Journal of Hematology* **72**, 234–242.

Lottenberg, R., Broder, C. C., Boyle, M. D., Kain, S. J., Schroeder, B. L., and Curtiss, R. (1992). Cloning, sequence analysis, and expression in *Escherichia coli* of a streptococcal plasmin receptor. *Journal of Bacteriology* **174**, 5204–5210.

Maguire, M., Coates, A. R. M., and Henderson, B. (2002). Chaperonin 60 unfolds its secrets of cellular communication. *Cell Stress and Chaperones* **7**, 317–329.

Maguire, M., Coates, A. R. M., and Henderson, B. (2003). Cloning expression and purification of three chaperonin 60 homologues. *Journal of Chromatography* **786**, 117–125.

Mandik-Nayak, L., Wipke, B. T., Shih, F. F., Unanue, E. R., and Allen, P. M. (2002). Despite ubiquitous autoantigen expression, arthritogenic autoantibody response initiates in the local lymph node. *Proceedings of the National Academy of Sciences USA* **99**, 14368–14373.

Martin, W. and Muller, M. (1998). The hydrogen hypothesis for the first eukaryote. *Nature* **392**, 37–41.

Matzinger, P. (2002). The danger model: a renewed sense of self. *Science* **296**, 301–305.

McCarthy, J. S., Wieseman, M., Tropea, J., Kaslow, D., Abraham, D., Lustigman, S., Tuan, R., Guderian, R. H., and Nutman, T. B. (2002). *Onchocerca volvulus* glycolytic enzyme fructose-1,6-bisphosphate aldolase as a target for a protective immune response in humans. *Infection and Immunity* **70**, 851–858.

Medzhitov, R. and Janeway, C. A. (1998). Innate immune recognition and control of adaptive immune responses. *Seminars in Immunology* **10**, 351–353.

Meghji, S., Lillicrap, M., Maguire, M., Tabona, P., Gaston, J. S. H., Poole, S., and Henderson, B. (2003). Human chaperonin 60 (Hsp60) stimulates bone resorption: Structure/function relationships. *Bone* **33**, 419–425.

Melmed, G., Thomas, L. S., Lee, N., Tesfay, S. Y., Lukasek, K., Michelsen, K. S., Zhou. Y., Hu, B., Arditi, M., and Abreu, M. T. (2003). Human intestinal epithelial cells are broadly unresponsive to Toll-like receptor 2-dependent bacterial ligands: implications for host-microbial interactions in the gut. *Journal of Immunology* **170**, 1406–1415.

Modun, B. and Williams, P. (1999). The staphylococcal transferring-binding protein is a cell wall glyceraldehyde-3-phosphate dehydrogenase. *Infection and Immunity* **67**, 1086–1092.

Nelson, D., Goldstein, J. M., Boatright, K., Harty, D. W., Cook, S. L., Hickman, P. J., Potempa, J., Travis, J., and Mayo, J. A. (2001). pH-regulated secretion of a glyceraldehyde-3-phosphate dehydrogenase from *Streptococcus gordonii* FSS2: purification, characterization, and cloning of the gene encoding this enzyme. *Journal of Dental Research* **80**, 371–377.

Ohashi, K., Burkart, V., Flohe, S., and Kolb, H. (2000). Cutting edge: heat shock protein 60 is a putative endogenous ligand of the toll-like receptor-4 complex. *Journal of Immunology* **164**, 558–561.

Pancholi, V. and Fischetti, V. A. (1992). A major surface protein on group A streptococci is a glyceraldehyde-3-phosphate-dehydrogenase with multiple binding activity. *Journal of Experimental Medicine* **176**, 415–426.

Pancholi, V. and Fischetti, V. A. (1993). Glyceraldehyde-3-phosphate-dehydrogenase on the surface of group A streptococci is also an ADP-ribosylating enzyme. *Proceedings of the National Academy of Sciences USA* **90**, 8154–8158.

Pancholi, V. and Fischetti, V. A. (1997). Regulation of the phosphorylation of human pharyngeal cell proteins by group A streptococcal surface dehydrogenase: Signal transduction between streptococci and pharyngeal cells. *Journal of Experimental Medicine* **186**, 1633–1643.

Pancholi, V. and Fischetti, V. A. (1998). Alpha-enolase, a novel strong plasmin(ogen) binding protein on the surface of pathogenic streptococci. *Journal of Biological Chemistry* **273**, 14503–14515.

Pancholi, V. (2001). Multifunctional alpha-enolase: its role in disease. *Cellular and Molecular Life Sciences* **58**, 902–920.

Parker, A. E. and Bermudez, L. E. (2000). Sequence and characterization of the glyceraldehyde-3-phosphate dehydrogenase of *Mycobacterium avium*: correlation with an epidermal growth factor binding protein. *Microbial Pathogenesis* **28**, 135–144.

Pockley, A. G. (2003). Heat shock proteins as regulators of the immune response. *Lancet* **362**, 469–476.

Prasadarao, N. V., Srivastava, P. K., Rudrabhatla, R. S., Kim, K. S., Huang, S. H., and Sukumaran, S. K. (2003). Cloning and expression of the *Escherichia coli* K1 outer membrane protein A receptor, a gp96 homologue. *Infection and Immunity* **71**, 1680–1688.

Purcell, A. W., Todd, A., Kinoshita, G., Lynch, T. A., Keech, C. L., Gething, M. J., and Gordon, T. P. (2003). Association of stress proteins with autoantigens: a possible mechanism for triggering autoimmunity? *Clinical and Experimental Immunology* **132**, 193–200.

Randow, F. and Seed, B. (2001). Endoplasmic reticulum chaperone gp96 is required for innate immunity but not cell viability. *Nature Cell Biology* **3**, 891–896.

Ranford, J., Coates, A. R. M., and Henderson, B. (2000). Chaperonins are cell-signalling proteins: the unfolding biology of molecular chaperones. *Expert Reviews in Molecular Medicine* 15 September, http://www-ermm.cbcu.cam.ac.uk/00002015h.htm.

Reiss, J., Gross-Hardt, S., Christensen, E., Schmidt, P., Mendel, R. R., and Schwarz, G. (2001). A mutation in the gene for the neurotransmitter receptor-clustering protein gephyrin causes a novel form of molybdenum cofactor deficiency. *American Journal of Human Genetics* **68**, 208–213.

Retzlaff, C., Yamamoto, Y., Hoffman, P. S., Friedman, H., and Klein, T. W. (1994). Bacterial heat shock proteins directly induce cytokine mRNA and interleukin-1 secretion in macrophage cultures. *Infection and Immunity* **62**, 5689–5693.

Rha, Y. H., Taube, C., Haczku, A., Joetham, A., Takeda, K., Duez, C., Siegel, M., Aydintug, M. K., Born, W. K., Dakhama, A., and Gelfand, E. W. (2002). Effect of microbial heat shock proteins on airway inflammation and hyperresponsiveness. *Journal of Immunology* **169**, 5300–5307.

Riffo-Vasquez, Y., Spina, D., Page, C. P., Desel, C., Whelan, M., Tormay, P., Singh, M., Henderson, B., and Coates, A. R. M. (2004). Differential effects of *Mycobacterium tuberculosis* chaperonins on bronchial eosinophilia and hyperresponsiveness in a murine model of allergic inflammation *Clinical and Experimental Immunology*, 34, 712–719.

Schulz, L. C. and Bahr, J. M. (2003). Glucose-6-phosphate isomerase is necessary for embryo implantation in the domestic ferret. *Proceedings of the National Academy of Sciences USA* **100**, 8561–8566.

Seifert, K. N., McArthur, W. P., Bleiweis, A. S., and Brady, L. J. (2003). Characterization of group B streptococcal glyceraldehyde-3-phosphate dehydrogenase: surface localization, enzymatic activity, and protein-protein interactions. *Canadian Journal of Microbiology* **49**, 350–356.

Sha, J., Galindo, C. L., Pancholi, V., Popov, V. L., Zhao, Y., Houston, C. W., and Chopra, A. K. (2003). Differential expression of the enolase gene under *in vivo* versus *in vitro* growth conditions of *Aeromonas hydrophila. Microbial Pathogenesis* **34**, 195–204.

Shimizu, K., Tani, M., Watanabe, H., Nagamachi, Y., Niinaka, Y., Shiroishi, T., Ohwada, S., Raz, A., and Yokota, J. (1999). The autocrine motility factor receptor gene encodes a novel type of seven transmembrane protein. *FEBS Letters* **456**, 295–300.

Sirover, M. A. (1999). New insights into an old protein: the functional diversity of mammalian glyceraldehyde-3-phosphate dehydrogenase. *Biochimica Biophysica Acta* **1432**, 159–184.

Smith, T. F. and Morowitz, H. J. (1982). Between history and physics. *Journal of Molecular Evolution* **18**, 265–282.

Suau, A., Bonnet, R., Sutren, M., Godon, J.-J., Gibson, G. R., Collins, M. D., and Dore, J. (1999). Direct analysis of genes encoding 16S rRNA from complex communities reveals many novel molecular species within the human gut. *Applied and Environmental Microbiology* **65**, 4799–4807.

Sun, Y. J., Chou, C. C., Chen, W. S., Wu, R. T., Meng, M., and Hsiao, C. D. (1999). The crystal structure of a multifunctional protein: phosphoglucose isomerase/autocrine motility factor/neuroleukin. *Proceedings of the National Academy of Sciences USA* **96**, 5412–5417.

Swedo, S. E. (2001). Genetics of childhood disorders: XXXIII. Autoimmunity, part 6: poststreptococcal autoimmunity. *Journal of the American Academy of Child & Adolescent Psychiatry* **40**, 1479–1482.

Tanaka, N., Haga, A., Uemura, H., Akiyama, H., Funasaka, T., Nagase, H., Raz, A., and Nakamura, K. T. (2002). Inhibition mechanism of cytokine activity of human autocrine motility factor examined by crystal structure analyses and site-directed mutagenesis studies. *Journal of Molecular Biology* **318**, 985–997.

Taylor, J. M. and Heinrichs, D. E. (2002). Transferrin binding in *Staphylococcus aureus*: involvement of a cell wall-anchored protein. *Molecular Microbiology* **43**, 1603–1614.

Tomlinson, I. P., Alam, N. A., Rowan, A. J., Barclay, E., Jaeger, E. E., Kelsell, D., Leigh, I., Gorman, P., Lamlum, H., Rahman, S., Roylance, R. R., Olpin, S., Bevan, S., Barker, K., Hearle, N., Houlston, R. S., Kiuru, M., Lehtonen, R., Karhu, A., Vilkki, S., Laiho, P., Eklund, C., Vierimaa, O., Aittomaki, K., Hietala, M., Sistonen, P., Paetau, A., Salovaara, R., Herva, R., Launonen, V., Aaltonen, L. A.; Multiple Leiomyoma Consortium (2002). Germline mutations in FH predispose to dominantly inherited uterine fibroids, skin leiomyomata and papillary renal cell cancer. *Nature Genetics* **30**, 406–410.

Tormay, P., Coates, A. R. M., and Henderson, B. (2005). The intercellular signalling activity of the *Mycobacterium tuberculosis* chaperonin 60.1 protein resides in the equatorial domain. *Journal of Biological Chemistry* (in press).

Triantafilou, K., Triantafilou, M., and Dedrick, R. L. (2001). A CD14-independent LPS receptor cluster. *Nature Immunology* **2**, 338–345.

Troen, B. R. (2003). Molecular mechanisms underlying osteoclast formation and activation. *Experimental Gerontology* **38**, 605–614.

Tsutsumi, S., Gupta, S. K., Hogan, V., Tanaka, N., Nakamura, K. T., Nabi, I. R., and Raz, A. (2003). The enzymatic activity of phosphoglucose isomerase is not required for its cytokine function. *FEBS Letters* **534**, 49–53.

Vabulas, R. M., Ahmad-Nejad, P., Ghose, S., Kirschning, C. J., Issels, R. D., and Wagner, H. (2002a). HSP70 as endogenous stimulus of the Toll/interleukin-1 receptor signal pathway. *Journal of Biological Chemistry* **277**, 15107–15112.

Vabulas R. M., Braedel S., Hilf N., Singh-Jasuja H., Herter S., Ahmad-Nejad P., Kirschning C. J., Da Costa, C., Rammensee, H. G., Wagner, H., and Schild, H. (2002b). The endoplasmic reticulum-resident heat shock protein Gp96 activates dendritic cells via the Toll-like receptor 2/4 pathway. *Journal of Biological Chemistry* **277**, 20847–20853.

van Eden, W., van der Zee, R., Paul, A. G., Prakken, B. J., Wendling, U., Anderton, S. M., and Wauben, M. H. (1998). Do heat shock proteins control the balance of T-cell regulation in inflammatory diseases? *Immunology Today* **19**, 303–307.

van Gurp, M., Festjens, N., van Loo, G., Saelens, X., and Vandenabeele, P. (2003). Mitochondrial intermembrane proteins in cell death. *Biochemical Biophysical Research Communications* **304**, 487–497.

Wallin, R. P. A., Lundqvist, A., More, S. H., von Bonin, A., Kiessling, R., and Ljunggren, H.-G. (2002). Heat shock proteins as activators of the innate immune response. *Trends in Immunology* **23**, 130–135.

Walsh, R. C., Koukoulas, I., Garnham, A., Moseley, P. L., Hargreaves, M., and Febbraio, M. A. (2001). Exercise increases serum Hsp72 in humans. *Cell Stress Chaperones* **6**, 386–393.

Walter, S. and Buchner, J. (2002). Molecular chaperones – cellular machines for protein folding. *Angewandte Chemie International Edition* **41**, 1098–1113.

Wang, Y., Kelly, C. G., Karttunen, J. T., Whittall, T., Lehner, P. J., Duncan, L., MacAry, P., Younson, J. S., Singh, M., Oehlmann, W., Cheng, G., Bergmeier, L., and Lehner, T. (2001). CD40 is a cellular receptor mediating mycobacterial heat shock protein 70 stimulation of CC-chemokines. *Immunity* **15**, 971–983.

Watanabe, H., Takehana, K., Date, M., Shinozaki, T., and Raz, A. (1996). Tumor cell autocrine motility factor is the neuroleukin/phosphohexose isomerase polypeptide. *Cancer Research* **56**, 2960–2963.

Wilkins, J. C., Beighton, D., and Homer, K. A. (2003). Effect of acidic pH on expression of surface-associated proteins of *Streptococcus oralis*. *Applied and Environmental Microbiology* **69**, 5290–5296.

Wilson, M., McNab, R., and Henderson, B. (2002). *Bacterial Disease Mechanisms: An Introduction to Cellular Microbiology*. Cambridge: Cambridge University Press.

Winrow, V. R., Coates, A. R. M., Tormay, P., Henderson, B., Singh, M., Blake, D. R., and Morris, C. J. (2002). Chaperonin 60.1 prevents bone destruction in Wistar rats with adjuvant-induced arthritis. *Rheumatology* **41**, (abstr. suppl. 1) 47.

Xu, W., Seiter, K., Feldman, E., Ahmed, T., and Chiao, J. W. (1996). The differentiation and maturation mediator for human myeloid leukemia cells shares homology with neuroleukin or phosphoglucose isomerase. *Blood* **87**, 4502–4506.

Yamakami, K., Yoshizawa, N., Wakabayashi, K., Takeuchi, A., Tadakuma, T., and Boyle, M. D. (2000). The potential role for nephritis-associated plasmin receptor in acute poststreptococcal glomerulonephritis. *Methods* **21**, 185–197.

Yoshida, N., Oeda, K., Watanabe, E., Mikami, T., Fukita, Y., Nishimura, K., Komai, K., and Matsuda, K. (2001). Protein function. Chaperonin turned insect toxin. *Nature* **411**, 44.

Zhu, W., Rainville, I. R., Ding, M., Bolus, M., Heintz, N. H., and Pederson, D. S. (2002). Evidence that the pre-mRNA splicing factor Clf1p plays a role in DNA replication in *Saccharomyces cerevisiae*. *Genetics* **160**, 1319–1333.

CHAPTER 17

Cell signalling pathways as targets for bacterial evasion and pathology

Andrew S. Neish

17.1. INTRODUCTION

Eukaryotic organisms are in intimate and continuous contact with members of the prokaryotic kingdom. The implications of this seemingly obvious statement reflect an emerging theme in the life sciences that has only recently come to the forefront of our general view of multicellular plants and animals – that microbes may affect our biology in profound and, perhaps, previously unsuspected ways. Cooperative interactions between eukaryotes and prokaryotes are well known. In these symbiotic relationships, the microbe benefits by acquisition of a stable nutrient supply and immediate environment. Eukaryotic hosts may gain extended metabolic/digestive ability, competitive exclusion of less benignly predisposed microbes, or more exotic talents, such as bioluminescence. Examples include the gut-dwelling flora of vertebrates, an ecosystem of increasingly recognized medical significance, and nitrogen-fixing rhizobia that induce root nodules in leguminous plants, a relationship with great ecological and economic importance. Beyond these classic examples, microbial life co-exists with eukaryotes in virtually all aspects of the biome, with actual symbioses at cellular, organismal and ecological levels probably occurring but simply unrecognized by us. Of course, the relationship of eukaryotes and prokaryotes is not always benign. From the traditional perspective of the medical doctor, veterinarian, or plant pathologist, microbes are exogenous invaders, bent on no good and suitable only for elimination. Although there is increasing recognition in these quarters of the role of beneficial bacteria, the viewpoint of microbes as deleterious invaders is well taken; clearly a wide variety of infectious disease exists and will always represent a scourge to individual organisms.

This review discusses the mechanisms by which animal (and plant) cells (i) recognize the presence of bacteria, necessarily both potential pathogens and commensals; (ii) how that recognition is translated into signalling events; and (iii) the initial effector mechanisms cells and tissues use to respond to microbial interactions. It then goes on to discuss mechanisms by which bacteria can manipulate these responses and the potential consequences for both parties. A more detailed discussion of the host receptors for bacteria is presented in Chapter 15.

17.2. THE RECOGNITION PHASE: INTRUDER ALERT!

Eukaryotes have evolved mechanisms to constantly survey their surroundings for the telltale presence of microbes. To employ a popular metaphor, the animal is sailing upon potentially hostile seas, and must remain vigilant against potential threats (and opportunities), both on the horizon and from within. To accomplish this task, virtually all metazoans scan their environment with pattern recognition receptors (PRRs), an operational term for transmembrane or intracytoplasmic receptors that are defined by the ability to specifically bind distinctive microbial ligands, designated "pathogen-associated molecular patterns" (PAMPs) (Barton and Medzhitov 2003). PAMPs represent structural motifs that *are restricted to, and definitive of,* microbial organisms. They consist of complex macromolecules such as lipopolysaccharides (LPS), peptidoglycan (PGN), lipoproteins, and glycoproteins, and unmodified proteins (flagellin) and nucleic acids (CpG-rich DNA, double-stranded RNA) are also included in this grouping. PAMPs are characteristic of specific classes of microbe. For example, LPS is a component of Gram-negative cell walls, whereas PGNs are a major part of the Gram-positive cell wall structure and dsRNA is typical of nucleic acid intermediates that occur during viral replication. From the hosts' point of view, an optimal PAMP is a structure that is vital to the viability of the microbe, limiting the amount of structural variation that would be tolerable (Barton and Medzhitov 2003). Because of the efficiency of eukaryotic recognition of PAMPs, bacteria have predictably evolved significant ability to modify and conceal these structures, including mechanisms such as flagellar phase variation, LPS chain modification, and encapsulation (Hornef et al. 2002b). Further discussion of PAMPs is found in Chapters 2, 3, 10, 15, and 16, and the suggestion that the term should be *microbial-associated molecular patterns* (MAMPs) is found in Chapter 16.

The best-studied vertebrate PRRs are the Toll-like receptors (TLRs). The designation *Toll-like receptor* reflects the homology to the Toll receptor in

Table 17.1. *Pattern recognition receptors and established microbial ligands*

PRR	Ligand
TLR1	Lipopeptides
TLR2	Peptidoglycan, lipopoproteins, lipoteichoic acid, others
TLR3	dsRNA
TLR4	LPS, taxol, Hsp60
TLR5	Flagellin
TLR6	?
TLR7	ssRNA, anti-viral compounds
TLR8	ssRNA
TLR9	Unmethylated CpG DNA
TLR10	?
Nod1	Peptidoglycan
Nod2	Peptidoglycan

Drosophila melanogaster, a protein involved in early embryonic patterning and which also plays a critical role in insect innate immunity (Hoffman 2003). The human genome contains at least ten known TLRs (Barton and Medzhitov 2003; Takeda et al. 2003). The gene products are transmembrane receptors defined by the presence of two ancient and highly conserved structural motifs. The first is the leucine-rich repeat (LRR) in the extracellular portion of the molecule that functions in selective ligand (PAMP) recognition (see Table 17.1). LRR domains are common to many cellular proteins involved in the recognition of foreign proteins, and are important in mediating bacterial detection in plants (Dangl and Jones 2001) and in lower invertebrates (Pujol et al. 2001; see also Chapter 14). The cytoplasmic domain of the TLR is the TIR (Toll/IL-1R/ plant resistance) domain, which interacts with and activates soluble signal-transducing proteins that are discussed in the next section.

Toll-like receptors may be strategically deployed; clearly a circulating cell such as a vertebrate macrophage or an invertebrate haemocyte would be expected to express receptors circumferentially around the cell. In contrast, epithelial cells – the apical surfaces of which must interface with the environment (or gut lumen) and the basal aspect with the interior of the organism – may spatially restrict TLR expression. For example, TLR5, which recognises the bacterial protein flagellin, is present along basolateral epithelial cell membranes, sheltered from the lumenal contents and poised to scrutinise breaches in the cell–cell barriers (Gewirtz et al. 2001). TLRs may be restricted to intracellular vacuolar membranes (Hornef et al. 2002a). A vacuolar TLR, although topologically external to the cell, is insulated from PAMP exposure in the external environment and is positioned as a monitor of material incoming from endocytic or phagocytic processes.

Pattern recognition receptors are capable of detecting soluble intracytoplasmic PAMPs (Girardin et al. 2001). Many pathogenic (and endosymbiotic) bacteria exhibit an intracytoplasmic stage of their life cycle, not occupying a cytoplasmic vacuole. The intracellular LRR containing proteins of the Nod family may act to monitor the cytoplasm of cells for PAMPs present on intracellular pathogens, providing a means of detecting internalized bacteria not perceived by surface receptors. This form of bacterial surveillance is especially prevalent in plant immunity (Dangl and Jones 2001). The Nod proteins have a modular structure similar to TLRs. The LRR ligand binding motifs interact with specific PAMPs, to date identified as specific peptide components of larger PGN macromolecules. The C-terminus of Nod1 and 2 proteins feature a CARD, or caspase-recruitment domain, which is functionally analogous to the TIR domain of the TLRs (actual TIR domains are present on plant Nod proteins) that mediate second messenger activation necessary for subsequent signalling pathways (Inohara and Nunez 2003). Nod proteins include Nod2, certain mutant alleles of which are associated with the debilitating human inflammatory condition Crohn's disease (Hugot et al. 2001; Ogura et al. 2001). This disorder is thought to represent abnormal host inflammatory responses to normal intestinal flora. The mutant Nod alleles are an "experiment of nature" that illustrates the role of PRRs in monitoring a microbial presence generally recognized as commensal, rather than overtly pathogenic. Perhaps other PRRs play a significant role in negotiating cellular interactions with commensal/symbiotic organisms.

Different PAMP/PRR binding events very likely activate distinct signalling intermediates and result in responses tailored to the specific class of inciting microbe (Kopp and Medzhitov 2003). Expression profiling and proteomic approaches are likely to be informative in delineating these differences. Of course, in real-world conditions, an actual bacterium offers up at least several PAMPs to a given host cell. Bacterial communities possess both Gram-negative (LPS-TLR4) and Gram-positive (PGN-TLR2) members. Many are flagellated (TLR5), and all presumably possess CpG DNA (TLR9). These products are also likely shed and/or released by dead organisms, ending up in the immediate environment, even if microbially sterile.

In summary, eukaryotic cells possess PRRs capable of detecting microbial "signatures." These PRRs possess a wide range of specificity and manner of deployment to allow efficient monitoring of the local environment.

17.3. THE SIGNALLING PHASE; BATTLE STATIONS!

Upon detecting evidence of microbes, eukaryotic cells have evolved mechanisms to sound the alarm in anticipation of defensive responses. Ligation of a PRR with its cognate PAMP results in the activation of cytoplasmic signalling relays. As signalling events are usually transient and reversible, they are prime candidates for manipulation – both by bacteria seeking counter defenses and by humans seeking points of therapeutic intervention. Such relays may be conducted by the controlled, regulated transfer of covalent modifications along a series of cytoplasmic protein intermediates or by sequential proteolytic cascades. An example of the former is the proinflammatory Rel/NF-κB pathway. The terminal result of activating this pathway is the nuclear appearance of transcription factors of the Rel family, which bind specific promoter elements and stimulate de novo transcriptional activation of genes involved in responses to microbes. An example of a controlled, regulated series of proteolytic cleavages initiated by PRRs is the activation of "extrinsic" pro-apoptotic signalling. This signalling cascade culminates in the appearance of activated effector caspases that mediate the controlled physiological demolition of the infected, or otherwise damaged, cell.

Mammalian signalling responses to bacteria have an ancient lineage. Striking functional and structural conservation of the Rel/NF-κB pathway is present in arthropods; indeed *Drosophila* are a vital experimental organism in the study of antimicrobial responses (Hoffman 2003). Several components of the Toll pathway have been characterized in nematodes, which even at this level are involved in pathogen recognition (Pujol et al. 2001). Caspase activation and extrinsically induced apoptosis are well described in plants, nematodes, and flies (Meier et al. 2000).

The details of cellular signalling pathways that are activated by bacteria, particularly via TLRs and Nod proteins, are beginning to be unravelled (Fig. 17.1) (Silverman and Maniatis 2001; O'Neill 2003). As currently understood, binding of an appropriate ligand to the LRR of a TLR likely results in dimerization and the formation of a TIR domain competent to bind members of a family of adaptor proteins. The original member of this family is known as MyD88, and additional family members have been described in recent years, including TRAM, MAL/TIRAP, and TRIF/TICAM (Kopp and Medzhitov 2003). Different adaptor proteins (or combinations of them) may

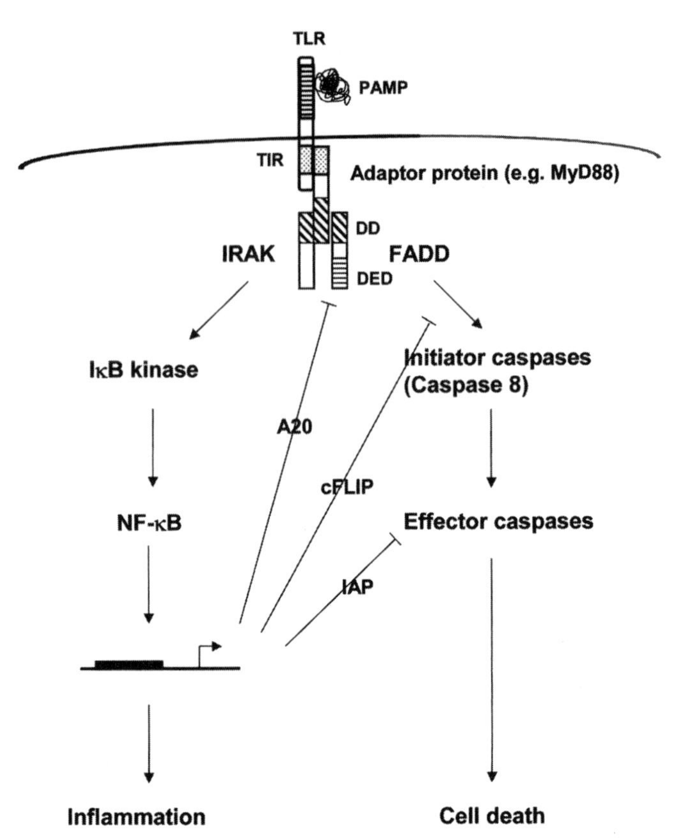

Figure 17.1. Schematic of proinflammatory and pro-apoptotic pathways. See text. PRRs bind PAMPs and transmit signals via cytoplasmic intermediates that can have proinflammatory or pro-apoptotic outcomes. Activating interactions are indicated with arrows, whereas inhibitory interactions are indicated with bars.

preferentially interact with specific TLRs or groups of TLRs and presumably have a role in orchestrating the most appropriate signalling pathways for a given PAMP/TLR interaction. Chapter 15 also describes the intracellular signalling induced by TLRs.

With the binding of an adaptor protein to the TIR of a PAMP-bound PRR, a bifurcation of signalling can occur along proinflammatory and/or pro-apoptotic pathways. The pro-apoptotic pathway will be addressed first. The MyD88 adaptor and other signalling adaptors/receptors, especially of the TNF receptor or Nod family, possess a protein–protein interaction motif termed the death domain (DD), so called because it is common to proteins involved in pro-apoptotic signalling. The death domain of these adaptor molecules can interact with other DD containing proteins that also encode a death effector domain (DED). The DED can directly interact with the inactive zymogen forms of initiator caspases, such as caspase 8, setting into motion the proteolytic cascades eventuating in programmed cell death (Adams 2003). For example, MyD88 can directly interact with the DED-containing adaptor FADD and subsequently activate procaspase 8. This pro-apoptotic action has been demonstrated for TLR2, but presumably other TLRs that bind MyD88 could mediate a similar outcome (Aliprantis et al. 2000). The intracytoplasmic Nod receptors are defined by their CARD domain (caspase activation and recruitment domains). Ligand-induced dimerization of Nod receptors may result in the similar recruitment of DED-bearing intermediates and caspase activation (Inohara et al. 1999). Thus, the PRRs are likely capable of directly activating the extrinsic apoptotic pathway, acting as "death receptors." The consequences of pro-apoptotic activation are discussed in the next section.

The other branch of the TLR signalling pathway is the initiation of proinflammatory signalling, largely mediated by activation of the NF-κB pathway. *NF-κB* is a collective term for members of the Rel family of DNA-binding transcription factors (Karin et al. 2004). Active NF-κB is a dimer that recognizes characteristic sequence motifs present in the promoters of many genes involved in immune, inflammatory, and anti-apoptotic responses. During unstimulated conditions, NF-κB is sequestered in the cytoplasm in an inactive state by the action of a third protein, IκB, or inhibitor of κB. NF-κB activation occurs by a rapid post-translational pathway that requires loss of the IκB (Karin and Ben-Neriah 2000) (Fig. 17.2). Free NF-κB dimer can then translocate to the nucleus and activate transcription. IκB degradation is preceded by an inducible phosphorylation event mediated by an *Iκ B kinase* (IKK), a large multi-subunit complex that is generally viewed as the key rate-limiting step in the activation pathway.

Activation of the IKK via PRR signalling occurs when the DD of MyD88 interacts with a DD in a second adaptor molecule, IRAK, of which several related family members are known. IRAK, a serine kinase, then activates the cytoplasmic signalling intermediate *TRAF6*. The activation of TRAF6 is dependant on ubiquitination (Deng et al. 2000). Ubiquitin is a highly conserved seventy-six–amino acid peptide employed as a common covalent modification of cellular proteins, usually on lysine residues (Schwartz and Hochstrasser 2003). Originally, ubiquitination of proteins was assumed to target the modified protein for regulated destruction by the cellular proteasome organelle. Recent discoveries of alternative ubiquitin chain linkages, as well as families of related molecules, such as SUMO (sentrin) and NEDD8, have revealed that these modifications play a role in diverse processes such as intracellular

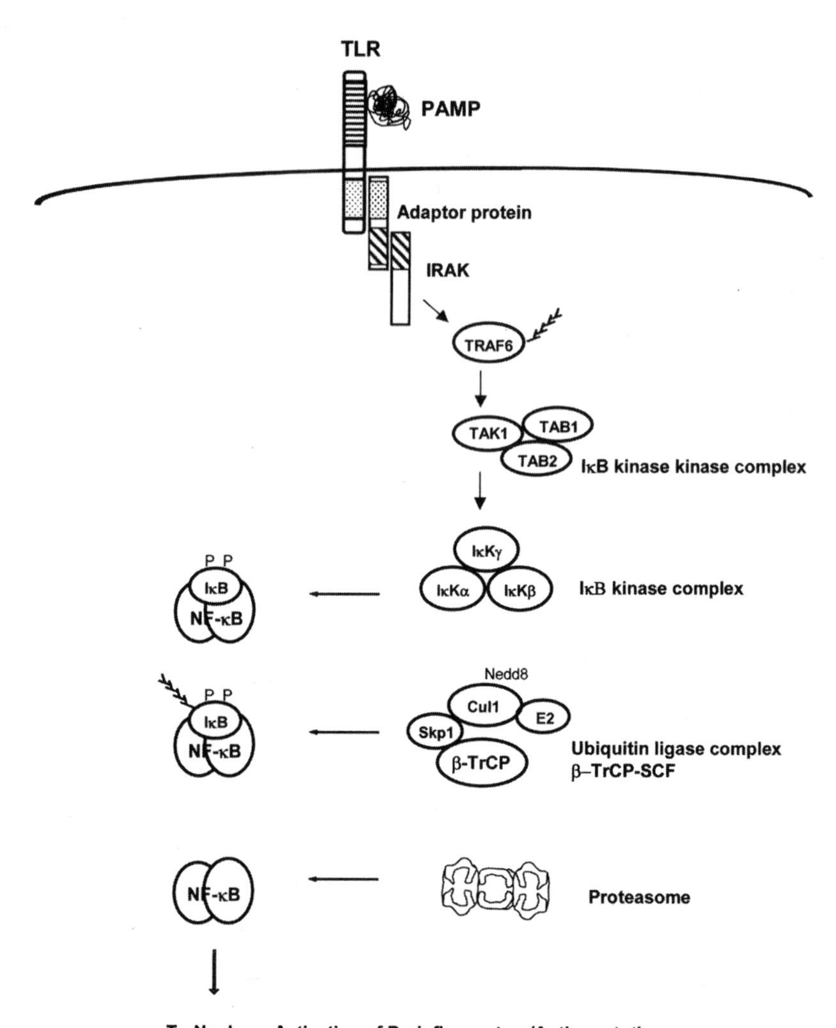

Figure 17.2. Diagram of the NF-κB activation pathway. See text. Phosphorylation is indicated with a P and ubiquitination is depicted as a brush.

trafficking, and, as in the case of TRAF6, enzymatic activation (Yeh et al. 2000). In any event, once activated by ubiquitination, TRAF6 in turn activates *TAK*1, in complex with *TAB*1 and *TAB*2. The TAK1/TAB1/TAB2 complex functions as an IKK kinase (Wang et al. 2001).

IKK is the signalling nexus, receiving and integrating signals from multiple proinflammatory signal transduction pathways, of which the Toll pathway is only one example (May and Ghosh 1999). The catalytic subunits of the IκB-kinase complex have been identified as IKK-α and IKK-β, which recognize a conserved 6 AA motif present on IκB and phosphorylate two serine residues within the motif (Karin 1999). The serine phosphorylated IκB isoforms are subjected to polyubiquitination by another multi-component enzymatic complex designated β-TrCP-SCF. The β-TrCP is the protein that physically interacts with the phospho domain of IκB-α, as well as a highly similar domain on the transcriptional activator/structural protein β-catenin (Maniatis 1999). Several other components comprise the complex, including the actual ubiquitin ligase enzyme (termed E2), and a subunit, *Cul1*, which is itself regulated by covalent modification of the ubiquitin-like molecule *Nedd8* (Read et al. 2000). IκB polyubiquitination results in the recognition of the modified IκB by the 18s regulatory subunit of the proteasome, followed by the proteolysis of the IκB molecule. Following IκB degradation, the NF-κB nuclear localization signal is exposed, allowing regulated translocation across the nuclear membrane, with subsequent site-specific binding to DNA of relevant promoters, transcriptional activation, and new mRNA synthesis.

In summary, eukaryotic cells have independently evolved, or co-opted from other biochemical circuits, two pathways for the transmittal of alarm signals: the proinflammatory NF-κB pathway and the pro-apoptotic extrinsic caspase pathway. Both pathways seem to be activated concurrently and in parallel. As we shall see, the end result of either pathway is very different at the cellular level (though not always at the tissue level). We now turn our attention to the effector phase.

17.4. THE EFFECTOR PHASE: CALL IN THE FLEET . . . , OR ABANDON SHIP!

The result of signalling pathways is the activation of effector molecules. As we have implied, animal cells seem to have two options to respond to a perceived threat. Firstly, activation of NF-κB and associated pathways result in inflammation. Broadly defined, inflammation is a cellular influx of immune effector cells that serve to eliminate the inciting microbial threat, generally by phagocytosis and subsequent intracellular lysis. Alternatively, or in parallel, apoptosis is initiated. Again broadly defined, apoptosis is a controlled, programmed autodigestion of a selected cell. In the sense of responding to microbial threats, apoptosis of an infected cell can accomplish precisely the same thing as phagocytosis. Loss of individual infected cells, especially in a population of interchangeable or highly turned over cells such as epithelia would have no deleterious effects on the whole organism.

17.4.1. Inflammation

Details of the inflammatory response in vertebrates have been well discussed elsewhere (Cotran et al. 1999). To briefly review, the usual paradigm holds that activation of NF-κB and other proinflammatory signal transduction pathways results in new transcription of a battery of effector molecules. Studies utilizing large-scale expression profiling techniques indicate these molecules include antibacterial peptides, metabolic enzymes (with roles in bacterial killing and wound healing processes), chemotactic messengers, anti-apoptotic proteins, cytokines, adhesion molecules, and mediators of adaptive immunity (MHC and costimulatory receptors) (Boldrick et al. 2002; Nau et al. 2002; Zeng et al. 2003). Similarly, in invertebrates, inflammatory responses include chemotactic activation of phagocytic cells (Hoffman 2003), also in concert with the transcriptional activation of a specific battery of genes (Irving et al. 2001). The up-regulation of soluble anti-microbial peptides appears to be a major effector mechanism in invertebrates.

Typically, inflamed vertebrate tissue exhibits increased numbers of phagocytic and immunomodulatory leukocytes, which busily phagocytose microbes and cellular debris. The process can involve any tissue and appears temporally as distinct acute and chronic phases, which usually resolve with complete restitution of function. It is only when overly intense, prolonged, or ill-situated that clinically recognized disease results. Inflammation is a classic two-edged sword in that the oxidants and proteases released by neutrophils can cause lasting tissue damage and likely plays a role in the early development of neoplasia. Therefore it is extremely important for the host to activate the inflammatory process sparingly.

17.4.2. Apoptosis

Apoptosis, or programmed cell death (PCD), is a morphologically distinct, genetically defined intrinsic mechanism by which individual cells can eliminate themselves while largely preserving the surrounding cells and overall tissue architecture (Meier et al. 2000; Adams 2003). Apoptosis, unlike necrosis, is not necessarily a pathological process. Necrosis is also a morphologically recognizable pattern of cell death generally induced by exogenous influences (physical and chemical injury, hypoxia – see Chapter 11). A key difference in necrosis and apoptosis is that the regulated sequence of events that mediates apoptosis results in the orderly elimination of the afflicted cell, without elicitation of acute inflammation and its potentially deleterious sequelae that typically accompanies necrotic cell death. Accordingly, apoptosis mediates a variety of physiological and adaptive events, such as embryological tissue remodelling, elimination of self-reactive lymphocytes or senescent neutrophils, and culling of continually proliferating cell populations (e.g., intestinal epithelial crypts – see Chapter 12).

The process of apoptosis is mediated by an arsenal of effector cysteinyl aspartate–specific proteases (caspases) that in the active state carry out limited proteolysis on apparently dozens of cellular structural and regulatory proteins, thus effectively and harmlessly (for surrounding tissues) dismantling the cell and accounting for the morphologic changes characteristic of apoptosis (Adam 2003). Understandably, this process is tightly regulated. The *effector* caspases exist in inactive zymogen forms until processed by an amplifying cascade of upstream initiator caspases. Commencing caspase activation can occur by several mechanisms. In the "intrinsic" pathway, a variety of cellular stressors (physical injury, DNA damage, withdrawal of growth factors/hormones) results in the leakage of mitochondrial cytochrome C into the cytosol, eliciting the formation of a scaffold complex, termed the apoptosome, which serves to activate the initiator caspase 9 and the subsequent effector caspases. Alternatively, and as discussed, in the extrinsic pathway, extracellular ligand-binding events result in activation of receptor complexes mediating assembly of proteins bearing DEDs and ultimately activation of initiator caspases, such as caspase 8.

Although inflammation is generally considered as the major effector process of innate immunity, apoptosis clearly also can function as a form of "immunity." Colonial or simple multicellular eukaryotes such as nematodes apparently utilize a form of PCD to eliminate infected cells without lethal effects to the whole organism, as they lack the luxury of having specialized phagocytic cells to do the dirty work for them (Aballay and Ausubel 2001). This strategy of microbial control may be prevalent in a variety of eukaryotic-prokaryotic interactions. It is common in certain forms of plant immunity (Dangl and Jones 2001; Staskawicz et al. 2001), in which detection of a microbial presence (by intracellular LRR-CARD-bearing proteins – more than 100 in the *Arabidopsis* genome) leads to activation of apoptotic pathways, resulting in PCD of the infected cells. This process is termed a hypersensitivity response (HR), and serves to effectively isolate the infection and protect the plant from systemic spread of the pathogen. Similar events can be observed in higher animals. For example, when mice deficient in the CD95 ligand (a common death receptor ligand) are infected with a respiratory epithelial pathogen, *Pseudomonas aeruginosa*, the epithelium fails to undergo local apoptosis as would normally occur in the wild-type tissue, and the mouse succumbs to systemic infection (Grassme et al. 2000).

The primordial function of PRRs (both cytoplasmic and transmembrane) may have been to activate cellular self-destruction. Just as highly turned-over or interchangeable tissue, such as epithelia or individual plant cells, can tolerate individual cell death without significant compromise of physiological tissue function, early metazoans with a colonial (or at least multipotential) cellular organization may have employed PCD to eliminate individual infected cells and thus spare the entire organism from overwhelming infection.

17.4.3. Which Response?

Given the parallel and concurrent activation of the proinflammatory and pro-apoptotic pathways, the individual cell (of a higher metazoan) is given a choice to attempt to recruit inflammatory cells or undergo controlled demolition, or use our naval metaphor, "call in the fleet or abandon ship." Recent work has demonstrated a complex and interrelated cross talk of checks and balances during the execution phases of either inflammation and/or apoptosis.

Proinflammatory Inhibition of Apoptosis

The generation of mouse mutants deficient in components of the NF-κB pathway uncovered a vital role for this system as an anti-apoptotic control. Mice null in the p65 subunit of NF-κB show embryonic lethality secondary to massive apoptosis in the liver (Beg et al. 1995). A mouse strain harboring null alleles of IKK-β solely in intestinal enterocytes exhibited no abnormalities under normal conditions, but responded to a systemic stress (transient intestinal ischemia) with massive apoptosis of the enterocytes (Chen et al. 2003). Expression profiling studies of eukaryotic/prokaryotic interactions have identified a subset of anti-apoptotic effectors consistently induced by bacterial (and other proinflammatory) stimuli (Karin and Lin 2002; Burstein and Duckett 2003). These proteins include the inducible NF-κB-dependant IAP (inhibitor of apoptosis) family (cIAP1, cIAP2, XIAP). These anti-apoptotic effectors act by binding to and inhibiting the activity of individual caspases, thus inactivating the apoptotic executioners induced by either intrinsic or extrinsic pathways. Interestingly, these proteins also possess C-terminal RING domains, indicating they may function as ubiquitin ligases, suggesting that a potential mode of action is the selective targeting of caspases for proteosomal-mediated degradation (Jesenberger and Jentsch 2002).

Other NF-κB-dependent, highly inducible anti-apoptotic proteins include *A20*, a zinc finger protein with Cys-protease activity that inhibits adaptor protein recruitment to the DD of death receptors (He and Ting 2002). Yet another similarly regulated protein is c*FLIP*. This apoptotic inhibitor contains both a DED and a catalytically inactive caspase-like domain. It interacts with the DED of FADD and/or procaspase 8, apparently acting as a dominant negative inhibitor of caspase 8 activation (Irmler et al. 1997). Thus, a complex network of inducible anti-apoptotic genes is present in vertebrates (and in flies – the IAP are key regulators of apoptosis in *Drosophila*) (Fig. 17.1). It has been commented that these proteins interrupt multiple steps of the pro-apoptotic pathway, acting as a multiply redundant "fail safe" system to allow proinflammatory signalling to occur without cell death (Karin and Lin 2002). Thus, activation of the NF-κB pathway, or perhaps more accurately, increased activation of the NF-κB pathway relative to proapoptotic pathways, serves to squelch whatever pro-apoptotic activation had been induced by threatening stimuli.

Both pro-apoptotic and proinflammatory effectors are activated by the perception of microbes by PRRs. It is intriguing that these sentinel receptors utilize much of the same signalling circuitry to initiate both pathways. The upstream components of the proinflammatory NF-κB pathway may have co-evolved with the apoptotic machinery to ensure that activation of NF-κB proceeds in parallel with initiation of the apoptotic death program. We have suggested the PRR-mediated stimulation of PCD programs may have represented a primordial "immune" reaction. Once a PRR-dependant inducible system for activating an early version of inflammation appeared later in animal evolution (in animals physically larger and with more specialised histological organization), the organism must have required a negative feedback to neutralize PCD pathways. This arrangement may also serve to protect the whole organism from microbial-mediated inhibition of proinflammatory pathways by committing individual infected cells to default apoptotic destruction. Such a hypothesis also suggests that microbes with immunomodulatory functions could have wide-ranging effects on the host.

Pro-apoptotic Inhibition of Proinflammatory Signalling

Conversely, activation of apoptotic pathways can inactivate NF-κB signalling, by caspase-mediated destruction of NF-κB signalling intermediates. IKKβ, p65, and IκB have all been identified as physiological substrates of activated effector caspases (Levkau et al. 1999; Tang et al. 2001). Partial cleavage products of some of these proteins have also been shown to act as dominant negative inhibitors of NF-κB signalling (Karin and Lin 2002). Other proteins act as inhibitors of the IAPs, providing another biochemical activity augmenting pro-apoptotic activity (Wu et al. 2000; Bergmann et al. 2003). These proteins, *Smac/Diablo*, seem to specifically target IAPs for proteasomal degradation either by stimulating an IAP autoubiquitination function or acting as ubiquitin ligases in their own right. Overall, proteolytic mediated inhibition of proinflammatory and anti-apoptotic components apparently ensures that caspase-mediated dismantling of the cell can occur unimpeded.

It is evident that eukaryotic cells possess parallel and heavily intermixed detection receptors and signalling pathways that can terminate in cellular proinflammatory responses or in PCD. Natural infection can vary in multiplicity of infection, duration, associated organisms, concurrent state of innate or adaptive immune activation, and undoubtedly many other variables. It is probably accurate to say that both proinflammatory and pro-apoptotic activation both occur rapidly and reliably during the initial interaction of a bacterial organism with a potential host. The cellular endpoints of these biochemical events is the result of whether proinflammatory or pro-apoptotic signalling "gains the upper hand" and is able to abort the sequence of events that mediates the contrary pathway.

17.5. A COUNTER STRATEGY: MICROBIAL INFLUENCES ON EUKARYOTIC SIGNALLING

It should come as no surprise that microbes have evolved myriad mechanisms to evade, inhibit, or usurp the various detection, signalling, and effector mechanisms deployed against them, a topic well reviewed elsewhere (Henderson and Oyston 2003). One form of microbial management of eukaryotic defenses that is gaining increasing attention is active repression of innate immune signalling (Neish 2003). A spectrum of bacterial pathogens exhibits mechanisms to directly inhibit the NF-κB pathway or reduce synthesis of inflammatory mediators in infected host cells. *Mycobacterium avium* blocks IL-8 secretion in cultured epithelia exposed to *Salmonella typhimurium* (Sangari et al. 1999). *Bordetella bronchoseptica* colonization of cultured respiratory epithelial cells inhibits the cytoplasmic to nuclear translocation of the NF-κB subunit p65 in response to TNF, potently activating apoptosis (Yuk et al. 2000). Members of the genus *Yersinia* repress MAPK and NF-κB in in vitro co-culture systems at the level of IκB phosphorylation, and are potently pro-apoptotic to the macrophages they infect (Ruckdeschel et al. 1998). Our laboratory has described several strains of non-pathogenic *Salmonella* that not only inhibit the NF-κB pathway in infected epithelial cells, but further block even subsequent proinflammatory responses elicited by endogenous cytokines (Neish et al. 2000). In contrast to the inhibitory mechanisms associated with *Yersinia* infection, non-pathogenic *Salmonella* block the polyubiquitination of IκB without interfering with phosphorylation. These bacteria are also pro-apoptotic (Collier-Hyams et al. 2002).

These observations suggest that microbially mediated active blockade of proinflammatory pathways could act at the cellular level by blunting cellular innate immune responses, essentially damping antimicrobial responses, and/or by activation of apoptosis. An example of the former, long-term intracellular parasites such as *Rickettsia* and *Chlamydia* do not elicit an inflammatory response and apparently repress apoptotic effects from even potent exogenous signalling (Weinrauch and Zychlinsky 1999). Conversely, highly virulent pathogens such as *Bacillus anthracis* and *Yersina pestis* strongly augment apoptosis by interference with proinflammatory signalling, which in infected macrophages results in elimination of key immunomodulatory cells ("dead men tell no tales!"), permitting rapid systemic spread (Monack et al. 1998; Weinrauch and Zychlinsky 1999). Thus, although the host can choose whether a desired antimicrobial response skews toward the proinflammatory or pro-apoptotic, bacteria with the ability to influence eukaryotic signalling pathways are apparently afforded the same options.

Bacteria have evolved multiple biochemical mechanisms to influence eukaryotic signalling. For example, many bacteria that have intimate relationships with eukaryotic organisms possess a type III secretion system (TTSS), a transmembrane conduit in the bacterial (Gram-negative) cell wall that allows the secretion of preformed soluble effector proteins from the interior of the bacterium into the environment or actually into the cytoplasm of a contacted eukaryotic cell (Galan 2001). A detailed discussion of the *Bordetella* TTSS is provided in Chapter 11. Type III secretion is characteristic of the family Enterobacteriaceae, which includes human pathogens such as *Salmonella*, *Yersinia*, *Shigella*, and the spectrum of pathogenic *Escherichia coli*. TTSSs and batteries of translocated effector proteins have also been characterized in many Gram-negative phytopathogenic bacteria (Staskawicz et al. 2001). Interestingly, these secretion systems have also been found to play a role in establishment of endosymbiotic relationships in plants and insects. Infection/colonization of leguminous plants by *Rhizobia* induces a defensive "nodulation" reaction by the plant, wherein the microbe thrives. Symbiotic *Rhizobia* establish their symbiosis by the action of proteins secreted via type III secretion apparatus (Viprey et al. 1998). *Sodalis glossinidius* is an endosymbiont of tsetse flies that requires a TTSS to invade insect cells and establish a stable symbiotic state (Dale et al. 2001). These later observations establish that the TTSS is not solely a virulence factor, but in essence a device employed by the prokaryotic kingdom as a "communication system" with eukaryotic cells (Dale et al. 2002). Further discussion of TTSS is to be found in Chapters 3, 10, and 11.

The TTSS functions to introduce effector proteins into the host cell, where they presumably influence some aspect of host biology. Needless to say, the biochemical characterization of various prokaryotic effector proteins is a topic of intense interest. In *Yersinia*, the anti-NF-κB activity is TTSS dependant and is mediated by the secreted effector molecule *YopP/J* (Ruckdeschel et al. 2001). Mutation of this gene renders *Yersinia* less virulent in in vivo murine models, and blocks the ability to induce apoptosis in macrophages in vitro. Orth and others showed that YopJ blocks the NF-κB pathway by physically associating with the IKK-β subunit and inhibiting the subsequent phosphorylation of IκB-α (Orth et al. 1999). Homologues of YopJ are present in several enteric pathogens; an example is *AvrA* in *Salmonella* sp, a protein which has potent inhibitory activity toward NF-κB activation. *Salmonella*-bearing mutations in the AvrA open reading frame have a reduced ability to elicit apoptosis in naturally infected epithelial cells (Collier-Hyams et al. 2002), though effects in vivo have not been demonstrated. As mentioned, TTSS-secreted factors are found in phytopathogens. Several members of the YopJ/Avr family have been identified in plant pathogens (e.g., AvrBsT in the plant pathogen *Xanthomonas campestris*). Similarly, in infected host plants, these effectors elicit the apoptotic hypersensitivity response. Bacteria without the effector molecule are not perceived by the plant cells' defensive machinery and are able to disseminate systemically and kill the plant. Thus, bacterial effectors that elicit the HR are commonly known as "avirulence factors," hence the designation "Avr" (Staskawicz et al. 2001). As we have discussed, the plant symbiont *Rhizobium* requires a TTSS to establish its environment within plant root nodules. Intriguingly, *Rhizobium* spp carry a homologue of the YopJ/AvrA family, designated YL4O (Viprey et al. 1998). Potentially, this translocated effector may function analogously to elicit developmental PCD, which is presumably necessary for establishment of the symbiotic state (Kobayashi et al. 2001).

The Avr/YopJ proteins represent a class of cysteine proteases, termed "ubiquitin-like protein proteases," an increasingly recognized subset of enzymes first characterized in adenovirus and also showing homology to the eukaryotic Ulp1 from budding yeast (Orth et al. 2000). These enzymes cleave the isopeptide bond between polymerized ubiquitin or ubiquitin-like proteins (SUMO, Nedd8), resulting in a reversal of the covalent modification on the target protein. It is presumed that the cellular role of the broad class of enzymes is to reverse the effects of regulatory protein modification by ubiquitin and related molecules. Ubiquitin and ubiquitin-like protein modifications are involved at several regulatory points of the NF-κB pathway, including activation of TRAF6, the Cul subunit of the SCF ubiquitin ligase, and IκB itself.

Thus, bacterial proteins with ubiquitin-like protease activity are highly plausible candidates for having negative effects on NF-κB activation pathways. Furthermore, given the emerging roles for this type of covalent modification in myriad cellular functions, not just in protein degradation but also in signal transduction, endocytic trafficking, and cytoplasmic/nuclear transport (Schwartz and Hochstrasser 2003), bacterial proteins with a proteolytic activity toward ubiquitin and related molecules could influence many aspects of host cell biology.

At the biochemical level, bacterially encoded effector proteins can inhibit proinflammatory/anti-apoptotic pathways, evidently in some cases, by interfering with aspects of ubiquitination and related processes. At the cellular level, these biochemical events may result in dampening of inflammatory pathways or activation of apoptotic pathways; whereas at the organismal level, the end results may be traditionally recognized pathogenesis (*Yersinia*) or symbiosis (*Rhizobium*). An intermediate organismal level outcome may be the eukaryotic host counterexploiting the bacterial effector to elicit a defensive reaction, either by detection of the foreign effector by PRR (with the effector representing a PAMP – a situation commonly recognized in plants) or by the pro-apoptotic default that occurs with inhibition of NF-κB activation. This concept is well illustrated by the plant hypersensitivity response, and it has been speculated that features of human enteric infections (diarrhoea) that serve to eliminate infection represent a functional parallel to the plant HR (Galan 1998). The recent demonstration of the anti-inflammatory/pro-apoptotic activity of YopJ and AvrA is consistent with these proteins mediating a cellular function analogous to their role in plants. Furthermore, the AvrA gene is found in virtually all human enteropathogenic strains of *Salmonella* but is invariably absent in all known strains of human systemic *Salmonella*, including *S. typhi* and *S. paratyphi* (Chan et al. 2003). Characteristically, these virulent human pathogens fail to stimulate proinflammatory effectors and result in systemic infection. As a matter of speculation, the AvrA gene product may augment pro-apoptotic pathways by blockade of NF-κB activation in epithelial cells in vivo (or simply may be recognized as foreign). PCD in infected epithelial cells, like plant cells (and unlike macrophages), are expendable without prohibitive damage to the organism as a whole. In the absence of an apoptotic or inflammatory response, bacteria could easily disseminate systemically.

As mentioned, apoptotic PCD acts as a mechanism in many normal developmental processes across a spectrum of life (Meier et al. 2000). As bacteria can be potent inducers of apoptosis, is there any evidence of bacterially mediated PCD influencing animal growth and development? Intriguingly, Foster and McFall-Ngai described the characteristic morphological changes of PCD occurring during the establishment of a symbiotic interaction between the marine squid *Euprymna scolopes* and the luminous bacterium *Vibro fischeri* (Foster et al. 2000). During the initial colonization of the light organ of the squid, the *Vibrio* elicted morphological and biochemical evidence of apoptosis in the superficial epithelium of this specialized tissue, ultimately leading to marked remodelling of the squid tissue to form the mature light organ stably colonized by the bacteria. This process is mediated by the *Vibrio* LPS, plausibly suggesting that a PRR is involved. This observation is the first description of a metazoan developmental (histogenic) process controlled by the interaction with a prokaryote. This process is discussed in detail in Chapter 11.

Apoptotic processes are involved in the physiologic control of cell number in the epithelium of the vertebrate intestinal lining. This tissue is known to turn itself over completely every two to three days, with new cells emerging from the stem cell compartment of the epithelial "crypts" while senescent cells are eliminated via PCD at a rate that equals their proliferation (Hall et al. 1994). Mammalian intestinal commensal bacteria have been observed to inhibit inflammatory pathways. In vitro coculture experiments with lactic acid bacteria have shown dampening of inflammatory responses (Wallace et al. 2003). *Lactobacillus* sp. can prevent colitis in spontaneous murine models (Masden et al. 1999). The mammalian intestinal commensal *Bacteroides thetaiotaomicron* has been demonstrated to inhibit NF-κB pathways by the novel mechanism of regulating cytoplasmic to nuclear translocation (Kelly et al. 2004). Plausibly, these immunomodulatory and pro-apoptotic effects may shape many aspects of intestinal function and it may not be an overstatement to conclude our own bacterial flora may affect our biology and health in ways we have only begun to consider.

17.6. CONCLUSIONS

It is obvious that bacteria including frank pathogens and well-recognized commensals have evolved a range of strategies for evading the consequences of PAMP binding to host cells. It is not clear if commensal organisms have developed clear-blue-water between their abilities to inhibit the NF-κB/Relish proinflammatory signalling system and that of pathogens such as *Yersinia*. What is fascinating is that we are discovering the ability of bacteria to interfere with the IKK/NF-κB system just at the time that the pharmaceutical industry is viewing this cellular signalling nexus as a key target for drug development (Karin et al. 2004). As usual, *Homo sapiens* is getting there just a few hundred million years too late.

REFERENCES

Aballay, A. and Ausubel, F. M. (2001). Programmed cell death mediate by ced-3 and ced-4 protects *Caenorhabditis elegans* from *Salmonella typhimurium*-mediated killing. *Proceedings of the National Academy of Sciences USA*, **98**, 2735–2739.

Adams, J. M. (2003). Ways of dying: multiple pathways to apoptosis. *Genes and Development* **17**, 2481–2495.

Aliprantis, A. O., Yang, R.-B., Weiss, D. S., Godowski, P., and Zychlinsky, A. (2000). The apoptotic signalling pathway activated by Toll-like receptor-2. *EMBO Journal* **19**, 3325–3336.

Barton, G. M. and Medzhitov, R. (2003). Toll-like receptors and their ligands. In *Toll-like Receptor Family Members and Their Ligands*, eds. B. Beutler and H. Wagner, pp. 81–92. New York: Springer.

Beg, A. A., Sha, W. C., Bronson, R. T., Ghosh, S., and Baltimore, D. (1995). Embryonic lethality and liver degeneration in mice lacking the RelA component of NF-κB. *Nature* **376**, 167–170.

Bergmann, A., Yang, A., Y.-P., and Srivastava, M. (2003). Regulators of IAP function: coming to grips with the grim reaper. *Current Opinion in Cell Biology* **15**, 717–724.

Boldrick, J. C., Alizadeh, A. A., Diehn, M., Dudoit, S., Liu, C. L., Belcher, C. E., Botstein, D., Staudt, L. M., Brown, P. O., and Relman, D. A. (2002). Stereotyped and specific gene expression programs in human innate immune responses to bacteria. *Proceedings of the National Academy of Sciences USA* **99**, 972–977.

Burstein, E. and Duckett, C. S. (2003). Dying for NF-κB? Control of cell death by transcriptional regulation of the apoptotic machinery. *Current Opinion in Cell Biology* **15**, 732–737.

Chan, K., Baker, S., Kim, C. C., Detweiler, C. S., Dougan, G., and Falkow, S. (2003). Genomic comparison of *Salmonella enterica* serovars and *Salmonella bongori* by use of an *S. enterica* serovar *typhimurium* DNA array. *Journal of Bacteriology* **185**, 553–563.

Chen, L.-W., Egan, L., Li, Z.-W., Greten, F. R., Kagnoff, M. F., and Karin, M. (2003). The two faces of IKK and NF-κB inhibition: prevention of systemic inflammation but increased local injury following intestinal ischemia-reperfusion. *Nature Medicine* **9**, 575–581.

Collier-Hyams, L. S., Zeng, H., Sun, J., Tomlinson, A. D., Bao, Z. Q., Chen, H., Madara, J. L., Orth, K., and Neish, A. S. (2002), *Salmonella* AvrA effector inhibits the key proinflammatory, anti-apoptotic NF-κB pathway. *Journal of Immunology* **169**, 2846–2850.

Cotran, R., Kumar, V., and Collins, T. (1999). *The Pathologic Basis of Disease.* Philadelphia: W. B. Saunders.

Dale, C., Plague, G. R., Wang, B., Ochman, H., and Moran, N. A. (2002). Type III secretion systems and the evolution of mutualistic endosymbiosis. *Proceedings of the National Academy of Sciences USA* **99**, 12397–12402.

Dale, C., Young, S. A., Haydon, D. T., and Welburn, S. C. (2001). The insect endosymbiont *Sodalis glossinidius* utilizes a type III secretion system for cell invasion. *Proceedings of the National Academy of Sciences USA* **98**, 1883–1888.

Dangl, J. L. and Jones, J. D. (2001). Plant pathogens and integrated defence responses to infection. *Nature* **411**, 826–833.

Deng, L., Wang, C., Spencer, E., Yang, L., Braun, A., You, J., Slaughter, C., Pickart, C., and Chen, Z. J. (2000). Activation of the IκB kinase complex by TRAF6 requires a dimeric ubiquitin-conjugating enzyme complex and a unique polyubiquitin chain. *Cell* **103**, 351–361.

Foster, J. S., Apicella, M. A., and McFall-Ngai, M. J. (2000). *Vibrio fischeri* lipopolysaccharide induces developmental apoptosis, but not complete morphogenesis, of the *Euprymna scolopes* symbiotic light organ. *Developmental Biology* **226**, 242–254.

Galan, J. E. (1998). 'Avirulence genes' in animal pathogens? *Trends in Microbiology* **6**, 3–6.

Galan, J. E. (2001). *Salmonella* interactions with host cells: type III secretion at work. *Annual Review of Cell and Developmental Biology* **17**, 53–86.

Gewirtz, A. T., Simon, P. O., Jr., Schmitt, C. K., Taylor, L. J., Hagedorn, C. H., O'Brien, A. D., Neish, A. S., and Madara, J. L. (2001). *Salmonella typhimurium* translocates flagellin across intestinal epithelia, inducing a proinflammatory response. *Journal of Clinical Investigation* **107**, 99–109.

Girardin, S. E., Tournebize, R., Mavris, M., Page, A.-L., Li, X., Stark, G. R., Bertin, J., DiStefano, P. S., Yaniv, M., Sansonetti, P., and Philpott, D. J. (2001). CARD4/Nod1 mediates NF-κB and JNK activation by invasive *Shigella flexner. EMBO Reports* **21**, 736–742.

Grassme, H., Kirschnek, S., Riethmueller, J., Riehle, A., von Kurthy, G., Lang, F., Weller, M., and Gulbins, E. (2000). CD95/CD95 ligand interactions on epithelial cells in host defense to *Pseudomonas aeruginosa. Science* **290**, 527–530.

Hall, P. A., Coates, P. J., Ansari, B., and Hopwood, D. (1994). Regulation of cell number in the mammalian gastrointestinal tract: the importance of apoptosis. *Journal of Cell Science* **107**, 3569–3577.

He, K.-I. and Ting, A. T. (2002). A20 inhibits tumor necrosis factor (TNF) alpha-induced apoptosis by disrupting recruitment of TRADD and RIP to the TNF

Receptor 1 complex in Jurkat T cells. *Molecular and Cellular Biology* **22**, 6034–0645.

Henderson, B. and Oyston, C. F. (eds.) (2003). *Bacterial Evasion of Host Immune Responses.* Cambridge: Cambridge University Press.

Hoffman, J. A. (2003). The immune response of *Drosophila. Nature* **426**, 33–37.

Hornef, M. W., Frisan, T., Vandewalle, A., Normark, S., and Richter-Dahlfors, A. (2002a). Toll-like receptor 4 resides in the Golgi apparatus and colocalizes with internalized lipopolysaccharide in intestinal epithelial cells. *Journal of Experimental Medicine* **195**, 559–570.

Hornef, M. W., Wick, M. J., Rhen, M., and Normark, S. (2002b). Bacterial strategies for overcoming host innate and adaptive immune responses. *Nature Immunology* **3**, 1033–1040.

Hugot, J.-P., Chamaillard, M., Zouall, H., Lesage, S., Cezard, J.-P., Belaiche, J., Almer, S., Tysk, C., O'Morain, C. A., Gassull, M., Binder, V., Finkel, Y., Cortot, A., Modigliani, R., Laurent-Puig, P., Gower-Rousseau, C., Macry, J., Colombel, J.-F., Sahbatou, M., and Thomas, G. (2001). Association of NOD2 leucine-rich repeat varients with susceptibility to Crohn's disease. *Nature* **411**, 599–603.

Inohara, N., Koseki, T., del Peso, L., Hu, Y., Yee, C., Chen, S., Carrio, R., Merino, J., Liu, D., Ni, J., and Nunez, G. (1999). Nod1, and Apaf-1-like activator of caspase-9 and nuclear factor-κB. *Journal of Biological Chemistry* **21**, 14560–14567.

Inohara, N. and Nunez, G. (2003). Nods: Intracellular proteins involved in inflammation and apoptosis. *Nature Reviews Immunology* **3**, 371–382.

Irmler, M., Thome, M., Hahne, M., Schneider, P., Hofmann, K., Steiner, V., Bodmer, J. L., Schroter, M., Burns, K., Mattmann, C., Rimoldi, D., French, L. E., and Tschopp, J. (1997). Inhibition of death receptor signals by cellular FLIP. *Nature* **388**, 190–195.

Irving, P., Troxler, L., Heuer, T. S., Belvin, M., Kopczynski, C., Reichart, J.-M., Hoffman, J. A., and Hetru, C. (2001). A genome-wide analysis of immune responses in *Drosophila. Proceedings of the National Academy of Sciences USA* **98**, 15119–15124.

Jesenberger, V. and Jentsch, S. (2002). Deadly encounter: ubiquitin meets apoptosis. *Nature Reviews Molecular and Cell Biology* **3**, 112–121.

Karin, M. (1999). The beginning of the end: IκB kinase (IKK) and NF-κB activation. *Journal of Biological Chemistry* **274**, 27339–27342.

Karin, M. and Ben-Neriah, Y. (2000). Phosphorylation meets ubiquitination: the control of NF-κB activity. *Annual Review of Immunology* **18**, 621–663.

Karin, M. and Lin, A. (2002). NF-κB at the crossroads of life and death. *Nature Immunology* **3**, 221–227.

Karin, M., Yamamoto, Y., and Wang, Q. M. (2004). The IKK-NF-κB system: A treasure trove for drug development. *Nature Reviews Drug Discovery* **3**, 17–26.

Kelly, D., Campbell, J. I., King, T. P., Grant, G., Jansson, E. A., Coutts, A. G. P., Pettersson, S., and Conway, S. (2004). Commensal anaerobic gut bacteria attenuate inflammation by regulating nuclear-cytoplasmic shuttling of PPAR-gamma and RelA. *Nature Immunology* **5**, 104–112.

Kobayashi, K., Sunako, M., Hayashi, M., and Murooka, Y. (2001). DNA synthesis and fragmentation in bacteroids during *Astragalus sinicus* root nodule development. *Biosciences, Biotechnology and Biochemistry* **65**, 510–515.

Kopp, E. and Medzhitov, R. (2003). Recognition of microbial infection by Toll-like receptors. *Current Opinion in Immunology* **15**, 1–6.

Levkau, B., Scatena, M., Giachelli, C. M., Ross, R., and Raines, E. W. (1999). Apoptosis overrides survival signals through a caspase-mediated dominant-negative NF-κB loop. *Nature Cell Biology* **1**, 227–233.

Maniatis, T. (1999). A ubiquitin ligase complex essential for the NF-κB, Wnt/Wingless, and Hedgehog signalling pathways. *Genes and Development* **13**, 505–510.

Masden, K., Doyle, J. S., Jewell, L. D., Tavernini, M. M., and Fedorak, R. N. (1999). *Lactobacillus* sp. prevents colitis in interleukin-10 deficient mice. *Gastroenterology* **116**, 1107–1114.

May, M. J. and Ghosh, S. (1999). IkappaB kinases: kinsmen with different crafts. *Science* **284**, 271–273.

Meier, P., Finch, A., and Evan, G. (2000). Apoptosis in development. *Nature* **407**, 796–801.

Monack, D. M., Mecsas, J., Bouley, D., and Falkow, S. (1998). *Yersinia*-induced apoptosis in vivo aids in the establishment of a systemic infection of mice. *Journal of Experimental Medicine* **188**, 2127–2137.

Nau, G. J., Richmond, J. L. F., Schlesinger, A., Jennings, E. G., Lander, E. S., and Young, R. A. (2002). Human macrophage activation programs induced by bacterial pathogens. *Proceedings of the National Academy of Sciences USA* **99**, 1503–1508.

Neish, A. S. (2003). Microbial interference with host inflammatory responses. In *Microbial Pathogenesis and the Intestinal Epithelial Cell*, ed. G. Hecht, pp. 175–189. Washington DC: ASM Press.

Neish, A. S., Gewirtz, A. T., Zeng, H., Young, A. N., Hobert, M. E., Karmali, V., Rao, A. S., and Madara, J. L. (2000). Prokaryotic regulation of epithelial responses by inhibition of IκB-α ubiquitination, *Science* **289**, 1560–1563.

Ogura, Y., Bonen, D. K., Inohara, N., Nicolae, D. L., Chen, F. F., Ramos, R., Britton, H., Moran, T., Karaliuskas, R., Duerr, R. H., Achkar, J. P., Brant, S. R.,

Bayless, T. M., Kirschner, B. S., Hanauer, S. B., Nunez, G., and Cho, J. H. (2001). A frameshift mutation in NOD2 associated with susceptibility to Crohn's disease. *Nature* **411**, 537–539.

O'Neill, L. A. J. (2003). Signal transduction pathways activated by the IL-1 receptor/Toll-like receptor superfamily. In *Toll-like Receptor Family Members and Their Ligands,* eds. B. Beutler and H. Wagner, pp. 47–62. New York: Springer.

Orth, K., Palmer, L. E., Bao, Z. Q., Stewart, S., Rudolph, A. E., Bliska, J. B., and Dixon, J. E. (1999). Inhibition of the mitogen-activated protein kinase kinase superfamily by a *Yersinia* effector. *Science* **285**, 1920–1923.

Orth, K., Xu, Z., Mudgett, M. B., Bao, Z. Q., Palmer, L. E., Bliska, J. B., Mangel, W. F., Staskawicz, B., and Dixon, J. E. (2000). Disruption of signalling by *Yersinia* effector YopJ, a ubiquitin-like protein protease. *Science* **290**, 1594–1597.

Pujol, N., Link, E. M., Liu, L. X., Kurz, C. L., Alloing, G., Tan, M.-W., Ray, K. P., Solari, R., Johnson, C. D., and Ewbank, J. J. (2001). A reverse genetic analysis of components of the Toll signalling pathway in *Caenorhabditis elegans. Current Biology* **11**, 809–821.

Read, M. A., Brownell, J. E., Gladysheva, T. B., Hottelet, M., Parent, L. A., Coggins, M. B., Pierce, J. W., Podust, V. N., Luo, R. S., Chau, V., and Palombella, V. J. (2000). Nedd8 modification of cul-1 activates SCF-bTrCP-dependent ubiquitination of IkappaBalpha. *Molecular and Cellular Biology* **20**, 2326–2333.

Ruckdeschel, K., Harb, S., Roggenkamp, A., Hornef, M., Zumbihl, R., Kohler, S., Heesemann, J., and Rouot, B. (1998). *Yersinia enterocolitica* impairs activation of transcription factor NF-κB: involvement in the induction of programmed cell death and in the suppression of the macrophage tumor necrosis factor alpha production. *Journal of Experimental Medicine* **187**, 1069–1079.

Ruckdeschel, K., Mannel, O., Richter, K., Jacobi, C. A., Trulzsch, K., Rouot, B., and Heesemann, J. (2001). *Yersinia* outer protein P of *Yersinia enterocolitica* simultaneously blocks the nuclear factor-kappa B pathway and exploits lipopolysaccharide signalling to trigger apoptosis in macrophages. *Journal of Immunology* **166**, 1823–1831.

Sangari, F. J., Petrofsky, M., and Bermudez, L. E. (1999). *Mycobacterium avium* infection of epithelial cells results in inhibition or delay in the release of interleukin-8 and RANTES. *Infection and Immunity* **67**, 5069–5075.

Schwartz, D. C. and Hochstrasser, M. (2003). A superfamily of protein tags: Ubiquitin, SUMO and related modifiers. *Trends in Biochemical Sciences* **28**, 321–328.

Silverman, N. and Maniatis, T. (2001). NF-κB signalling pathways in mammalian and insect immunity, *Genes and Development* **15**, 2321–2343.

Staskawicz, B. J., Mudgett, M. B., Dangl, J. L., and Galan, J. E. (2001). Common and contrasting themes of plant and animal diseases. *Science* **292**, 2285–2289.

Takeda, K., Kaisho, T., and Akira, S. (2003). Toll-like receptors, *Annual Review of Immunology* **21**, 335–376.

Tang, G., Yang, J., Minemoto, Y., and Lin, A. (2001). Blocking caspase-3-mediated proteolysis of IKKbeta suppresses TNF-alpha-induced apoptosis, *Molecular Cell* **8**, 1005–1016.

Viprey, V., Del Greco, A., Golinowski, W., Broughton, W. J., and Perret, X. (1998). Symbiotic implications of type III protein secretion machinery in *Rhizobium*. *Molecular Microbiology* **28**, 1381–1389.

Wallace, T. D., Bradley, S., Buckley, N. D., and Green-Johnson, J. M. (2003). Interactions with lactic acid bacteria with human intestinal epithelial cells: effects on cytokine production. *Journal of Food Protection* **66**, 466–472.

Wang, C., Deng, L., Hong, M., Akkaraju, G. R., Inoue, J., and Chen, Z. J. (2001). TAK1 is a ubiquitin-dependent kinase of MKK and IKK. *Nature* **412**, 346–351.

Weinrauch, Y. and Zychlinsky, A. (1999). The induction of apoptosis by bacterial pathogens, *Annual Review of Microbiology* **53**, 155–187.

Wu, G., Chai, J., Suber, T. L., Wu, J. W., Du, C., Wang, X., and Shi, Y. (2000). Structural basis of IAP recognition by Smac/DIABLO. *Nature* **408**, 1008–1012.

Yeh, E., Gong, L., and Kamitani, T. (2000). Ubiquitin-like proteins: new wines in new bottles, *Gene* **248**, 1–14.

Yuk, M. H., Harvill, E. T., Cotter, P. A., and Miller, J. F. (2000). Modulation of host immune responses, induction of apoptosis and inhibition of NF-κB activation by the Bordetella type III secretion system, *Molecular Microbiology* **35**, 991–1004.

Zeng, H., Carlson, A. Q., Guo, Y., Yu, Y., Collier-Hyams, L. S., Madara, J. L., Gewirtz, A. M., and Neish, A. S. (2003). Flagellin is the major proinflammatory determinant of enteropathogenic *Salmonella*. *Journal of Immunology* **171**, 3668–3674.

CHAPTER 18

Shaping of the bacterial world by human intervention

Rino Rappuoli

18.1. INTRODUCTION

Of the more than 1 million species of bacteria that are estimated to populate the world, a minority (approximately 500 species) belong to the human gut flora and are somehow an essential part of every human being. An even smaller portion are human pathogens (approximately fifty bacterial species or less than 0.0005 percent of the total bacterial species). In spite of the small number of bacteria that interact with *Homo sapiens*, our species is now having a profound impact on the bacterial world. The abuse of tons of antibiotics for animal farming and human treatment has soaked the world in a solution of antibacterials, which, during the last 50 years, introduced an unprecedented selection pressure on the microbial world. For example, the 800 tons of fluoroquinolones used annually for human treatment and the 120 tons for animals have recently selected resistant lineages of *Campylobacter jejuni*, *Salmonella*, *Yersinia*, and *Escherichia coli* O157 (Falkow and Kennedy 2001; Byrd et al. 2001). This theme is the focus of the discussion in Chapter 4. In this chapter, the focus is on the consequences of other human interventions, such as hygiene and vaccination, on bacterial pathogens.

18.2. VIRTUAL ELIMINATION OF A BACTERIAL VIRUS BY VACCINATION AGAINST *CORYNEBACTERIUM DIPHTHERIAE*

Corynebacterium diphtheriae is a ubiquitous Gram-positive bacterium that is abundant in the environment. When the bacterium is infected by the phage that carries the gene for diphtheria toxin, the bacterium becomes a very dangerous pathogen that is responsible for diphtheria, a disease that

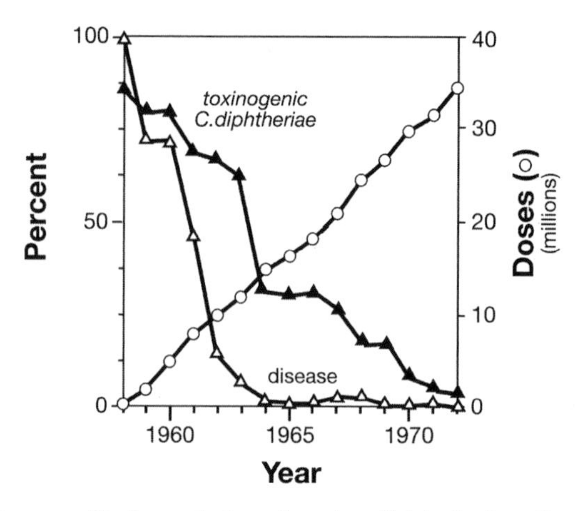

Figure 18.1. Decrease of toxinogenic *Corynebacterium diphtheriae* from Romania following vaccination against diphtheria toxin.

before the introduction of general vaccination, was a killer. The vaccine, composed of formaldehyde-inactivated diphtheria toxin (diphtheria toxoid), was developed by Ramon (1924) and introduced for mass immunization a few decades later. In a study initiated in Romania in 1958 (Pappenheimer 1984), it was shown that mass immunization with the toxoid (30 million doses were used between 1958 and 1972) decreased the incidence of disease from 600 per 100,000 population to less than 1 per 100,000 population. In 1958, ninety-five percent of the strains isolated carried the toxinogenic phage. By 1972, the proportion of human isolates carrying the gene for diphtheria toxin had declined to five percent (Fig. 18.1). The toxinogenic strains disappeared not only from the human population but also from horses. The disappearance of the phage followed several years after the elimination of the disease.

Interestingly, the disappearance of the isolates carrying the toxinogenic phage had no impact on the nontoxinogenic population of *C. diphtheriae,* the carrier rate of which remained unchanged. Today diphtheria vaccination is universal in both the developing and developed worlds, and toxinogenic strains are only rarely isolated, mostly from skin infections. However, they are not eradicated, and when the immunity level in the population decreases because of inadequate vaccination, as recently happened in the former Soviet Union, toxinogenic strains become prevalent again and cause devastating disease (Vitek and Wharton 1998).

18.3. *HELICOBACTER PYLORI* AND ITS DISAPPEARANCE FROM THE WESTERN WORLD

Helicobacter pylori is a Gram-negative bacterium that colonizes the extreme acidic environment of the human stomach. The infection causes inflammation and peptic ulcer and, in the long term, is responsible for most gastric cancer. The bacterium is usually transmitted by a member of the family to children during the first years of life, and once established, it remains in this environment for the rest of the individual's life. The selection of a niche with no competition and the ability to establish a chronic infection make *H. pylori* one of the most successful human bacterial parasites, which colonizes more than half of the human population (Montecucco and Rappuoli 2001). Given the vertical transmission of the infection and the life-long persistence of the bacterium, the evolution of this organism occurs within the stomach of the same individual, and the evolved bacterium is transmitted vertically to a member of the family. As a consequence, each individual is colonized by an *H. pylori* with a unique DNA signature, which is closely related to that of the family members and completely different from unrelated individuals (Fig. 18.2). Closer analysis of the DNA sequence has revealed that isolates can

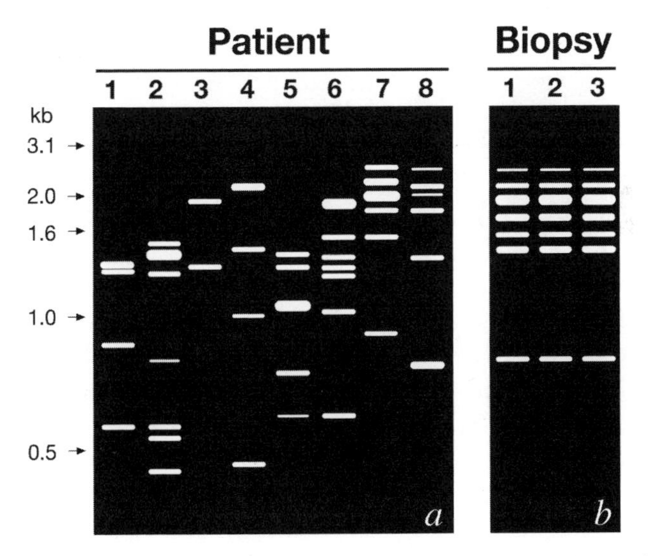

Figure 18.2. PCR-based DNA fingerprinting (RAPD) of (a) clinical *Helicobacter pylori* isolates deriving from different patients (patients 1 to 8), and (b) biopsies of the same patients carried out at different times (biopsies 1 to 3). Members of the same family tend to have a similar profile as in (b).

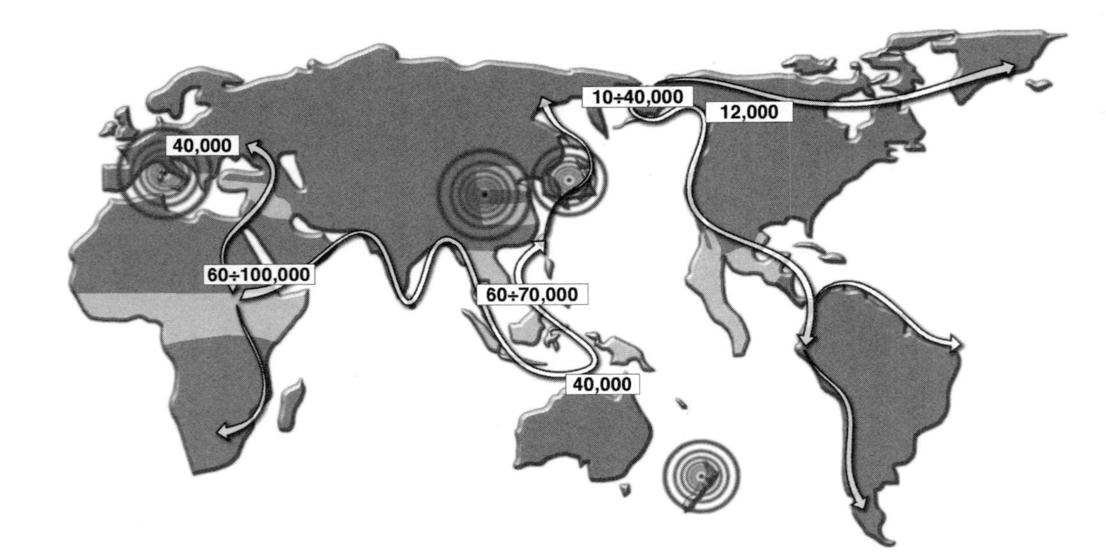

Figure 18.3. *Helicobacter pylori* genotypes, indicated by concentric circles, derive from the evolution of the bacterium, which migrated out of Africa in the stomach of primitive humans, and followed them during the colonization of the world (numbers indicate the years of the human migration).

be grouped according to their geographic origin. For instance, the analysis of DNA sequence of small fragments of DNA is sufficient to establish whether the bacterium comes from Asia, from Europe and North America, or from Africa. Within these regions, the bacterium can be further divided into subgroups. For example, Eastern European strains cluster in a subgroup that is different from one of the strains from Western Europe, especially Spain and Portugal. On the other hand, strains from South America are closely related to those of Spain and Portugal (Falush et al. 2003). The signature of the bacterium is very stable. For instance, Japanese populations that migrated to South America a century ago still carry largely strains of Asian origin.

The analysis of the genetic profile of *H. pylori* strains from different populations supports the hypothesis that the bacterium colonized primitive humans before they left Africa and since then, followed mankind during the migrations that generated today's "races" and populations (Covacci et al. 1999) (Fig. 18.3). Therefore, the bacterium co-evolved with humans and is a unique identifier of populations, and we can learn about human evolution and migrations by studying the genetics of *H. pylori* populations. In other words, *H. pylori* can be used to trace human history in a manner similar to languages, a tool widely used for the same purpose.

After more than 100,000 years' coexistence and co-evolution with humans, *H. pylori* is now on its way to extinction, at least in developed countries (Fig. 18.4). In fact, since World War II, the bacterium started to disappear from the European and North American populations. Increased hygiene,

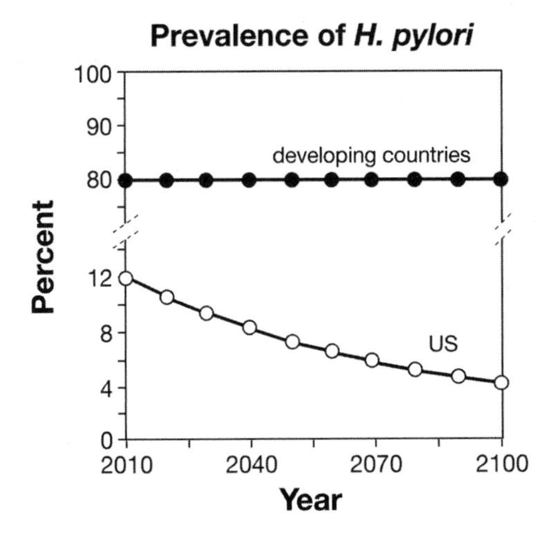

Figure 18.4. Present and predicted future prevalence of *Helicobacter pylori* in the United States and developing countries.

antibiotic use, and other unknown factors have decreased the ability of this bacterium to colonize and survive within the human stomach, and today the new generations have very low frequency of infection. Colonization of younger generations has decreased from more than ninety percent to ten to fifteen percent. It has been calculated that within the next century, the bacterial infection will be eradicated from the European and North American populations (Rupnow et al. 2001). In developing countries on the other hand, infection is still very high. *H. pylori* is a unique example of the impact that modern human lifestyle is having on a bacterial population that has stably co-existed with humans from the very beginning of human evolution.

18.4. MENINGOCOCCUS C ELIMINATION BY VACCINATION

Neisseria meningitidis, which colonizes the upper respiratory tract of ten percent of the human population is a major cause of meningitis and septicemia in humans (Rosenstein et al. 2001; Granoff et al. 2004). Both diseases are very severe, and in spite of the most sophisticated modern medicine, they are associated with five to fifteen percent mortality and up to twenty-five percent permanent sequelae. Fortunately, only one out of every 10,000 infections ends up in disease, which occurs with a frequency of one case per 10^5 population. Therefore, *N. meningitidis* should be considered a commensal, which, on rare occasions, becomes a very dangerous pathogen. The reasons why this meningococcus can be a commensal and a pathogen at the same time are poorly understood. The bacterium can be isolated only from humans, and the isolates may derive either from patients or from healthy people. The latter are usually classified as "carriers."

Overall, meningococci are classified in serogroups based on the chemical composition of their polysaccharide capsule. Although thirteen chemically defined serogroups have been described, only serogroups A, B, C, Y, and W-135, and to a very minor extent X and Z, have been associated with disease. A very effective vaccine was recently developed against serogroup C meningococcus. The vaccine was used for mass immunization in the United Kingdom in 2000. In less than a year, the entire United Kingdom population aged from 2 months to 18 years, was immunized with the vaccine. The effect on the disease was dramatic. As shown in Figure 18.5, within a year from the beginning of vaccination, the devastating disease had virtually disappeared in the United Kingdom (Bose et al. 2003).

Similarly, the carriage of the bacterium within the population has decreased. Although this is a unique example of how an effective vaccine can eliminate, in a short period of time, a dreadful disease and increase the quality

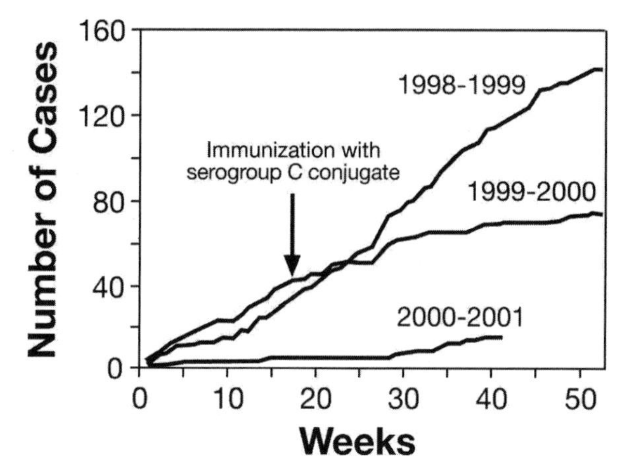

Figure 18.5. Cumulative cases of meningococcus C disease in the United Kingdom before vaccination with conjugate vaccine against meningococcus C (1998 to 1999), the year the vaccination was introduced (1999 to 2000), and the year after the introduction of vaccination (2000 to 2001).

of life, this is also another example of how human intervention can dramatically affect the ecology of the microbial population. The questions that may require a long time to answer are whether the ecological niche, which has been left empty by serogroup C meningococcus is going to remain empty or will be colonized by other bacteria.

A similar story is going on in New Zealand. Since the beginning of the 1990s, an epidemic of meningitis, caused by meningococcus B, has affected the country. Every year, approximately 600 cases are reported. Given the small population of the country, this number is extremely high. Although the disease is present in all populations, the native Maori and Pacific Islanders have a rate of disease that is two to four times higher than that of Caucasians. The epidemic is caused by a serogroup B strain for which a vaccine can be developed, using a technology originally developed in Norway by the National Institute of Public Health. Given the small size of the New Zealand population, development of a vaccine is not economically viable for a private company. To overcome this obstacle, the New Zealand government established a partnership with a private vaccine manufacturer to develop a vaccine dedicated to the New Zealand population. To date, the vaccine has been developed and successfully tested in several phase I and phase II clinical studies. Large-scale immunization for the entire population of the country is ongoing. Such immunization is expected to have an effect on this disease similar to that observed in the United Kingdom: immediate disappearance of the disease and dramatic effect on the meningococcal population colonizing the New Zealand population.

18.5. PNEUMOCOCCUS, A BACTERIUM FIGHTING FOR AN ECOLOGICAL NICHE

Streptococcus pneumoniae causes pneumonia, meningitis, and otitis media. More than ninety serotypes are known that are differentiated by the composition of the capsular polysaccharide. In the United States, more than eighty percent of the invasive disease is caused by seven serotypes (Feiken and Klugman 2002). Recently, a vaccine has been introduced against the seven serotypes that are responsible for most of the disease in the United States. In a like manner to the meningococcus, this pneumococcal vaccine, has been very successful in eliminating the invasive disease caused by strains included in the vaccine, and these serotypes have disappeared from the population. Unlike the case of meningococcus and *H. pylori*, in which the ecological niche left empty by the bacterium has not been filled by other bacteria, in the case of pneumococcus, other serotypes have taken the place left empty by the bacteria present in the vaccine. It is not yet well understood why serotype replacement occurs in pneumococcus and does not occur with meningococcus, and *H. pylori*. A possible hypothesis is that *H. pylori* colonizes the acidic environment of the stomach and other bacteria cannot survive there. In the case of meningococcus, the answer is less clear. Perhaps five pathogenic serogroups of meningococcus are not as efficient as more than ninety serotypes of pneumococcus in filling an empty ecological niche. Whatever the reason, the vaccination with pneumococcus although clearly beneficial in the short term, raises the question of whether in the long term we will see novel serotypes emerging and causing disease. To address this question, research efforts are in place to develop a protein-based universal vaccine that can protect against all serotypes.

18.6. CONCLUSIONS

It is ironic that although we know so little about the cooperative bacteria that populate animals, in particular the animal we care so much about, *Homo sapiens*, information is becoming available to suggest that we are already altering the cooperative bonds with members of our own microbiota. Although ecological research shows the possible downside to the creation of gaps in an ecological network, our 200 years of experience with vaccines to eliminate toxic pathogens has enhanced the quality of life. It is hoped that future work on vaccine development can be done against a background of precise knowledge of the human microbiota and the consequences of removing a bacteria species or strain from this integrated network.

REFERENCES

Bose, A., Coen, P., Tully, J., Viner, R., and Booy, R. (2003). Effectiveness of meningococcal C conjugate vaccine in teenagers in England. *Lancet* **361**, 675–676.

Covacci, A., Telford, J. L., Del Giudice, G., Parsonnet, J., and Rappuoli R. (1999). *Helicobacter pylori*, virulence, and genetic geography. *Science* **284**, 1328–1333.

Falkow, S. and Kennedy, D. (2001). Antibiotics, animals, and people . . . again! *Science* **291**, 397.

Falkow, S. and Kennedy, D. (2001). Response letter [Byrd, D. M. 3rd, Cox, L. A. Jr, and Wilson, J. D. (2001)]. Tracking antibiotics up the food chain. *Science* **291**, 2550a.

Falush, D., Wirth, T., Linz, B., Pritchard, J. K., Stephens, M., Kidd, M., Blaser, M. J., Graham, D. Y., Vacher, S., Perez-Perez, G. I., Yamaoka, Y., Megraud, I., Otto, K., Reichard, U., Katzowitsch, E., Wang, X., Achtman, M., and Suerbaum, S. (2003). Traces of human migrations in *Helicobacter pylori* population. *Science* **299**, 1582–1585.

Feiken, D. R. and Klugman, K. P. (2002). Historical changes in pneumococcal serogroup distribution: implications for the era of pneumococcal conjugate vaccines. *Clinical Infectious Diseases* **35**, 547–555.

Granoff, D. M., Feavers, I. M., and Borrow, R. (2004). Meningococcal vaccines. In *Vaccines*, ed. S. A. Plotkin and W. A. Orenstein, pp. 959–987. Philadelphia: Saunders.

Montecucco, C. and Rappuoli, R. (2001). Living dangerously: how *Helicobacter pylori* survives in the human stomach. *Nature Reviews Molecular Cell Biology* **2**, 457–466.

Pappenheimer, A. M. Jr. (1984). Diphtheria. In *Bacterial Vaccines*, ed. R. Germanier, pp. 1–32. Boca Raton, Florida: Academic Press, Inc.

Ramon, G. (1924). Sur la toxine et surranatoxine diphtheriques. *Ann Inst Pasteur Paris* **38**, 1–106.

Rosenstein, N., Perkins, B. A., Stephens, D. S., Popovic, T., and Hughes, J. M. (2001). Meningococcal diseases. *The New England Journal of Medicine* **344**, 1378–1388.

Rupnow, M. F., Shachter, R. D., Owens, D. K., and Parsonnet J. (2001). Quantifying the population impact of a prophylactic *Helicobacter pylori* vaccine. *Vaccine* **20**, 879–885.

Vitek, C. R. and Wharton, M. (1998). Diphtheria in the former Soviet Union: reemergence of a pandemic disease. *Emerging Infectious Diseases* **4**, 539–550.

Index